# Radiation

E. B. Manoukian

# Radiation

## The Photon in the Classical Context

Springer

E. B. Manoukian
Institute for Fundamental Study
Naresuan University
Phitsanulok, Thailand

Center for Theoretical Physics and Natural
Philosophy, Nakhonsawan Studiorum for
Advanced Studies—NAS
Mahidol University
Salaya, Thailand

ISBN 978-3-032-21505-5 ISBN 978-3-032-21506-2 (eBook)
https://doi.org/10.1007/978-3-032-21506-2



This Springer imprint is published by the registered company Springer Nature Switzerland AG
The registered company address is: Gewerbestrasse 11, 6330 Cham, Switzerland

*My greatest desire in writing this book is to show that classical electrodynamics may be described in terms of the photon the same way as our present field theories of all the fundamental particles of nature, in a much simpler way. At some stage, this description, as a consequence of gauge invariance, joins the classic one which originated from the monumental work of the great James Clerk Maxwell of 1865. From this, emerges quite a rich appealing clear language in which, in particular, a dynamical process in the theory may be described, in a time evolutionary manner, in which a photon is emitted and then detected as in experiments. Moreover, it opens a way for further applications, and certainly contributes in providing a unity in our understanding and explaining nature.*

*—E. B. Manoukian*

*This book is dedicated to all the girls I loved before*

# Preface

The writing of this book spread over three continents: North America (Canada), Europe (Ireland), and Asia (Thailand). It has reached a point at which every "i" has been dotted and every "t" has been crossed. The book covers all of fundamental classical electrodynamics, and encompasses a wide range of applications.

The theory is developed starting from the concept of a photon using methods of field theory. In particular, the so-called derived celebrated vacuum-to-vacuum transition amplitude, well known to field theorists, provides a tool for carrying out computations to study radiation and describes the interactions of the theory. While the propagator, monitoring the photon, describes the entire dynamical process, in a time evolutionary manner, from the time a photon is emitted to the time it is detected in an experiment. As a consequence of the latter, one avoids using plane waves, as is invariably done. One learns from this approach that what one calls electric and magnetic fields simply emerge from the formalism as a consequence of gauge invariance and, in passing, Maxwell's equations, in turn, simply emerge, thus recovering the Maxwellian theory of classical electrodynamics.

The power of the vacuum-to-vacuum transition amplitude, as a tool for studying radiation and basic interactions, as well as the propagator describing the time evolution of underlying processes, together with these fields give rise to a language rich to study electrodynamics in a more complete and, certainly, more appealing physically as one monitors the photon in a time evolutionary process.

Electrodynamics, over the years, was invariably developed again and again beginning with Maxwell's equations, and a reader will find it refreshing to experience the present treatment emphasizing physical clarity, simplicity of derivations, and elegance of formalism. This book is intended for physicists, all those in the sciences and engineering, for graduate/advanced undergraduate students and professionals alike. The prerequisites are basic introductions to electrodynamics, quantum mechanics, and undergraduate mathematical methods.

The Introduction, briefly, and rather effectively, discusses how the theory of classical electrodynamics, as we know it today, originated from the monumental work of James Clerk Maxwell in 1865. It will also describe the content of the book and guide the reader on how to go through it.

During the writing of this book in its final form, I discussed, almost daily, the development of the content of this book with my Ph.D. students who already had my course on classical electrodynamics, several years before, starting from Maxwell's equations. This was done in order that they experience, first hand, the simplicity in studying electrodynamics from the point of view of the photon, and the emerging clear and elegant formalism.

Without further ado, let us find out "where were we, and where are we now".

Phitsanulok/Nakhon Sawan, Thailand
January 2026

Prof. Dr. E. B. Manoukian

# Acknowledgements

Almost daily, I have enjoyed discussing the development of the content of this book with some of my former graduate students who, several years before, have taken my course in Classical Electrodynamics given by the traditional approach. These include Chaiyapoj Muthaporn, Seckson Sukhasena, Siri Sirininlakul, Tukkamon Vijaktanawudhi (aka Kanchana Limboonsong), Prasopchai Viriyasrisuwattana, Pornrad Srisawad, Songvudhi Chimchinda, who are now well established physicists in their own rights.

I am intended to all of them, as well as to the late Suppiya Siranan, and the late Moragote Buddhakala, for partly justifying my existence by attending and listening to my endless lectures on physics at the graduate level, as well as to many of them for their collaborations in our research.

I want to thank Seckson Sukhasena who asked me so many non-trivial questions all along, from the time he was a graduate student, that always persuaded me to go into more details, both, in the physics and the mathematics in writing my books. I am also grateful to him for all the help he gave me on computer and LATEX intricacies.

Although I have typed the entire manuscript myself, and drew the figures as well, Chaiyapoj Muthaporn prepared the LATEX input files, and saved the figures in eps form. Without his constant help in LATEX throughout the development, this work would never have been completed. I applaud him, thank him and will always remember how helpful he was. He was and still is a very disciplined hard worker.

I would like to give my special thanks to colleague Sikarin Yoo-Kong of the "Institute for Fundamental Study" of Naresuan University, and colleague Burin Gumjudpai of the "Center for Theoretical Physics and Natural Philosophy, Nakhonsawan Studiorum for Advanced Studies—NAS" of Mahidol University for their support.

Over the years, I have benefited much from lectures and writings of Julian Schwinger who had a unique way of doing physics with clarity and much elegance, and developed most powerful formalisms for carrying out computations. He certainly had one of the greatest minds in theoretical physics of our time. Attending a lecture by Schwinger was considered as a very special event, perhaps, comparable to a Mozart symphony, so to speak. Regarding Schwinger's lectures, Richard Feynman states[1]: "You can imagine for a lecturer like me, what a sensation it was to see such a thing. I was supposed to be a good lecturer according to some people, but this was really a masterpiece". Schwinger was literally able to develop such powerful methods for computations in physics, as this book attests, that once you master them, you wouldn't want to use, even accept, other methods of computations. In one lecture, he stated that "if you have a variable in your theory and you don't quite understand its role in the theory, vary it and see what happens". Needless to say, Feynman was also great, but don't ask him to compute things; just guess! He was brilliant at it. As a graduate student at McGill, I saw them together for the first time in a 1967 McGill-conference. At that time, I started looking at physics in a different way, particularly, because of those two legends who, just a couple of years before, were awarded the Nobel Prize for their work on QED.

It was such a pleasure and a privilege to work with my Editor Annelies Kersbergen of Springer, for her constant help, encouragement, and her professionalism which made the publication of this book possible. She was with me almost all the time during the writing of the book, exhibiting infinite patience and understanding. It was an unforgettable collaboration that I will always cherish. She is a remarkable human being.

This project would not have also been possible without the patience, encouragement, and understanding of my wife Tuenjai.

[1] R. P. Feynman (1996) *in*: Julian Schwinger-*The Physicist, the Teacher and the Man*, Editor Y. Jack Ng. Singapore: World Scientific, p. 91.

# Contents

# About the Author

**Prof. Dr. E. B. Manoukian** is Emeritus Professor of Physics at the "Institute for Fundamental Study" of Naresuan University, and Honorary Fellow of the "Center for Theoretical Physics and Natural Philosophy", Nakhonsawan Studiorum for Advanced Studies—NAS of Mahidol University, in Thailand. He carried out graduate studies at McGill and University of Toronto obtaining the M.Sc. and Ph.D. degrees, respectively, in 1968 and 1971. He held research positions at the "Theoretical Physics Institute" of the University of Alberta, the "Dublin Institute for Advanced Studies", and the "Centre de Recherches Mathématiques Appliquées" of the University of Montreal, and held Professorships as well at the "Royal Military College of Canada" of the Department of National Defence, at the "School of Physics" of Suranaree University of Technology in Thailand, and at the "Institute for Fundamental Study" of Naresuan University. He has authored over 190 refereed papers in most aspects of theoretical physics, as well as of 10 books, several of which are by Springer.

# Chapter 1
# Introduction

Classical electrodynamics may be described in terms of the photon, like all of our present field theories involving all of the fundamental particles (building blocs) of Nature, but, because of its limitations, in a much simpler way. The photon has been identified already about 120 years ago, thanks to Albert Einstein, with well know properties. As a matter of fact, by applying very basic field theory methods to the photon involving the so-called derived vacuum-to-vacuum transition amplitude and the photon propagator, classic electrodynamics may be simply developed as described in the book.

The vacuum-to-vacuum transition amplitude is a tool for studying radiation and basic interactions, while the propagator, monitoring the photon, describes a given process, in a time evolutionary manner, from the time a photon is emitted to the time it is detected in an experiment. It turns up what one calls the electric and magnetic fields emerge from the formalism as a consequence of gauge invariance, and, in passing, Maxwell's equation just simply pop up. Accordingly, from the formalism involving the photon, one eventually recovers the classic theory that originated from the work of legendary James Clerk Maxwell of 1865.[1]

This classic theory, as we know it today, given in terms of *four* equations using *vector* calculus, rather than of the many more equations derived by Maxwell in 1865, is thanks due to the contributions of several people in the 1880s. These mainly include Oliver Heaviside,[2] Joseph W. Gibbs and Edwin B.Wilson,[3] as well as Heinrich Hertz, [4] who simplified Maxwell's equations. In his work, Maxwell unified electricity and magnetism and predicted that light itself was an electromagnetic wave. The key result in Maxwell's equations is that they were formulated in terms of *fields*, avoiding, in the process, action at a distance, as we will explicitly see in Eq. (3.5.27) in

[1] Maxwell (1865) [18].

[2] Heaviside, (2011) [11].

[3] Gibbs and Wilson (1901) [10].

[4] Hertz (1884) [13].

 
E. B. Manoukian, *Radiation*,
https://doi.org/10.1007/978-3-032-21506-2_1

Chap. 3, consistent with Einstein's special relativity as given in Eqs. (5.1.19)/(5.1.20) in Chap. 5. The idea of fields originated from the experimental work of Michael Faraday, between the 1830s and 1850s,[5] who considered the electromagnetic forces as *lines* of forces.

The remarkable aspect of Maxwell's work, as mentioned above, that it is described in terms of fields avoiding *action at a distance*, was unlike Newton's formulation of the gravitational interaction which didn't. Although Isaac Newton did not use this term, he was troubled[6] by the implication of one body may act upon another in empty space, and thus was unable to explain how gravity operates at a distance.[7]

That light is a form of electromagnetic radiation obeying Maxwell's equations was established by Heinrich Hertz through his experimental work in 1880s,[8] confirming Maxwell's prediction. On the other hand, in earlier years, about 1818, Augustin Jean Fresnel[9] worked out the mathematical framework of the wave nature of light, which was observed earlier in the classic experimental work of Thomas Young's[10] experiments on the double slit experiment predicting a wave nature of light.

The building bloc of light, the photon concept was born by Albert Einstein's description of the photoelectric effect in 1905,[11] and named as such by Gilbert N. Lewis in 1926.[12]

What do we know about the photon?. Well it is a massless particle, i.e., it is an ultra-relativistic particle of, by convention, *exactly* given speed $c = 299,792,458$ meters per second, in vacuum, with the meter defined as the distance light travels per $10^{-8}/2.99792458$ second. It lives in Minkowski spacetime (Chap. 2), and has two polarization vectors perpendicular to its momentum.

Photons do not discriminate between positive and negative charges and they interact with both charges, they are unbiased. Differently said, a photon is a *bipartisan*, so to speak. They do not carry a charge and because of this, they essentially don't see one another, i.e., they do not interact directly with one another; they are essentially blind. More appropriately, we may call a photon as a *blind bipartisan*. Or, perhaps even better, one may call a photon a *xenophile*, so to speak. Imagine how deformed an object will be seen, if they interact with one another before reaching your eyes. That is, they are readily transmitted across space while being "indifferent" to one another.

In the quantum region, the probability that they interact is insignificant. For example the probability that two photons scatter of each other, the process of which is

---

[5] Faraday (2004) [6].

[6] Cohen, Whitman, and Budenz (translators of the Principia) (1999) [4]. See also Cohen (1962) [3].

[7] To understand the simple idea how Einstein was able to solve Newton's action at a distance problem and how gravity may operate at a distance, see the pedagogical treatment of this in Chap. 4 of Manoukian (2024) [17], for the interested reader.

[8] E.g., Hertz and Jones (1893) [12].

[9] Fresnel et al. (1866, 1868) [9].

[10] Young (1802, 1804) [22, 23].

[11] Einstein (1905) [5] including English translation by Arons and Peppard (1965).

[12] Lewis (1926) [14].

denoted by $\gamma + \gamma \;\rightarrow\; \gamma + \gamma$, where $\gamma$ denotes a photon, is of the order $\alpha^4$, as will be estimated at the end of Chap. 2, where $\alpha \simeq 1/137$ is the fine-structure constant. Another process $\gamma + \gamma \;\rightarrow\; e^+ + e^-$, with $e^\mp$ denoting the electron/positron, the corresponding probability of occurrence is of the order $\alpha^2$. They are not only small as predicted by explicit computations in the quantum region (QED), is that, experimentally, they are verified under very restricted conditions for which the detection of the outgoing photons is done indirectly by their interactions with other systems. Detailed experimental studies of these processes are given by the two critical, highly recommended, references given below.[13]

It is precisely because photons are indifferent to one another, that classical electrodynamics works perfectly well described in terms of photons. On the other hand, we will learn, through propagator theory, how a photon scatters off obstacles and different media. When we say the probability that a photon is reflected is greater than it is transmitted through a glass, this means that the intensity of reflected light is higher than the intensity of transmitted light. Not only that, we can compute power of radiation in terms of photons from these probabilities.

The power of the vacuum-to-vacuum transition amplitude, and the propagator, together with the emerging fields, give rise to a rich language to study classical electrodynamics, providing an elegant, clear, physically appealing formalism and opens the way for various applications, as the present book asserts.

The vacuum-to-vacuum transition amplitude is due to Julian Schwinger, and has been his basic tool[14] of research during his entire career. He also used the concept of the propagator extensively. The propagator, in a *scattering* process,[15] as developed by the author, is a reminder of Richard Feynman's,[16] well publicized, easy to read, approach of adding phase factors of different amplitudes in describing the scattering of light off obstacles. This will be particularly clear in Chap. 15 on X-rays of adding up phase factors with different amplitudes.

Chapter 2 is involved with Minkowski spacetime, i.e., with special relativity. The latter has been around, now, for 121 years, at the time of writing this book. If you haven't study it yet, here it is an easy quick way for you to learn the subject. In Chap. 3, the vacuum-to-vacuum transition amplitude is derived from scratch and the concept of the propagator is introduced, as well as the basic interactions of electrodynamics between currents are derived directly from the phase of the vacuum-to-vacuum transition amplitude. No electric or magnetic fields are used. In this chapter, the reader will encounter an extensive treatment of the Dirac delta distribution of paramount importance in all types of applications.

In Chap. 4, the power of radiation off an accelerating relativistic charged particle is derived *directly* from the vacuum-to-vacuum transition amplitude, where neither electric nor magnetic fields are used.

---

[13] Aad et al. ATLAS Collaboration (2019) [1]; and Adams et al. STAR Collaboration (2021) [2].

[14] In particular, see Schwinger (1970, 1973, 2003) [19–21].

[15] See, e.g., Manoukian (2013, 2016) [15, 16], and to earlier references to the author therein.

[16] Feynman (1985) [8].

The emergence of the electric and magnetic fields as a consequence of gauge invariance and the derivation of Maxwell's equations are the subject matter of Chap. 5, where the Maxwellian theory is recovered. In this chapter the phenomenological parameters of electrodynamics are also introduced and the *rationalized CGS unis* used in this book are defined in Sect. 5.4. The appropriate boundary conditions arising in applications are worked out by using the powerful method involving Dirac delta distributions. The chapter also deals with electromagnetic waves. Chapter 6 deals rather extensively with the underlying theory of synchrotron radiation.

In Chap. 7, the vacuum-to-vacuum transition amplitude is expressed in terms of the phenomenological parameters by simple scaling transformations. In Chap. 8, propagator theory is studied in detail emphasizing, in the process, polarization aspects of radiation. Various applications of propagators are, particularly, given in Chaps. 9–12.

Chapter 10 deals with the application of propagator theory, to study the fates of red and blue photons impinging on a piece of glass and see which has a better chance of being transmitted through or reflected by the glass. I was inspired by Feynman's[17] treatment of the problem to actually work out and *derive* the essential results. Chapter 11 deals with the important problem of radiation seeping through a metal with application to silver. In particular it is shown that the textbook exponential damping of radiation, as radiation seeps through a metal, is not always valid and bounds of the accuracy of it to hold are derived.

Chapter 13 deals with a subject matter that interests most people who want to understand as to why the sky is blue, and sometimes red, as well as to why clouds are white. We provide a detailed treatment of the subject matter for the color of the sky and of clouds. Realizing, that water droplets in clouds are much larger than molecules in the atmosphere necessitates to use a rather sophisticated method referred to as Mie radiation, also covered in detail.

Chapter 14 deals with the classic problem of radiation off a perfectly conducting plane screen through a circular hole using a powerful method involving Green's Theorem in terms of vector fields generated from tensor ones–a treatment most often omitted in the literature.

Radiation of X rays as they scatter off crystals, is investigated at length in Chap. 15. The interest in this problem is that X rays having small wavelengths, in comparison to light in the visible spectrum, small enough to unravel the structure of a crystal, that their scattering off electrons in atoms of the crystal may be directly studied. As a consequence of this, and most importantly, *no* phenomenological parameters need to be introduced here. As Feynman points out that as a consequence of this, a "chemistry problem" becomes a problem of pure physics.[18]

Chapter 16 is a fascinating chapter involving the cosmic background radiation originated when the Universe was about 380,000 years old and cooled down enough so that neutral atoms were formed and photons became mere spectators free to move, and are right here around us as we observe them today. Here the reader

---

[17] Feynman (1985) [8].

[18] Feynman (1967) [7].

will encounter important technical aspects of the underlying theory involving spin-weighted spherical harmonics to deal with light.

There is much emphasis on polarizations aspects of radiation throughout the volume. The book is self contained with detailed derivations. The prerequisites of the book are very basic introductions to quantum mechanics, to elements of electrodynamics, and to undergraduate mathematical methods.

My earlier book on "How energy considerations have shaped our fundamental modern theories of physics: Theory and key historical moments",[19] together with the present one provide a unity for studying fundamental theoretical physics as one interconnected theory, and give rise to a method for a student to master it *all*. I believe future generations will teach physics in phases, in such an integrated manner, as one theory. I might still be around to see this. Several medical schools have been teaching their students this way.

In physics, we have stories to tell, most of them quite entertaining. We look at one another, or we look at an object, and we understand that photons are indifferent to one another, that is, they hardly interact with one another when they are transmitted in space. We look at the sky, we understand why it is blue. We look at the clouds and we understand why they are white. We know how to estimate how big is our visible Universe, and determine its age, as well determine when these photons were liberated and became free after neutral atoms were formed, and so on. We understand why when we irradiate a material with X-rays, we do not have to introduce phenomenological parameters to do the underlying physics. The stories are endless. Are physicists privileged or not?; you tell me.

Although it will be a smooth ride to live through such stories and much more, you better fasten your seat belts.

## References


1. Aad, G. et al. ATLAS Collaboration (2019). Observation of light-by-light scattering in ultraperipheral Pb+Pb collisions with the ATLAS detector. *Physical Review Letters, 123*, 052001.
2. Adams, J. et al. STAR Collaboration (2021). Measurement of $e^+ e^-$ momentum and angular distributions from linearly polarized photon collisions. *Physical Review Letters, 127*, 052302.
3. Cohen, I. B. (1962). The first English version of Newton's hypotheses non fingo. *Isis, University of Chicago Press Journals, 53 (3)*, 379–388.
4. Cohen, I. B., Whitman, A, & Budenz, J. (Transls.) (1999): *Isaac Newton, The Principia: Mathematical principles of natural philosophy*, 916-917 Book III. California: University of California Press.
5. Einstein, A. (1905). Über einen die Erzeugung und Verwandlung des Lichtes betreffenden heuristischen Gesichtspunkt. *Annalen der Physik, 17 (6)*, 132-148. [English Translation: Arons, A. B., & Peppard, M. B. (1965). Einstein's proposal of the photon concept. *American Journal of Physics, 33 (5)*, 367-374.
6. Faraday, M. (2004). *Experimental Researches in Electricity, published in three volumes.* Mineola, N Y: Dover Publications.


[19] Manoukian (2024) [17].

7. Feynman, R. (1967). *The character of physical law*. Cambridge, MA: MIT Press.
8. Feynman, F (1985). *QED: The strange theory of light and matter*. Princeton, New Jersey: Princeton University Press.
9. Fresnel, A. J., Fresnel, L, Fresnel, F. & Lecauchez, W. Vol 1.(1866), Vol. 2 (1868). *Oeuvres complètes d'Augustin Fresnel*. Paris: Imprimerie Impériale.
10. Gibbs, J. W., & Wilson, E. B. (1901).*Vector Analysis*. New Haven: Yale University Press.
11. Heaviside, O. (2011). *Electromagnetic theory*. Cambridge. Cambridge. UK.
12. Hertz, H., & Jones, D. E. (1893). *Electric Waves: Being Researches on the Propagation of Electric Action with Finite Velocity Through Space*. London: Macmillan and Co.
13. Hertz, H. (1884). Über die Gleichungen, welche die electrodynamimischen Vorgänge in ruhenden Körpern darstellen, *Annalen der Physik und Chemie, 31*, 421-448.
14. Lewis, G. N. (1926). The conservation of photons. Nature 118, 874-875.
15. Manoukian, E.B. (2013). Red light, blue light, Feynman and a layer of glass. The causal propagator, *Journal of Electromagnetic Waves and Applications, 27. No.15*, 1986-1996.
16. Manoukian, E. B. (2016). Electromagnetic propagation across a metal: The causal propagator. *Journal of Electromagnetic Waves and Applications, 30 (5)*, 677-688.
17. Manoukian, E. B. (2024). *How energy considerations have shaped our fundamental modern theories of physics: Theory and key historical moments*. Switzerland: Springer.
18. Maxwell, J. C. (1865), A Dynamical Theory of the Electromagnetic Field. *Philosophical Transactions of the Royal Society of London 155*, 459–512.
19. Schwinger, J. (1970). *Particles, sources, and fields*. Reading, Massachusetts: Addison-Wesley.
20. Schwinger, J. (1973). A report on quantum electrodynamics. In. J. Mehra (Ed.), *The physicist's conception of nature*. Dordrecht/Holland: D. Reidel Publishing Company.
21. Schwinger, J. (Editor) (2003). *Selected papers on quantum electrodynamics*, Dover Publications, Mineola, N Y.
22. Young, T. (1802). On the theory of light and colours. *Philosophical Transactions of the Royal Society of London, 92*, 12–48.
23. Young, T. (1804). Experiments and calculations relative to physical optics. *Philosophical Transactions of the Royal Society of London, 94*, 1–16.

# Chapter 2
# The Photon in Minkowski Spacetime

*The photon started it all, and all the things that Einstein had to do for it, and remodel space and time into a unified description of spacetime*

The quantum of light, the photon concept was born by Albert Einstein's description of the photoelectric effect in 1905[1] and named as such by Gilbert N. Lewis in 1926.[2]

In order to deal with particles that may travel at speeds comparable to the speed of light and of light itself, and develop a proper description for their dynamics at sufficiently high energies, it becomes essential to have a closer look at the description of space and time in which such particles live, and in which such processes occur. This inescapably leads to a unified description of space and time together as a pseudo-Euclidian four-dimensional space referred to as *Minkowski spacetime*.

In this chapter, we learn how such a spacetime description, unifying space and time, naturally arises, and gives rise to the special theory relativity, as well as of the equivalence of matter and energy, and quite importantly how a concept of *relativistic invariance*, to be discussed below, guides us in setting up dynamical equations of critical importance in the study of electrodynamics.[3]

We consider only those aspects of the subject which are relevant to our description of classical electrodynamics and radiation theory, in the simplest possible case.[4] The subject is essential, and I will spell it out for you.

To the above end, and see how such a spacetime description naturally arises, we first introduce the concept of an inertial frame.

---

[1] Einstein (1905) [8], with English translation.

[2] Lewis (1926) [14].

[3] It is rather surprising that after the subject was introduced 121 years ago, at the time of writing this book, even some senior physicists have not mastered the subject. The answer I always get, regarding this, is: "We never use it".

[4] The reader who is interested in a complete and detailed study of the special theory of relativity and of the concept of Minkowski spacetime, may consult my earlier textbook treatments of the subject: Manoukian (2020, 2024) [15, 16].



E. B. Manoukian, *Radiation*,
https://doi.org/10.1007/978-3-032-21506-2_2

A frame in which a particle which is initially at rest in it, remains at rest, or a particle in it which is initially in uniform motion continues in uniform motion, unless acted upon by an external force, is referred to as an *inertial frame* (IF) of reference. An IF in uniform motion w.r.t. an IF is also an IF. Taken as a physical law, IFs should be all equivalent in describing physical laws, referred to as the *principle of relativity*, and that the speed of light, in vacuum, is the same in all IFs, independent of the speed of its source and independent of its direction of propagation. The verifications of these properties of light were carried out by various people over the years and, in particular, by G. C. Babcock, and T. G. Bergman,[5] T. Alväger et al.[6] in 1964, R. O. Waddoups, W. F. Edwards, and J. J. Merrill,[7] P. Beckmann and P. Mandics[8] in 1965, and more recently by K. Brecher[9] in 1977, and V. V. Ragulsky[10] in 1997.

The principle of relativity and the universality of the speed of light in IFs constitute the postulates of the theory of *special relativity* SR, introduced by Albert Einstein in 1905.[11] If the speed of light determined in two different IFs were different this would allow one to distinguish between the two IFs thus violating the principle of relativity.

The modern definition is that the speed of light c is an absolute constant $c = 2.99792458 \times 10^8\,\mathrm{m/s}$ with the meter defined as the distance traveled by light in $10^{-8}/2.99792458$ of a second.

Spacetime consists of all possible events. An event $E$, such as an explosion, in spacetime occurs independently of any coordinate system that one may set up to investigate it.

In a given coordinate system, associated with a given inertial frame, an event $E$ is labeled by coordinates $(t, \mathbf{r})$ stating when and where the event has occurred *in* the given coordinate system. Different inertial frames give different labeling for the same event in spacetime. The relationship which exists between such different labeling in these different coordinate systems, is provided by the so-called *Lorentz transformations* (LT).[12]

We will see that the Galilean transformations are recovered from the LTs in the limit as c is taken to infinity. An IF is also referred to as a *Lorentz frame* in which, by definition, the SR theory holds.

The different realization of so-called *four*-vectors in Minkowski spacetime, as described in a *four*-dimensional spacetime, in different LF, are again related by LT. This allows us to readily satisfy the first postulate of ST in constructing physical

---

[5] Babcock and Bergman (1964) [4].

[6] Alväger et al. (1964) [3].

[7] Waddoups, Edwards and Merrill (1965) [19].

[8] Beckmann and Mandics (1965) [5].

[9] Brecher (1977) [6].

[10] Ragulsky (1997) [18].

[11] Einstein (1905) [9], see also Einstein et al. (1952) [10], Chapter III: On the Electrodynamics of Moving Bodies. pp. 35–65, as well as Pauli (1958) [17] and French (1968) [12].

[12] For the possible contribution of Einstein's wife Mileva, to the theory of SR, see Esterson et al. (2020) [11].

laws which have the same form in every LF resulting simply by merely relabeling of the underlying variables involved. The set of all possible such four-vectors, as just described, generate a vector space.

The set of all events together with the generated vector space is what one calls the *Minkowski spacetime* which is equipped with a *scalar product* $UV$ for any given two four-vectors $U$ and $V$. This scalar product will be spelled out subsequently as it emerges from the theory. With the above brief introduction, let us spell out the underlying LT of SR first.

## 2.1 The Lorentz Transformations and Concept of Invariance

In this section, we study the Lorentz transformations LT and discuss the important concept of invariance. The importance of this concept is that the invariance of an expression means it has the *same* value in .every inertial frame. For example the speed of light is an invariant as it has the same numerical value in every inertial frame. The derivation of the LT is given in the following section. Although the derivation is not important in the rest of the book, we strongly advise the reader to go through it.

Consider the simplest possible case of two inertial frames F and F′ such that initially the coordinate axes $(x, y, z)$ and $(x', y', z')$, respectively, and their origins 0, 0′, coincide. Now suppose, the latter frame moves with a constant velocity $\mathbf{v}$ relative to the first.

A particle's position is being investigated in the two inertial frames, and if the time/space coordinate label determined in F is $(t, \mathbf{r})$ in the corresponding coordinate system, then in the frame F′ it is denoted this by $(t', \mathbf{r}')$ , when two experimentalists, in question,stationary in their respective frames, are both right at the particle.[13] The basic problem now is to find the relationship between these two (time,space) coordinate labels.

If the particle moves, the time space coordinate labels, on the respective coordinate systems, for infinitesimal changes, will be labelled as

$$\mathrm{F}: \ \big(t, \mathbf{r}\big), \ \big(t + \mathrm{d}t, \mathbf{r} + \mathrm{d}\mathbf{r}\big), \qquad \mathrm{F}': \ \big(t', \mathbf{r}'\big), \ \big(t' + \mathrm{d}t', \mathbf{r}' + \mathrm{d}\mathbf{r}'\big). \quad (2.1.1)$$

[13] In relativity, it is essential to assume that there is an experimenter (recording device) at every point in a coordinate system all the time, so that an experimenter, stationary in one frame, will always meet, with the passage of time, a new, i.e., different, experimenter, stationary in another inertial frame, that is moving relative to the other, all the time.

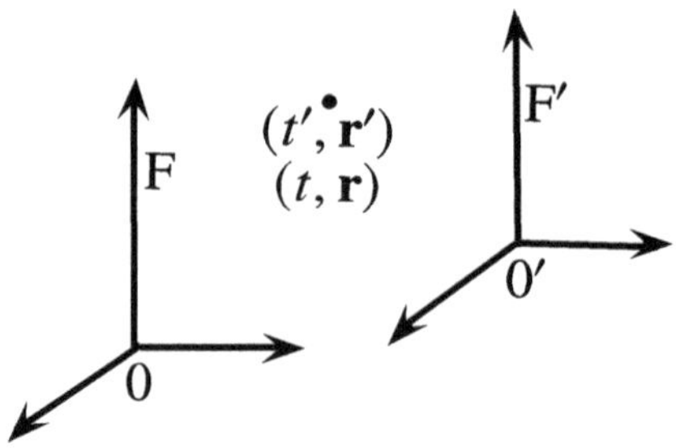


We assume that a functional relationship $\mathbf{r}' = \mathbf{r}'(\mathbf{r}, t),\ t' = t'(\mathbf{r}, t)$ exists between the variables in question. The chain rule of calculus gives

$$\mathrm{d}\mathbf{r}' = \sum_{j=1}^{3} \frac{\partial \mathbf{r}'}{\partial x^j}\,\mathrm{d}x^j + \frac{\partial \mathbf{r}'}{\partial t}\,\mathrm{d}t,\quad \mathrm{d}t' = \sum_{j=1}^{3} \frac{\partial t'}{\partial x^j}\,\mathrm{d}x^j + \frac{\partial t'}{\partial t}\,\mathrm{d}t,\qquad \mathbf{r} = (x^1, x^2, x^3). \tag{2.1.2}$$

Quite generally, we may write

$$\mathrm{d}\mathbf{r}' = A\,\mathrm{d}\mathbf{r} + \big(B\,\mathrm{d}t + C\,\mathbf{v}\cdot\mathrm{d}\mathbf{r}\big)\mathbf{v},\qquad \mathrm{d}t' = D\,\mathrm{d}t + E\,\mathbf{v}\cdot\mathrm{d}\mathbf{r}, \tag{2.1.3}$$

where $\mathbf{r}$, $\mathbf{v}$, $\mathrm{d}\mathbf{r}$ are the only available 3-vectors at hand in F, with $\mathrm{d}\mathbf{r}$ and $\mathrm{d}t$ the only infinitesimal quantities. In deriving the LT of SR, one assumes spacetime is homogeneous, referred to as the *homogeneity of spacetime*. That is, one assumes that all points in spacetime are equivalent to carry out an experiment. This means, in turn, that the coefficients $A$ to $E$ must be independent of $\mathbf{r}$ and $t$. The directional dependence of $\mathrm{d}\mathbf{r}'$ in Eq. (2.1.3) has already been taken into account, and the assumed property of spacetime taken as *isotropic* means that the coefficients in question cannot depend on a particular direction of the velocity $\mathbf{v}$ vector.[14]

That is, these coefficients, in general, may depend only on the speed v and may be expressed as functions of $\mathrm{v}^2$. The inverse transformation to Eq. (2.1.3) may be simply written by replacing $\mathbf{v}$ by $-\mathbf{v}$ and using the same coefficients $A$ to $E$ since they are invariant under the replacement $\mathbf{v}$ by $-\mathbf{v}$. The speed of light, of course, is considered as a fixed number. That is,

$$\mathrm{d}\mathbf{r} = A\,\mathrm{d}\mathbf{r}' - \big(B\,\mathrm{d}t' - C\,\mathbf{v}\cdot\mathrm{d}\mathbf{r}'\big)\mathbf{v},\qquad \mathrm{d}t = D\,\mathrm{d}t' - E\,\mathbf{v}\cdot\mathrm{d}\mathbf{r}', \tag{2.1.4}$$

By applying the postulates of SR, mentioned in the introductory section, gives the following transformation rules, i.e., the Lorentz transformations, in question, as derived in the Appendix in the next section:

$$\mathrm{d}\mathbf{r}' = \mathrm{d}\mathbf{r} - \gamma\,\mathbf{v}\,\mathrm{d}t + \frac{(\gamma - 1)}{\mathrm{v}^2}\big(\mathbf{v}\cdot\mathrm{d}\mathbf{r}\big)\,\mathbf{v},\qquad \gamma = \frac{1}{\sqrt{1 - \mathrm{v}^2/\mathrm{c}^2}} \tag{2.1.5}$$

---

[14] For an experiment on the isotropy of space, see, e.g., Chupp et al. (1989) [7].

$$dt' = \gamma\Big(dt - \frac{\mathbf{v}\cdot d\mathbf{r}}{c^2}\Big), \tag{2.1.6}$$

where $\gamma$ is referred to as the *Lorentz factor*.

If the origins of both coordinate systems coincide at $t = 0,\ t' = 0$, as mentioned above, the above equations are readily integrated to give,

$$\mathbf{r}' = \mathbf{r} - \gamma\,\mathbf{v}\,t + \frac{(\gamma-1)}{v^2}\big(\mathbf{v}\cdot\mathbf{r}\big)\,\mathbf{v}, \qquad t' = \gamma\Big(t - \frac{\mathbf{v}\cdot\mathbf{r}}{c^2}\Big). \tag{2.1.7}$$

Note that for $v^2/c^2 \to 0,\ \gamma \to 1$, and we recover the so-called *Galilean transformations* of Newtonian mechanics: $\mathbf{r}' = \mathbf{r} - \mathbf{v}t\ \ t' = t$.

An important invariance property is given below in the third equation which follows from the first two equations above it

$$(\mathbf{r}')^2 = (\mathbf{r})^2 + (\gamma^2 - 1)\frac{(\mathbf{v}\cdot\mathbf{r})^2}{v^2} - 2\gamma^2\,\mathbf{v}\cdot\mathbf{r}\,t + \gamma^2\,v^2 t^2, \tag{2.1.8}$$

$$(ct')^2 = \gamma^2 (ct)^2 - 2\gamma^2\,\mathbf{v}\cdot\mathbf{r}\,t + \gamma^2\,\frac{(\mathbf{v}\cdot\mathbf{r})^2}{c^2}, \tag{2.1.9}$$

$$(\mathbf{r}')^2 - (ct')^2 = (\mathbf{r})^2 - (ct)^2, \qquad \Big(\frac{\gamma^2 - 1}{v^2} \equiv \frac{\gamma^2}{c^2}\Big), \tag{2.1.10}$$

which, of course, has the fundamental built in property of the constancy of the speed light in different inertial frames.[15] This invariance property generalizes invariance under a rotation in Euclidean space for which $\mathbf{r}'^2 = \mathbf{r}^2,\ t' = t$.

To make the invariance property of a given equation evident, we introduce the following standard notation for time and space variables: $\big(\mathbf{r} = (x^1, x^2, x^3)\big)$

$$x^0 = c\,t, \quad x^i \ \text{for the space components, with Latin index}\ i = 1, 2, 3, \tag{2.1.11}$$

and due to the minus sign in Eq. (2.1.10), one may introduce a diagonal matrix $[\eta_{\mu\nu}] = \text{diag}\,[-1, 1, 1, 1]$, with the Greek indices $\mu, \nu$ running over $\{0, 1, 2, 3\}$, and rewrite (2.1.10) in the compact form:

$$\eta_{\mu\nu}\, x'^{\mu} x'^{\nu} = \eta_{\mu\nu}\, x^{\mu} x^{\nu}, \tag{2.1.12}$$

using the Einstein summation convention to sum over upper and lower identical indices. The inverse of the matrix $[\eta_{\mu\nu}]$ is given by $[\eta^{\nu\sigma}]$, which coincides with $[\eta_{\nu\sigma}]$, such that $\eta_{\mu\nu}\eta^{\nu\sigma} = \delta_{\mu}{}^{\sigma}$, $[\delta_{\mu}{}^{\sigma}] = \text{diag}[1, 1, 1, 1]$. The matrix $[\eta_{\mu\nu}]$ is called the *Minkowski metric* and generalizes the Euclidean metric $[\delta_{ij}]$, with the Latin indices $i, j$ going over $1, 2, 3$, with invariance under coordinate rotation given by $\delta_{ij}\, x'^i x'^j = \delta_{ij}\, x^i x^j$.

---

[15] Some textbooks, start from this last invariant equality to discuss LT.

We note the following key equations that follow upon setting $\mu = 0$ and $\mu = i$, in turn, in the metric $\eta_{\mu\nu}$, and then sum over the remaining index

$$\eta_{0\nu}x^{\nu} = \eta_{00}x^{0} = -x^{0} \equiv x_0, \qquad \eta_{i\,\nu}x^{\nu} = \eta_{ij}\,x^{j} = \delta_{ij}\,x^{j} = +x^{i} \equiv x_i. \tag{2.1.13}$$

That is, the metric may lower an upper index changing an upper index to a lower index: $\eta_{\mu\nu}x^{\nu} = x_{\mu}$. These, in turn, imply that we may simply write $x^0 = -x_0$, and $x^i = x_i$ for the space (Latin) indices. The Lorentz invariance statement in Eq. (2.1.12) may be then also simply rewritten as $x'_{\mu}x'^{\mu} = x_{\sigma}x^{\sigma}$. On the other hand, one may raise a lower index via the inverse matrix: $\eta^{\mu\nu}x_{\nu} = x^{\mu}$.

Summarising then, an event in spacetime that occurs at $\mathbf{r}$ at time $t$, in a given inertial frame F, with a corresponding coordinate system, in relativity it is labeled by:

$$x^{\mu} = (x^0, x^i),\ x^0 = \mathrm{c}t,\ x^i = \mathbf{r}^i \equiv x^i,\ i = 1, 2, 3,\ x_0 = -x^0,\ x_i = x^i, \tag{2.1.14}$$

From Eq. (2.1.7), the Lorentz transformations may be then conveniently rewritten as

$$x'^{\mu} = \Lambda^{\mu}{}_{\nu}\,x^{\nu}, \tag{2.1.15}$$

$$\Lambda^{i}{}_{j} = \delta^{i}{}_{j} + (\gamma - 1)\frac{\mathrm{v}^i\,\mathrm{v}_j}{\mathrm{v}^2},\quad \Lambda^{0}{}_{0} = \gamma,\ \ \Lambda^{0}{}_{i} = -\gamma\,\mathrm{v}_i/\mathrm{c},\ \ \Lambda^{i}{}_{0} = -\gamma\,\mathrm{v}^i/\mathrm{c}, \tag{2.1.16}$$

$$\gamma = (1 - \mathrm{v}^2/\mathrm{c}^2)^{-1/2},\ \mathrm{v}^i = \mathrm{v}_i. \tag{2.1.17}$$

Moreover, the scalar product in Minkowski spacetime, in a given inertial frame, with corresponding coordinate system, is defined by

$$x^{\mu}x_{\mu} \equiv x^i x^i - x^0 x^0 \equiv \mathbf{r}\cdot\mathbf{r} - \mathbf{c}^2 t^2, \tag{2.1.18}$$

with the Einstein convention of a summation over upper and lower repeated indices, understood, where $i = 1, 2, 3$, and the scalar product $x^{\mu}x_{\mu}$ is *invariant*, i.e.,

$$\mathrm{d}x'^{\mu}\mathrm{d}x'_{\mu} = \mathrm{d}x^{\mu}\mathrm{d}x_{\mu}, \tag{2.1.19}$$

i.e., it has the same value in every inertial frame, with corresponding coordinate system as discussed above.

The recorded time $t$ corresponding to an event in spacetime in a coordinate system, associated with a given frame F, is referred to as a *coordinate time*. On the other hand, the clock reading in the object's rest frame, is referred to as proper time denoted by $\tau$. Accordingly, from Eq. (2.1.19) for infinitesimals $\mathrm{d}x^{\mu}$, $\mathrm{d}x'^{\mu}$,

$$(\mathrm{d}t')^2\bigg(\Big(\frac{\mathrm{d}\mathbf{r}'}{\mathrm{d}t'}\Big)^2 - \mathrm{c}^2\bigg) = (\mathrm{d}t)^2\bigg(\Big(\frac{\mathrm{d}\mathbf{r}}{\mathrm{d}t}\Big)^2 - \mathrm{c}^2\bigg), \tag{2.1.20}$$

where F′ is the proper frame of the particle, i.e., the frame moving relative F in which it is at rest. Hence by setting $\mathrm{d}\mathbf{r}'/\mathrm{d}t' = 0$, $\mathrm{d}t' = \mathrm{d}\tau$, we have

$$-\,\mathrm{c}^2\mathrm{d}\tau^2 = \mathrm{d}x^\mu \mathrm{d}x_\mu, \qquad \Rightarrow \mathrm{d}\tau = \mathrm{d}t\sqrt{1-\Big(\frac{\mathrm{d}\mathbf{r}}{\mathrm{c}\mathrm{d}t}\Big)^2}, \tag{2.1.21}$$

showing that $\mathrm{d}\tau < \mathrm{d}t$, i.e., the time elapsed in the rest frame of the moving object is shorter than in the frame F. This is referred to as *time dilation*.[16] Note that $|\mathrm{d}\mathbf{r}/\mathrm{d}t|$ denotes the speed of the object in F. Accordingly, for $|\mathrm{d}\mathbf{r}/\mathrm{d}t| \to \mathrm{c}$, time stays still in such a moving frame as for the photon itself.

One may also consider measuring the length of a moving object in a given inertial frame F.

To the above end, consider an object stationary in a frame F′ with the latter moving relative to F, as described before. When one makes a measurement of, say, a moving stick one has to locate both end points of the stick *simultaneously*, otherwise the stick will move in the mean time and a recording of the other end will not provide the actual length measured in the frame F. Accordingly, consider the expression obtained from Eq. (2.1.7),

$$(\mathbf{r}'_2 - \mathbf{r}'_1) = (\mathbf{r}_2 - \mathbf{r}_1) - \gamma\,\mathbf{v}\,(t_2 - t_1) + \frac{(\gamma - 1)}{\mathrm{v}^2}\big(\mathbf{v}\cdot(\mathbf{r}_2 - \mathbf{r}_1)\big)\,\mathbf{v}. \tag{2.1.22}$$

Therefore, consider a stick with $y_\parallel$ denoting the length measured in F along the direction of $\mathbf{v}$. Accordingly, the above equation, *with* $t_2 = t_1$, gives

$$\begin{aligned} y'_\parallel &= y_\parallel + +\frac{(\gamma-1)}{\mathrm{v}^2}(y_\parallel \mathrm{v})\mathrm{v} \\ &= \gamma\, y_\parallel. \end{aligned} \tag{2.1.23}$$

$y'_\parallel \equiv L_0$ denotes the length of the stick attached to the moving frame, hence it is stationary in it, and is referred to as its *proper* length, while $L = y_\parallel$ denotes the length measured in F in which the stick is in motion. From the above equation,

$$L_0 = \frac{1}{\sqrt{1-\mathrm{v}^2/\mathrm{c}^2}}\,L, \quad \text{that is} \quad L = \sqrt{1-\mathrm{v}^2/\mathrm{c}^2}\,L_0 < L_0, \tag{2.1.24}$$

showing a length contraction in the measurement of a stick in motion.

Clearly there is no length contraction in the length measured in a direction perpendicular to $\mathbf{v}$, because $\mathbf{v}\cdot(\mathbf{r}_2 - \mathbf{r}_1)$ will be zero in Eq. (2.1.22).

The interest in this observation, is that from Eq. (2.1.23), by conveniently defining a 3 dimensional volume element in F by $\mathrm{d}y_\parallel \mathrm{d}y^1_\perp \mathrm{d}y^2_\perp$, i.e.,

---

[16] In a well known classic experiment (Frisch and Smith (1963) [13]) involving the muon decay $\mu^- \to e^- + \nu_\mu + \tilde{\nu}_e$, fewer decays were observed with the moving muons than expected.

$$\mathrm{d}y'_{\|}\mathrm{d}y'^{1}_{\perp}\mathrm{d}y'^{2}_{\perp} = \mathrm{d}V_0, \qquad \mathrm{d}y_{\|}\mathrm{d}y^{1}_{\perp}\mathrm{d}y^{2}_{\perp} = \mathrm{d}V, \tag{2.1.25}$$

we obtain

$$\mathrm{d}V = \sqrt{1 - \mathrm{v}^2/\mathrm{c}^2}\,\mathrm{d}V_0, \tag{2.1.26}$$

and providing an expression for the three-dimensional volume element defined relative to one given in a moving frame–a fundamental result that will be used in electrodynamics. An application of this will be given in Sect. 2.4, Eq. (2.4.7).

Most importantly, what happens to the four-dimensional volume element, defined below with *round* brackets, from a Lorentz transformation?

$$(\mathrm{d}x) \equiv \mathrm{d}x^0\mathrm{d}x^1\mathrm{d}x^2\mathrm{d}x^3. \tag{2.1.27}$$

From the definition of the Jabobian of a transformation

$$(\mathrm{d}x') = \left|\det\left[\frac{\partial x'^{\mu}}{\partial x^{\nu}}\right]\right|(\mathrm{d}x), \tag{2.1.28}$$

and Eq. (2.1.15),

$$\frac{\partial x'^{\mu}}{\partial x^{\nu}} = \Lambda^{\mu}{}_{\nu}, \qquad |\det[\Lambda^{\mu}{}_{\nu}]| = +1, \tag{2.1.29}$$

with the latter equality established in the Appendix, in the following section in Eq. (2.2.25), we obtain the invariance property of the four-dimensional volume element in Minkowski spacetime:

$$(\mathrm{d}x') = (\mathrm{d}x). \tag{2.1.30}$$

One may, in turn, define four-vectors, e.g., $(U, V)$, such that in *given* coordinate systems may have coordinates $(U'^{\mu}, V'^{\mu})$, $(U^{\mu}, V^{\mu})$ such that;

$$U'^{\mu} = \Lambda^{\mu}{}_{\nu}\, U^{\nu}, \ \ V'^{\mu} = \Lambda^{\mu}{}_{\nu}\, V^{\nu}, \ \ UV \equiv U^i V^i - U^0 V^0, \ \ U'V' = UV. \tag{2.1.31}$$

with the last equality stating the invariance of a scalar product. Examples of such four-vectors which will be given in Sect. 2.3.

The above Lorentz transformations are referred to as "Homogeneous Lorentz Transformations". One may consider further generalizations such as having two initial frames with corresponding coordinate systems initially not coinciding, with displaced origins, and even with relative rotated axes. Such generalizations will not be needed in the sequel.

## 2.2 Appendix: Derivation of the Lorentz Transformations

This Appendix may be omitted in a first reading, and the derivation of the Lorentz transformations is optional, but we strongly advise the reader to go through it eventually.

As described in the previous section, consider two inertial frames F and F′ such that the latter moves with a velocity **v** relative to the first frame. Two experimenters, stationary in their respective inertial frames, both right where a given particle is, label their respective clocks reading and the position of the particle, in their respective coordinate systems, by $(t, \mathbf{r})$, $(t', \mathbf{r}')$.

Suppose the particle travels from one spacetime time point to another spacetime point. By invoking homogeneity and isotropy of spacetime, meaning that an experiment carried out anywhere in space and at any time is equivalent, we have seen in Eqs. (2.1.3), (2.1.4), that for infinitesimal changes in the particle's labelings, we may write

$$\mathrm{d}\mathbf{r}' = A\,\mathrm{d}\mathbf{r} + \big(B\,\mathrm{d}t + C\,\mathbf{v}\cdot\mathrm{d}\mathbf{r}\big)\mathbf{v}, \qquad \mathrm{d}t' = D\,\mathrm{d}t + E\,\mathbf{v}\cdot\mathrm{d}\mathbf{r}, \tag{2.2.1}$$

$$\mathrm{d}\mathbf{r} = A\,\mathrm{d}\mathbf{r}' - \big(B\,\mathrm{d}t' - C\,\mathbf{v}\cdot\mathrm{d}\mathbf{r}'\big)\mathbf{v}, \qquad \mathrm{d}t = D\,\mathrm{d}t' - E\,\mathbf{v}\cdot\mathrm{d}\mathbf{r}', \tag{2.2.2}$$

where $A$, $B$, $C$, $D$ may, in general, be functions of $\mathrm{v}^2$.

Upon considering first the components of (2.2.1), (2.2.2) perpendicular to **v**, we obtain

$$\mathrm{d}\mathbf{r}'_\perp = A(\mathrm{v}^2)\,\mathrm{d}\mathbf{r}_\perp, \quad \mathrm{d}\mathbf{r}_\perp = A(\mathrm{v}^2)\,\mathrm{d}\mathbf{r}'_\perp, \tag{2.2.3}$$

from which $A^2(\mathrm{v}^2) = 1$, or $A(\mathrm{v}^2) = +1$, and by using the boundary condition that for $\mathbf{v} = 0$, we have $A(0) = +1$. Thus we may write

$$\mathrm{d}\mathbf{r}' = \mathrm{d}\mathbf{r} + \big(B\,\mathrm{d}t + C\,\mathbf{v}\cdot\mathrm{d}\mathbf{r}\big)\mathbf{v}, \qquad \mathrm{d}t' = D\,\mathrm{d}t + E\,\mathbf{v}\cdot\mathrm{d}\mathbf{r}, \tag{2.2.4}$$

$$\mathrm{d}\mathbf{r} = \mathrm{d}\mathbf{r}' - \big(B\,\mathrm{d}t' - C\,\mathbf{v}\cdot\mathrm{d}\mathbf{r}'\big)\mathbf{v}, \qquad \mathrm{d}t = D\,\mathrm{d}t' - E\,\mathbf{v}\cdot\mathrm{d}\mathbf{r}', \tag{2.2.5}$$

These equations should *better* give us the infinitesimal changes in space and time variables corresponding to particles as determined in the two frames.

For a particle at rest in F, $\mathrm{d}\mathbf{r} = \mathbf{0}$, and the particle would move with velocity $\mathrm{d}\mathbf{r}'/\mathrm{d}t' = -\mathbf{v}$ in F′. Accordingly, from the two equations in Eq. (2.2.4), we, respectively, have

$$-\,\mathbf{v} = \frac{\mathrm{d}\mathbf{r}'}{\mathrm{d}t'} = B\,\mathbf{v}\frac{\mathrm{d}t}{\mathrm{d}t'}, \quad \mathrm{d}t' = D\,\mathrm{d}t, \quad \Rightarrow \quad D = -\,B. \tag{2.2.6}$$

On the other hand, for a particle at rest in F′, $\mathrm{d}\mathbf{r}' = \mathbf{0}$, and the first equation in Eq. (2.2.4) gives

$$\mathbf{0} = \frac{\mathrm{d}\mathbf{r}}{\mathrm{d}t} + \Big(B + C\,\mathbf{v}\cdot\frac{\mathrm{d}\mathbf{r}}{\mathrm{d}t}\Big)\mathbf{v} = \mathbf{v} + \big(B + C\,\mathrm{v}^2\big)\mathbf{v}, \ \Rightarrow\ C = -\frac{(1+B)}{\mathrm{v}^2}. \tag{2.2.7}$$

Moreover, in this latter case, the second equations in Eqs. (2.2.5) and (2.2.4) together, in turn, give

$$\mathrm{d}t = D\,\mathrm{d}t' = D\,\mathrm{d}t\big(D + E\,\mathrm{v}^2\big), \ \Rightarrow\ D\big(D + E\,\mathrm{v}^2\big) = 1, \tag{2.2.8}$$

$$\Rightarrow\ B\big(B - E\,\mathrm{v}^2\big) = 1, \qquad E = \frac{B^2 - 1}{B\,\mathrm{v}^2}. \tag{2.2.9}$$

Therefore, from Eqs. (2.2.4), (2.2.7), (2.2.9), we may express everything in terms of $B$ as follows

$$\mathrm{d}\mathbf{r}' = \mathrm{d}\mathbf{r} + \Bigg(B\,\mathrm{d}t - \frac{(1+B)}{\mathrm{v}^2}\,\mathbf{v}\cdot\mathrm{d}\mathbf{r}\Bigg)\mathbf{v}, \ \ \mathrm{d}t' = -B\,\mathrm{d}t + \Big(\frac{B^2-1}{B\,\mathrm{v}^2}\Big)\,\mathbf{v}\cdot\mathrm{d}\mathbf{r}. \tag{2.2.10}$$

Two problems remain. We have to invoke the equality of the speed of light as determined in both frames. This will allow us to determine the only unknown $B$ and, in turn, show that the B.C. at $\mathbf{v} = \mathbf{0}$ is satisfied.

We have the explicit expression following from Eq. (2.2.10)

$$\frac{\mathrm{d}\mathbf{r}'}{\mathrm{d}t'} = \frac{\mathrm{d}\mathbf{r}/\mathrm{d}t + \Bigg(B - (1+B)/\mathrm{v}^2\,\mathbf{v}\cdot\mathrm{d}\mathbf{r}/\mathrm{d}t\Bigg)\mathbf{v}}{-B + \Big(B^2 - 1/B\,\mathrm{v}^2\Big)\,\mathbf{v}\cdot\mathrm{d}\mathbf{r}/\mathrm{d}t}. \tag{2.2.11}$$

Now consider light traveling in a direction perpendicular to $\mathbf{v}$ in F, with $\mathrm{d}\mathbf{r}/\mathrm{d}t = \mathbf{c}$. This gives from

$$\frac{\mathrm{d}\mathbf{r}'}{\mathrm{d}t'} = -\frac{\mathrm{d}\mathbf{r}/\mathrm{d}t + B\,\mathbf{v}}{B}, \tag{2.2.12}$$

for the components of the speed of light in frame F′ perpendicular and parallel, respectively, to $\mathbf{v}$, the expressions

$$\mathbf{c}'_{\perp} = -\frac{\mathbf{c}}{B}, \ \ \mathbf{c}'_{\parallel} = -\mathbf{v}, \ \Rightarrow\ (\mathbf{c}'_{\perp})^2 + (\mathbf{c}'_{\parallel})^2 = \frac{\mathrm{c}^2}{B^2} + \mathrm{v}^2 \equiv \mathrm{c}^2, \tag{2.2.13}$$

$$\Rightarrow B^2 = \frac{1}{(1 - \mathrm{v}^2/\mathrm{c}^2)}, \qquad B = -\frac{1}{\sqrt{1 - \mathrm{v}^2/\mathrm{c}^2}}, \tag{2.2.14}$$

with a minus sign, so that for $\mathrm{v} \to 0$,

$$\frac{(1+B)}{\mathrm{v}^2} \rightarrow -\frac{1}{2\mathrm{c}^2}, \qquad \left(\frac{1}{\sqrt{1-\mathrm{v}^2/\mathrm{c}^2}} \simeq 1+\frac{\mathrm{v}^2}{2\mathrm{c}^2}\right), \tag{2.2.15}$$

leads to a finite expression, and from Eq. (2.2.10), in this limit, $\mathbf{dr}' = \mathbf{dr}$.

All told, we have from Eq. (2.2.10),

$$\mathbf{dr}' = \mathbf{dr} - \gamma\,\mathbf{v}\,\mathrm{d}t + \frac{(\gamma-1)}{\mathrm{v}^2}\big(\mathbf{v}\cdot\mathbf{dr}\big)\,\mathbf{v}, \qquad \gamma = \frac{1}{\sqrt{1-\mathrm{v}^2/\mathrm{c}^2}} \tag{2.2.16}$$

$$\mathrm{d}t' = \gamma\Big(\mathrm{d}t - \frac{\mathbf{v}\cdot\mathbf{dr}}{\mathrm{c}^2}\Big), \tag{2.2.17}$$

where $\gamma$ is referred to as the *Lorentz factor*.

If the origins of both coordinate systems coincide at $t=0,\ t'=0$, then the above equations are readily integrated to give,

$$\mathbf{r}' = \mathbf{r} - \gamma\,\mathbf{v}\,t + \frac{(\gamma-1)}{\mathrm{v}^2}\big(\mathbf{v}\cdot\mathbf{r}\big)\,\mathbf{v}, \qquad t' = \gamma\Big(t - \frac{\mathbf{v}\cdot\mathbf{r}}{\mathrm{c}^2}\Big). \tag{2.2.18}$$

Before closing this Appendix, we derive, in particular, an important property of the transformation matrix $\Lambda$.

Using the convenient matrix multiplication notation, $\mathrm{d}x^\top = [\mathrm{d}x^0\, \mathrm{d}x^1, \mathrm{d}x^2, \mathrm{d}x^3]$, $\eta = [\eta_{\nu\mu}]$, where $\top$ denotes the transpose of a matrix, we may write

$$\mathrm{d}x' = \Lambda\,\mathrm{d}x, \tag{2.2.19}$$

$$\mathrm{d}x'_\mu\,\mathrm{d}x'^\mu = \mathrm{d}x'^\nu\eta_{\nu\mu}\,\mathrm{d}x'^\mu = \mathrm{d}x'^\top\eta\,\mathrm{d}x' = \mathrm{d}x^\top\big(\Lambda^\top\eta\,\Lambda\big)\,\mathrm{d}x. \tag{2.2.20}$$

Therefore, invoking Lorentz invariance, as given, in Eq. (2.1.19)

$$\mathrm{d}x'^\top\eta\,\mathrm{d}x' = \mathrm{d}x^\top\,\eta\,\mathrm{d}x, \tag{2.2.21}$$

we have from Eq. (2.2.20) the basic equation

$$\Lambda^\top\eta\,\Lambda = \eta, \tag{2.2.22}$$

Using the basic facts that for two square matrices $A$ and $B$,

$$\det A\,B = \det A \det B, \qquad \det A^\top = \det A, \tag{2.2.23}$$

and $\det\eta = -1$, we obtain from Eq. (2.2.22),

$$\det\Big(\Lambda^\top\eta\,\Lambda\Big) = \det\eta, \;\Rightarrow\; (\det\Lambda)^2 = 1 \tag{2.2.24}$$

and for the special Lorentz transformations considered, as may be explicitly verified,

$$\det \Lambda \;=\; +1. \tag{2.2.25}$$

The solution with a minus sign corresponds to time or space reflection, which is not being considered here.

For the keen reader who is interested in the transformation property of the four-dimensional derivative,

$$\frac{\partial}{\partial x^{\nu}} \;\equiv\; \partial_{\nu}. \tag{2.2.26}$$

it is readily established. Upon multiplying it by the inverse of the Minkowski metric $[\eta^{\nu\mu}] = \mathrm{diag}[-1, 1, 1, 1]$, introduced below Eq.(2.1.12) having the same matrix elements as $[\eta_{\nu\mu}]$, by $\partial'_{\nu} \equiv \partial/\partial x'^{\nu}$, used in another inertial frame F$'$, we obtain

$$\eta^{\mu\nu}\partial'_{\nu} \;=\; \partial'^{\mu}, \qquad \partial'^{\mu} \;=\; \Lambda^{\mu}{}_{\nu}\,\partial^{\nu}, \tag{2.2.27}$$

where the explicit elements of the matrix $[\Lambda^{\mu}{}_{\nu}]$ have been spelled out in Eqs. (2.1.16), (2.1.17). We may, in turn, multiply the above equation by $\eta_{\sigma\mu}$, use the identity $\eta_{\sigma\mu}\eta^{\mu\nu} = \delta_{\sigma}{}^{\nu}$, with the latter denoting the Kronecker delta $= 1$ for $\sigma = \nu$, and zero otherwise, to obtain the explicit transformation law of a derivative:

$$\partial'_{\sigma} \;=\; \big(\eta_{\sigma\mu}\Lambda^{\mu}{}_{\nu}\,\eta^{\nu\beta}\big)\partial_{\beta}, \tag{2.2.28}$$

involving matrix multiplication. Of course, in applications one does not have to explicit carry out these matrix multiplications as long as one is aware of the given transformation law.

## 2.3 Four-Momentum, Four-Velocity and $E = mc^2$

In relativity, the three dimensional velocity you are familiar with has to be generalized to a four-vector, whose interpretation will easily follow. With $\tau$ as an invariant, one may define the four velocity vector $U$, in spacetime, with components: $U^{\mu} = \mathrm{d}x^{\mu}/\mathrm{d}\tau$, $U'^{\mu} = \mathrm{d}x'^{\mu}/\mathrm{d}\tau, \ldots$, in different coordinate systems. The components are related by a Lorentz transformation $U'^{\mu} = \Lambda^{\mu}{}_{\nu}U^{\nu}$, such that

$$U^{\mu}U_{\mu} \;=\; U'^{\mu}U'_{\mu} \;=\; -\,\mathrm{c}^2, \tag{2.3.1}$$

as follows directly from Eq. (2.1.19) and the first equation in Eq. (2.1.21), and is an invariant, with $-\mathrm{c}^2$ just a numerical constant.

What is the significance of $U^0$ ? or more precisely what is the significance of the more relevant object $mU^0$ ?, where $m$ is the mass of a particle. From Eq. (2.1.14), it is explicitly given by

$$m\,U^0 = m\,\frac{\mathrm{d}x^0}{\mathrm{d}\tau} = m\mathrm{c}\frac{\mathrm{d}t}{\mathrm{d}\tau}. \tag{2.3.2}$$

In detail, with $\mathbf{dr}/\mathrm{d}t = \mathbf{v}$, denoting the speed of a particle, we have from the second equation in Eq. (2.1.21)

$$\frac{\mathrm{d}t}{\mathrm{d}\tau} = \frac{1}{\sqrt{1-\frac{\mathrm{v}^2}{\mathrm{c}^2}}} \equiv \gamma = 1 + \frac{1}{2}\frac{\mathrm{v}^2}{\mathrm{c}^2} + \cdots, \tag{2.3.3}$$

expanded in powers of v. Accordingly,

$$m\,U^0 = m\,\mathrm{c}\gamma = \frac{1}{\mathrm{c}}\Big[m\mathrm{c}^2 + \frac{1}{2}m\mathrm{v}^2 + \cdots\Big], \quad \frac{\mathbf{dr}}{\mathrm{d}t} = \mathbf{v}, \quad \gamma = \frac{1}{\sqrt{1-\mathrm{v}^2/\mathrm{c}^2}}. \tag{2.3.4}$$

We recognize $m\mathrm{v}^2/2$ as the non-relativistic kinetic energy of the particle, and hence $mc^2$, the first term within the square brackets above, is interpreted as a rest energy arising by putting $\mathrm{v} = 0$ in it, and thus $\mathrm{c}m\,U^0 \equiv E$ is the total energy of the particle which includes its rest energy $E_0 = m\,\mathrm{c}^2$ (the famous **Einstein equation**).

We may thus define the energy-momentum of a particle, or more conveniently the four-momentum, of a massive particle of mass $m$, by

$$p^\mu = (p^0, \mathbf{p}),\ p^0 = m\,U^0 = \frac{E}{\mathrm{c}},\ E = \gamma m \mathrm{c}^2, \quad (U^\mu) = \Big(\frac{\mathrm{d}x^\mu}{\mathrm{d}\tau}\Big) = (\gamma\mathrm{c}, \gamma\mathbf{v}), \tag{2.3.5}$$

$$\mathbf{p} = m\mathbf{U},\ p^2 \equiv p^\mu p_\mu = m^2 U^\mu U_\mu = -m^2\mathrm{c}^2, \tag{2.3.6}$$

$$\Rightarrow\ \mathbf{p}^2 - (p^0)^2 \equiv p^2 = -m^2\mathrm{c}^2, \quad E = \sqrt{\mathbf{p}^2\mathrm{c}^2 + m^2\mathrm{c}^4}, \tag{2.3.7}$$

and the mass $m$ of a particle is an invariant,[17] and where we have used Eq. (2.1.21), and the equalities in (2.3.1) in writing the last equality in Eq. (2.3.6).

We may then summarize the expressions for the four-momentum of a massive particle to be given by[18] ($U = (\gamma\,\mathrm{c}, \gamma\,\mathbf{v})$)

$$p = \Big(\frac{E}{\mathrm{c}}, \mathbf{p}\Big),\ E = \frac{m\mathrm{c}^2}{\sqrt{1-\frac{\mathrm{v}^2}{\mathrm{c}^2}}} = \sqrt{\mathbf{p}^2\mathrm{c}^2 + m^2\mathrm{c}^4},\ p^\mu p_\mu = -m^2\mathrm{c}^2, \tag{2.3.8}$$

What about the above expressions for a massless particle as for our star here, the *photon*?. In this case one may use the Einstein equation of a photon energy $E = \mathrm{h}\nu$,

---

[17] The old terminology of $\gamma m$ being the relativistic mass, and $m$ the rest one, has essentially disappeared in the modern literature.

[18] Recall $\mathbf{p} = m\mathbf{U} = m\,\mathbf{v}/\sqrt{1-\mathrm{v}^2/\mathrm{c}^2}$.

where $\nu$ is its frequency, and h is Planck's constant, to write

$$E = \mathrm{h}\nu, \quad \mathbf{p} = \frac{\mathrm{h}\nu}{\mathrm{c}}\mathbf{n}, \quad |\mathbf{n}| = 1, \;\; E = \mathrm{c}|\mathbf{p}|, \;\; p^{\mu} p_{\mu} = 0. \tag{2.3.9}$$

## 2.4 How to Make Use of Special Relativity

We have learnt in Eq. (2.1.31), that two vectors $U$, $V$, represented by components $U^{\mu}$, $V^{\mu}$, when in the inertial frame F, the scalar product $UV$ is Lorentz invariant, i.e., it has the same value in other inertial frames, as defined in Sect. 2.1, one has

$$UV \equiv U^i V^i - U^0 V^0, \;\; U'V' = UV. \tag{2.4.1}$$

In the same way as introducing a vector $V$, with components $V^{\mu}$, in a given inertial frame, with Lorentz transformation $V'^{\mu} = \Lambda^{\mu}{}_{\nu} V^{\nu}$, as given in Eq. (2.1.31), we may introduce an object, which in a given coordinate system has components $T^{\mu\nu}$, involving two indices, referred to as a second rank tensor, which transforms as the product of two four-vectors, i.e.,

$$T'^{\mu\nu} \;=\; \Lambda^{\mu}{}_{\lambda}\, \Lambda^{\nu}{}_{\sigma}\, T^{\lambda\sigma}, \tag{2.4.2}$$

with the Einstein convention of summing over upper and lower repeated indices understood. The indices of the tensor may be lowered with the Minkowski metric as before for vector components. Higher rank tensors may be similarly defined.

Such a typical tensor is the Faraday tensor $F^{\mu\nu}$, which will be encountered in Chap. 5, whose space components $F^{ij}$, with $i, j = 1, 2, 3$ are, most probably, familiar to the reader.

These vectors and tensors are extremely useful as they guide one to set up dynamical equations, such as, for a given second rank tensor and a vector,

$$\partial_{\mu} T^{\mu\nu} \;+\; V^{\nu} \;=\; 0, \tag{2.4.3}$$

where $\partial_{\mu}$ defined Eq. (2.2.26), is given by

$$\frac{\partial}{\partial x^{\nu}} \;\equiv\; \partial_{\nu}. \tag{2.4.4}$$

Equation (2.4.3) has exactly the same form in any other inertial frame, having just relabeled vectors and tensors related to the above ones by Lorentz transformations, i.e.,

$$\partial'_{\mu} T'^{\mu\nu} \;+\; V'^{\nu} \;=\; 0, \tag{2.4.5}$$

where, in particular, $\partial_\mu T^{\mu\nu}$ defines the components of a vector in a given inertial frame.

In Minkowski spacetime, we have learnt in Eq. (2.1.30) that the four-dimensional volume defined below, with round brackets, is invariant,

$$(\mathrm{d}x) \equiv \mathrm{d}x^0\mathrm{d}x^1\mathrm{d}x^2\mathrm{d}x^3. \qquad (\mathrm{d}x') = (\mathrm{d}x), \tag{2.4.6}$$

while we have learnt in Eq. (2.1.26) that a 3-dimensional volume element $\mathrm{d}V$ relative to one determined as $\mathrm{d}V_0$ in a frame moving relative to the former one, is given by

$$\mathrm{d}V = \sqrt{1-\mathrm{v}^2/\mathrm{c}^2}\,\mathrm{d}V_0, \tag{2.4.7}$$

A most fundamental 4-vector in electrodynamics is the current. A 3-current, i.e, with 3 components, $\mathbf{J}$ is defined such that $\mathbf{J}\cdot\mathbf{n}\,\mathrm{d}S$ is the rate at which charge passes through the element of surface element $\mathrm{d}S$, of the volume $V$, in a direction of the unit vector $\mathbf{n}$ perpendicular to it.

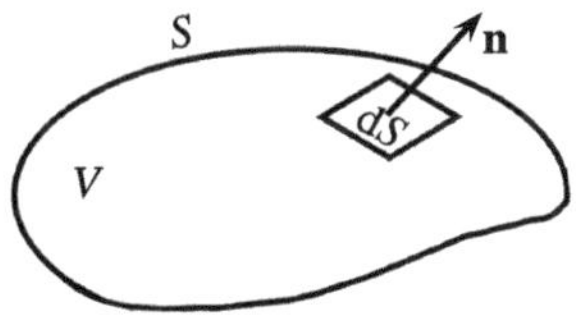


That is, if $\rho$ denotes the charge density, then the rate of charge that passes through the closed surface $S$ from a small volume $V$ is given by

$$\int_S \mathbf{J}\cdot\mathbf{n}\,\mathrm{d}S = -\frac{\partial}{\partial t}\int_V \mathrm{d}^3\mathbf{x}\,\rho, \tag{2.4.8}$$

emphasizing that a decrease of charge inside the volume $V$, with time, corresponds to the leakage of charge through the surface $S$.

Using Gauss' Theorem which states that

$$\int_S \mathbf{J}\cdot\mathbf{n}\,\mathrm{d}S = \int_V \mathrm{d}^3\mathbf{x}\,\nabla\cdot\mathbf{J}, \tag{2.4.9}$$

gives

$$\int_V \mathrm{d}^3\mathbf{x}\Big[\nabla\cdot\mathbf{J} + \frac{\partial}{\partial t}\rho\Big] = 0. \tag{2.4.10}$$

In the limit of $V \to 0$, about a point $\mathbf{x}$, we obtain the continuity equation

$$\frac{\partial}{\partial t}\rho + \nabla\cdot\mathbf{J} = 0. \tag{2.4.11}$$

We have to recast this equation into a "relativistic form", referred to as in a covariant form, ensuring its validity in every inertial frame, and, in particular, find the expression of the 4-current in electrodynamics.

To the above end, consider an infinitesimal portion of charge $\mathrm{d}q = \rho_0 \mathrm{d}V_0$, within a volume $\mathrm{d}V_0$, instantaneously at rest, with corresponding charge density $\rho_0$. An argument by the legendary Wolfgang Pauli[19] is that there is no reason that such an infinitesimal charge leaks outside the corresponding volume $\mathrm{d}V$ when the infinitesimal charge $\mathrm{d}q$ is set in motion (with velocity $\mathbf{v}$), implying the invariance of the infinitesimal charge

$$\rho_0 \mathrm{d}V_0 \;=\; \rho \mathrm{d}V \;\equiv\; \mathrm{d}q, \qquad \mathrm{d}V = \sqrt{1 - \frac{\mathrm{v}^2}{\mathrm{c}^2}}\,\mathrm{d}V_0, \qquad \Rightarrow \qquad \rho = \gamma \rho_0, \tag{2.4.12}$$

by using, in the process, Eq. (2.4.7). Moreover, by definition $\mathbf{J} = \rho \mathbf{v}$. We may thus write

$$\rho \;=\; \rho_0 \frac{U^0}{\mathrm{c}}, \quad \mathbf{J} = \rho_0\, \mathbf{U} = \rho_0 \gamma\, \mathbf{v}, \qquad U^\mu = (\gamma \mathrm{c}, \gamma \mathbf{v}), \quad \gamma = \frac{1}{\sqrt{(1 - \mathrm{v}^2/\mathrm{c}^2)}}, \tag{2.4.13}$$

$$\text{and set} \quad J^\mu \;=\; \big(\rho \mathrm{c}, \mathbf{J}\big). \tag{2.4.14}$$

where the expression for the four-velocity $U$, with coordinate components $(\mathrm{c}\gamma, \gamma \mathbf{v})$, is given in Eq. (2.3.5).

Therefore, as before in Eq. (2.1.14), with $x^0 = ct$, $\partial_0 = \partial/\partial x^0 = \partial/(\partial(ct)$, $\partial^i = \partial/\partial x^i$, and from Eq. (2.4.14), $J^0 = c\rho$, Eq. (2.4.11) simply becomes

$$\partial_\mu J^\mu(x) \;=\; 0, \qquad x = (\mathrm{c}t, \mathbf{x}). \tag{2.4.15}$$

## 2.5 Photons: Their Transmission in Space and the Seeing Mechanism

Photons interact with both positive and negative charges, they are unbiased; bipartisans. They do not carry a charge and because of this, they do not interact directly with one another, i.e., they essentially don't see one another. [20] Imagine how deformed an object will be seen, if they interact with one another before reaching your eyes. Moreover, they are readily transmitted across space exhibiting, fortunately for signal transmission, this "indifference". These are excellent demonstrations of this particular property of photons.

[19] Pauli (1958) [17].

[20] One may call a photon a *xenophile*, so to speak.

As a matter of fact, in a quantum description, the probability that two photons would feel the presence of each other and, as a result, their momentum-directions be altered, by the exchange of an electron-positron pair, is of the order $(\alpha)^4 \sim (1/137)^4$, and is indeed very small, as inferred from the amplitude corresponding to figure (b) below, involving a coupling of order of $\sqrt{\alpha}$, at each of the four points indicated of exchange, where $\alpha \cong 1/137$ is the fine structure constant:

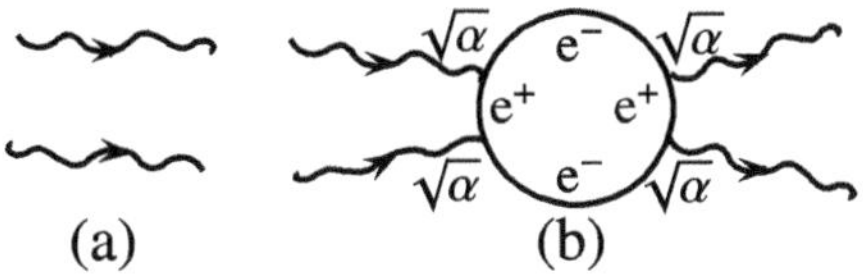


Another process $\gamma + \gamma \to \mathrm{e}^+ + \mathrm{e}^-$, occurs with probability of the order $\alpha^2$ according to QED. Both these processes are not only small as predicted by explicit computation in the quantum description, in that experimentally, they are verified under very restricted condition as the detection of the outgoing photons is done indirectly by their interactions with other systems. Detailed experimental studies of these processes are given by the two critical references given below,[21] and are highly recommended.

That is, as far as classical electrodynamics is concerned, photons are "indifferent" to one another. On the other hand, we will learn, through propagator theory, how a photon scatters off obstacles and different media.

## References

1. Aad, G. et al. ATLAS Collaboration (2019). Observation of light-by-light scattering in ultraperipheral Pb+Pb collisions with the ATLAS detector. *Physical Review Letters, 123*, 052001.
2. Adams, J. et al. STAR Collaboration (2021). Measurement of $\mathrm{e}^+\,\mathrm{e}^-$ momentum and angular distributions from linearly polarized photon collisions. *Physical Review Letters, 127*, 052302.
3. Alväger, T., et al. (1964). Measuring the velocity of light emitted by fast sources, using accelerated particles. *Physics Letters, 12*, 260–262.
4. Babcock, G. C., & Bergman, T. G. (1964). Determination of the constancy of the speed of light. *Journal of the Optical Society of America, 54*, 147–151.
5. Beckmann, P., & Mandics, P. (1965). Test of the constancy of the velocity of electromagnetic radiation in high vacuum. *Radio Science Journal of Research, 69D*, 623–628.
6. Brecher, K. (1977). Is the speed of light independent of the source ? *Physical Review Letters, 39*, 1051–1054. [Erratum: ibid., 1236.]
7. Chupp, T. E., et al. (1989). Results of a new test of local Lorentz invariance: A search for mass anisotropy in 21Ne. *Physical Review Letters, 63*, 1541–1545.
8. Einstein, A. (1905). Über einen die Erzeugung und Verwandlung des Lichtes betreffenden heuristischen Gesichtspunkt. *Annalen der Physik, 17 (6)*, 132-148. [English Translation: Arons, A. B., & Peppard, M. B. (1965). Einstein's proposal of the photon concept. *American Journal of Physics, 33 (5)*, 367–374.
9. Einstein, A. (1905). Zur elektrodynamik bewegter Körper. *Annalen der Physik, 17*, 891–921

[21] Aad et al. ATLAS Collaboration (2019) [1]; and Adams et al. STAR Collaboration (2021) [2].

10. Einstein, A., Lorentz, H. A. , Weyl, H., & Minkowski, H. (1952). The principle of relativity. New York: Dover Publications.
11. Esterson, A., Cassidy, D. C., & Sime, R. L. (2020). *Einstein's wife: The real story of Mileva Einstein-Maric*. Boston: The MIT Press.
12. French, A. P. (1968). *Special relativity* (M.I.T. Introductory Physics). Boston: The MIT Press.
13. Frisch, D. H., & Smith, J. H. (1963). Measurement of relativistic time dilation using $\mu$-mesons. *American Journal of Physics, 31*, 342–355.
14. Lewis, G. N. (1926). The conservation of photons. *Nature 118*, 874–875.
15. Manoukian, E. B. (2020). *100 years of fundamental theoretical physics in the palm of your hand: Integrated technical treatment*. Switzerland: Springer.
16. Manoukian, E. B. (2024). *How energy considerations have shaped our fundamental modern theories of physics: Theory and key historical moments*. Switzerland: Springer.
17. Pauli, W. (1958). *Theory of relativity*. London: Pergamon Press.
18. Ragulsky, V. V. (1997). Determination of light velocity dependence on direction of propagation, *Physics Letters A, 235*, 125–128.
19. Waddoups, R. O., Edwards, W. F., & Merrill, J. J. (1965). Experimental investigation of the second postulate of special relativity. *Journal of the Optical Society of America, 55*, 142–143.

# Chapter 3
# The Photon and the Vacuum-to-Vacuum Transition Amplitude $< 0_+ \mid 0_- >$

*The vavuum-to-vacuum transition amplitude and the photon propagator are derived while photons are emitted from and detected by current distributions. Their flow in spacetime is described, and current-current interactions of classical electrodynamics pop up*

The fine-structure constant $\alpha$ is a dimensionless quantity which provides a measure of the strength of the coupling of the interaction of the electron, or charged particles, in general, and the photon. According to the Particle Data Group (PDG),[1] (2024), a recommended numerical value of $\alpha$ is

$$\frac{1}{\alpha} = 137.035999084. \tag{3.1}$$

In rationalized CGS units, in which, in particular, $\hbar$, and the speed of light c (in vacuum), are set equal to one, which we adopt, the magnitude of the electron charge $|\mathrm{e}|$ is simply, given by

$$|\mathrm{e}| = \sqrt{4\pi\alpha}, \tag{3.2}$$

and is dimensionless. That is, charge is dimensionless in a units with $\hbar$ and c set equal to one. We will elaborate further on rationalized CGS units, when we encounter phenomenological parameters needed to describe radiation in matter in Chap. 5, Sect. 5.4.

The photon is a massless particle of spin one. That is, a photon of energy-momentum $(|\mathbf{k}|, \mathbf{k})$, may have associated with it two, mutually perpendicular, polarization vectors $\mathbf{e}_1$, $\mathbf{e}_2$ perpendicular to its 3-momentum $\mathbf{k}$:

$$\mathbf{e}_1 \cdot \mathbf{k} = 0 = \mathbf{e}_2 \cdot \mathbf{k}, \qquad \mathbf{e}_1 \cdot \mathbf{e}_2 = 0, \tag{3.3}$$

[1] Navas, et al. PDG [4].



E. B. Manoukian, *Radiation*,
https://doi.org/10.1007/978-3-032-21506-2_3

and for

$$\mathbf{k} = |\mathbf{k}|(\sin\theta\cos\phi, \sin\theta\sin\phi, \cos\theta), \tag{3.4}$$

we may take,

$$\mathbf{e}_1(k) = (\cos\theta\cos\phi, \cos\theta\sin\phi, -\sin\theta), \tag{3.5}$$

$$\mathbf{e}_2(k) = (-\sin\phi, \cos\phi, 0). \tag{3.6}$$

For the vector $\mathbf{k}$ along a $z-$ axis, we may, conveniently write,

$$\mathbf{k} = |\mathbf{k}|(0, 0, 1), \quad \mathbf{e}_1(k) = (1, 0, 0), \quad \mathbf{e}_2(k) = (0, 1, 0). \tag{3.7}$$

For a covariant description, we set $\mathrm{e}^{\mu}_{\lambda}(k) = \big(0, \mathbf{e}_\lambda(k)\big)$, with $\lambda$ denoting a given polarization state.

We have learnt in the previous chapter in Sect. 2.4, that one may introduce the four electromagnet current $J^\mu(x)$ in Eq. (2.4.14), as a function of the space time variable $x = \big(t, \mathbf{x}\big)$, with $x^0 = t$, $\partial_0 = \partial/\partial t$, such that

$$J^\mu(x) = (\rho(x), \mathbf{J}(x)), \quad \text{satisfying the conservation law,} \tag{3.8}$$

$$\partial_\mu J^\mu(x) = 0, \quad \Rightarrow \quad \frac{\partial\rho}{\partial t} + \nabla\cdot\mathbf{J} = 0, \tag{3.9}$$

as given in Eqs. (2.4.15), (2.4.11), where, in particular, $\rho$ defines the charge distribution in consideration.

For a photon, with $k = (k^0, \mathbf{k})$, $k^0 = |\mathbf{k}|$, one has the mass-shell restriction

$$k^2 \equiv \mathbf{k}^2 - (k^0)^2 = 0, \quad \text{with } k^0 > 0. \tag{3.10}$$

satisfied.

Accordingly, the infinitesimal volume element in the *energy-momentum* space description, with a given three momentum $\mathbf{k}$, may be defined by

$$\frac{(\mathrm{d}k)}{(2\pi)^3}\delta(k^2)\theta(k^0), \quad \text{where } (\mathrm{d}k) \equiv \mathrm{d}k^0\mathrm{d}^3\mathbf{k}, \tag{3.11}$$

$$\theta(k^0) = 1, \quad \text{for } k^0 > 0, \quad \text{and } 0 \text{ otherwise}, \tag{3.12}$$

and $\delta(k^2)$ is the Dirac delta distribution (see Appendix Sect. 3.6, Eq. (3.6.14), at the end of the chapter) which, in particular, satisfies the invariant relation

$$\delta\big(\mathbf{k}^2 - (k^0)^2\big) = \frac{1}{2|\mathbf{k}|}\Big[\delta\big(k^0 - |\mathbf{k}|\big) + \delta\big(k^0 + |\mathbf{k}|\big)\Big]. \tag{3.13}$$

Therefore, for $k^0 > 0$, only the first term above contributes, and the volume element in the energy-momentum space description, corresponding to a given three vector $\mathbf{k}$, may be now expressed as a 3-dimensional integration variable, after singling out the value for $k^0 = |\mathbf{k}|$, to be given simply by the expression recorded here:

$$\frac{\mathrm{d}^3\mathbf{k}}{(2\pi)^3}\frac{1}{2|\mathbf{k}|}. \tag{3.14}$$

Now we are ready to derive the expression of the vacuum-to-vacuum transition amplitude $< 0_+ \mid 0_- >$,[2] which is of a monumental importance in the present work.

## 3.1 $< 0_+ \mid 0_- >$: Emergence of the Poisson Distribution

To carry out our treatment in the energy-momentum space description, it is necessary to carry out a Fourier transform of the current $J^\mu(x)$ as follows

$$J^\mu(k^0,\mathbf{k}) \equiv J^\mu(k) = \int (\mathrm{d}x)\,\mathrm{e}^{-\mathrm{i}k_\mu x^\mu}\, J^\mu(x^0,\mathbf{x}), \quad \text{where} \quad (\mathrm{d}x) = \mathrm{d}t\,\mathrm{d}^3\mathbf{x}. \tag{3.1.1}$$

We recall that the three dimensional integral of $J^0(x) = \rho(x)$ gives the total charge which is dimensionless,[3] i.e., $\rho(x)$ has the dimension of $[length]^{-3}$, while $(\mathrm{d}x)$, in the integral above, has the dimension of $[length]^4$. That is, $J^\mu(k^0,\mathbf{k})$, in the energy-momentum description, on the left-hand side of the above equation has the dimension of *[length]*.

We may denote the state of a system under consideration before any current distribution is introduced into the system by $|0_->$ involving no photons. We have at our disposal three four vectors $k^\mu$, $J^\mu(k)$ and $\mathrm{e}^\mu_\lambda(k)$, where $k = (k^0,\mathbf{k})$, and the invariant measure $(\mathrm{d}k)\delta(k^2)\theta(k^0)/(2\pi)^3$, for a photon with energy-mentum $k = (|\mathbf{k}|,\mathbf{k})$ and polarization states specified by $\lambda = 1$ or $2$.

We make the following observations:

1. When there is no current $J^\mu(k)$, we have no photons.
2. The *invariant* product $\mathrm{e}^\mu_\lambda J_\mu(k)$ has the dimension of *[length]*.
3. The measure $\mathrm{d}^3\mathbf{k}/\big((2\pi)^3 2|\mathbf{k}|\big)$ corresponding to a photon of momentum $\mathbf{k}$ has the dimension of *[length]*$^{-2}$.
4. $k^\mu k_\mu = 0$, $k^\mu \mathbf{e}_{\lambda\mu} = 0$, $k^\mu J_\mu(k) = 0$.

In 4., the first relation is just $\mathbf{k}^2 - k^{02} = 0$, the second relation in is $\mathbf{k}\cdot\mathbf{e}_\lambda = 0$, and the last relation is just the conservation law $\partial^\mu J_\mu = 0$ in the energy-momentum description, i.e., obtained upon Fourier transformation.

Accordingly, from the points 1.–4. above, we may define the probability that a photon is created in a state $|\mathbf{k},\lambda>$, upon the introduction of the current distribution $J^\mu(k)$, to be given by

---

[2] Schwinger (1970) [6].

[3] Recall we are working in units for which $\hbar$ and c are set equal to one, see Eqs. (3.1), (3.2).

$$\big| < \mathbf{k}, \lambda | 0_- > \big|^2 \propto \big| \mathrm{e}^{\mu}_{\lambda}(k) J_\mu(k) \big|^2 \frac{\mathrm{d}^3\mathbf{k}}{(2\pi)^3} \frac{1}{2|\mathbf{k}|} \equiv \left| \mathrm{e}^{\mu}_{\lambda}(k) J_\mu(k) \sqrt{\frac{\mathrm{d}^3\mathbf{k}}{(2\pi)^3} \frac{1}{2|\mathbf{k}|}} \right|^2, \tag{3.1.2}$$

involving necessarily both the polarization vector and the current, up to a dimensionless proportionality constant to be determined, witnessing the power of relativity at work. It is convenient to introduce the following notation used frequently by Schwinger[4] which is especially useful for book-kipping purposes:[5]

$$\mathrm{i}\, \mathrm{e}^{\mu}_{\lambda}(k) J_\mu(k) \sqrt{\frac{\mathrm{d}^3\mathbf{k}}{(2\pi)^3} \frac{1}{2|\mathbf{k}|}} \equiv \mathrm{i}\, J_{k\lambda}, \tag{3.1.3}$$

as an amplitude of the creation of a photon, by the given current distribution, with attributes $\mathbf{k}$, $\lambda$, up to a proportionality constant to be determined. The i factor is introduced for convenience.

In a convenient discrete notation, we may enumerating all of possible momenta-polarization taken by a photon as follows:

$$(\mathbf{k}_1, \lambda_1), (\mathbf{k}_2, \lambda_2)(\mathbf{k}_3, \lambda_3), \ldots,$$

Using the Bose character of photons that any number of them may be put in the same state of a given momentum-polarization pair, we may infer that the probability that $n_1$ photons where created in the state $|\mathbf{k}_1, \lambda_1 >$, $n_2$ of them in the state $|\mathbf{k}_2, \lambda_2 >$, and so on is, up to a proportionality constant to be determined, is

$$| < n_1, n_2, \ldots | 0_- > |^2 \propto \frac{\big(|J_{k_1\lambda_1}|^2\big)^{n_1}}{n_1!} \frac{\big(|J_{k_2\lambda_2}|^2\big)^{n_2}}{n_2!} \cdots, \tag{3.1.4}$$

where we have taken into account of the indistinguishability of photons in the same state, giving rise to the divisions by the $n_j$ factorials. We may also infer that the corresponding amplitude to be given by

$$< n_1, n_2, \ldots | 0_- > \propto \frac{\big(\mathrm{i}\, J_{k_1\lambda_1}\big)^{n_1}}{\sqrt{n_1!}} \frac{\big(\mathrm{i}\, J_{k_2\lambda_2}\big)^{n_2}}{\sqrt{n_2!}} \cdots. \tag{3.1.5}$$

We know that if $n_1 = 0, n_2 = 0, \ldots$ the left-hand side of the above expression will be simply $< 0_+ \mid 0_- >$, where $|0_+ >$ is the vacuum, i.e., no photon, state after the current $J^\mu$ is introduced. Accordingly, with this boundary condition, we must have

---

[4] Schwinger (1970) [6].

[5] See, in particular, Eq. (3.1.9) to realize the importance of this notation.

$$< n_1, n_2, \ldots | 0_- > = < 0_+ \mid 0_- > \frac{\left(\mathrm{i}\, J_{k_1\lambda_1}\right)^{n_1}}{\sqrt{n_1!}} \frac{\left(\mathrm{i}\, J_{k_2\lambda_2}\right)^{n_2}}{\sqrt{n_2!}} \cdots . \tag{3.1.6}$$

Of particular interest is to obtain the probability that the current distribution $J^\mu$ generates $N = n = n_1 + n_2 + \ldots$ photons of arbitrary momenta and polarizations. From the above equation, this is simply given by

$$\mathrm{Prob}[N = n] = \Big| < 0_+ \mid 0_- > \Big|^2 \sum_{n_1+n_2+\ldots=n} \frac{\left(\left|J_{k_1\lambda_1}\right|^2\right)^{n_1}}{n_1!} \frac{\left(\left|J_{k_2\lambda_2}\right|^2\right)^{n_2}}{n_2!} \cdots . \tag{3.1.7}$$

The summation may be explicitly carried out by using the multinomial expansion formula:

$$\sum_{n_1+n_2+\ldots=n} \frac{(x_1)^{n_1}}{n_1!} \frac{(x_2)^{n_2}}{n_2!} \cdots = \frac{\left(x_1 + x_2 + \cdots\right)^n}{n!}. \tag{3.1.8}$$

Accordingly,

$$\begin{aligned} \mathrm{Prob}[N = n] &= \Big| < 0_+ \mid 0_- > \Big|^2 \frac{\left(\sum_{k\lambda} |J_{k\lambda}|^2\right)^n}{n!}. \\ &\equiv \Big| < 0_+ \mid 0_- > \Big|^2 \frac{\left(\int \frac{\mathrm{d}^3\mathbf{k}}{(2\pi)^3} \frac{1}{2|\mathbf{k}|} J^*_\mu(k) \big[\sum_\lambda \mathrm{e}^\mu_\lambda(k) \mathrm{e}^\nu_\lambda(k)\big] J_\nu(k)\right)^n}{n!}, \end{aligned} \tag{3.1.9}$$

where we have used Eq. (3.1.3). The reader, now, realizes the importance of introducing the notation used in Eq. (3.1.3) and for introducing a discrete set of momentum-polarization states so that the final step in the above may be explicitly carried out.

To evaluate the expression $\big[\sum_\lambda \mathrm{e}^\mu_\lambda(k) \mathrm{e}^\nu_\lambda(k)\big]$, note the following. The orthonormal vectors $\mathbf{e}_1, \mathbf{e}_2, \mathbf{k}/|\mathbf{k}|$ provide a complete set of vectors in 3D, i.e., we may write:

$$\delta^{ij} = \frac{\mathbf{k}^i}{|\mathbf{k}|} \frac{\mathbf{k}^j}{|\mathbf{k}|} + \sum_{\lambda=1,2} \mathbf{e}^i_\lambda \mathbf{e}^j_\lambda, \quad i, j = 1, 2, 3, \tag{3.1.10}$$

as a completeness relation. To find the covariant 4D corresponding to this completeness relation, we introduce, in the process, the four vectors: $k = (k^0, \mathbf{k})$, $\underline{k} = (k^0, -\mathbf{k})$, and note that

$$k^2 = \mathbf{k}^2 - (k^0)^2 = \underline{k}^2 = 0, \text{ and} \tag{3.1.11}$$

$$k\underline{k} \equiv -\mathbf{k}^2 - (k^0)^2 \equiv -2\,\mathbf{k}^2 \equiv -2\,(k^0)^2 \neq 0. \tag{3.1.12}$$

Also recall that, $\mathrm{e}^{\mu}_{\lambda} = (0, \mathbf{e}_{\lambda})$. We claim that the four dimensional (4D) completeness relation becomes

$$\eta^{\mu\nu} = \frac{k^{\mu}\underline{k}^{\nu} + \underline{k}^{\mu}k^{\nu}}{k\,\underline{k}} + \sum_{\lambda=1,2} \mathrm{e}^{\mu}_{\lambda}\mathrm{e}^{\nu}_{\lambda}, \tag{3.1.13}$$

involving, now, the Minkowski metric $\eta^{\mu\nu}$. To prove this, consider specific values of $\mu$ and $\nu$ as given below:

$$-1 = \eta^{00} = \frac{k^{02} + k^{02}}{-2k^{02}} + 0, \tag{3.1.14}$$

$$0 = \eta^{01} = \eta^{10} = 0 + 0, \tag{3.1.15}$$

$$\delta^{ij} = \eta^{ij} = \frac{-2k^i k^j}{-2\mathbf{k}^2} + \sum_{\lambda} \mathrm{e}^{i}_{\lambda}\mathrm{e}^{j}_{\lambda}, \tag{3.1.16}$$

with the latter coinciding with the 3D completeness relation already given in Eq. (3.1.10), thus establishing the proof of relation in Eq. (3.1.13).

From the complete relation Eq. (3.1.13), we get

$$\begin{aligned} J^{*}_{\mu}(k)\Big[\sum_{\lambda} \mathrm{e}^{\mu}_{\lambda}(k)\mathrm{e}^{\nu}_{\lambda}(k)\Big]J_{\nu}(k) &= J^{*}_{\mu}(k)\Big[\eta^{\mu\nu} - \frac{k^{\mu}\underline{k}^{\nu} + \underline{k}^{\mu}k^{\nu}}{k\,\underline{k}}\Big]J_{\nu}(k) \\ &= J^{*}_{\mu}(k)\eta^{\mu\nu}J_{\nu}(k) = J^{*}_{\mu}(k)J^{\nu}(k), \end{aligned} \tag{3.1.17}$$

where we have used the conservation law in the energy-momentum space description:

$$k^{\nu}J_{\nu}(k) = 0, \quad J^{*}_{\mu}(k)k^{\mu} = \big(k^{\mu}J_{\mu}(k)\big)^{*} = 0. \tag{3.1.18}$$

This ingenious way of expressing the 4D completeness relation is due to Schwinger.[6]

The probability in Eq. (3.1.9) becomes

$$\mathrm{Prob}[N = n] = \big|< 0_+ \mid 0_- >\big|^2 \frac{\Big(\int \frac{\mathrm{d}^3\mathbf{k}}{(2\pi)^3}\frac{1}{2|\mathbf{k}|}J^{*}_{\mu}(k)J^{\nu}(k)\Big)^n}{n!}, \tag{3.1.19}$$

Upon summing over all $n$ gives:

$$1 = \sum_{n=0}^{\infty} \mathrm{Prob}[N = n] = \big|< 0_+ \mid 0_- >\big|^2 \exp\Big[\int \frac{\mathrm{d}^3\mathbf{k}}{(2\pi)^3}\frac{1}{2|\mathbf{k}|}J^{*}_{\mu}(k)J^{\nu}(k)\Big], \tag{3.1.20}$$

[6] Schwinger [5].

where we have used the expansion formula $\sum_n (x)^n/n! = \exp(x)$. That is

$$\big| < 0_+ \mid 0_- > \big|^2 = \exp\Big[ - \int \frac{\mathrm{d}^3\mathbf{k}}{(2\pi)^3} \frac{1}{2|\mathbf{k}|} J^*_\mu(k) J^\nu(k) \Big], \tag{3.1.21}$$

and from Eq. (3.1.19)

$$\mathrm{Prob}[N = n] \; = \; \frac{\big( < N > \big)^n}{n!} \, \mathrm{e}^{-<N>}, \tag{3.1.22}$$

$$< N > \; = \; \int \frac{\mathrm{d}^3\mathbf{k}}{(2\pi)^3} \frac{1}{2|\mathbf{k}|} J^*_\mu(k) J^\nu(k), \tag{3.1.23}$$

giving rise to the celebrated Poisson distribution, with $< N >$ denoting the average number of photons created which arises as a consequence of the definition of an average as spelled out below,

$$\begin{aligned} \mathrm{Average}(N) \equiv \sum_{n=0}^{\infty} n\, \mathrm{Prob}[N = n] &= < N > \sum_{n=1}^{\infty} \frac{\big( < N > \big)^{(n-1)}}{(n-1)!} \, \mathrm{e}^{-<N>} \\ &= < N > \sum_{n=0}^{\infty} \frac{\big( < N > \big)^{n}}{n!} \, \mathrm{e}^{-<N>} \\ &= < N > \sum_{n=0}^{\infty} \mathrm{Prob}[N = n] \; = < N > . \end{aligned} \tag{3.1.24}$$

It is interesting to rewrite Eq. (3.1.21) in terms of $< N >$,

$$\big| < 0_+ \mid 0_- > \big|^2 \; = \; \mathrm{e}^{-<N>}. \tag{3.1.25}$$

It is remarkable, that upon taking the absolute value squared, we obtain in the exponent the average number of photons created by an arbitrary conserved current. We note that $< N >$ in Eq. (3.1.23) may be also re-written in configuration space by using directly the Fourier transform in Eq. (3.1.1) giving

$$< N > = \int (\mathrm{d}x) \int (\mathrm{d}x') J^\mu(x) J_\mu(x') \int \frac{\mathrm{d}^3\mathbf{k}}{(2\pi)^3} \frac{1}{2|\mathbf{k}|} \, \mathrm{e}^{\mathrm{i}k(x-x')}. \tag{3.1.26}$$

In the next section, we study the nature of $< 0_+ \mid 0_- >$ and derive its explicit expression.

## 3.2 The Explicit $< 0_+ \mid 0_- >$

As we are interested in deriving the expression for $< 0_+ \mid 0_- >$, ending with the situation that any photons that have been created are eventually re-absorbed by the current distribution, ending up again with the vacuum, we have to introduce the following amplitude of absorption:

$$< 0_+|n_1, n_2, \ldots > = \cdots \frac{\left(\mathrm{i}\, J^*_{k_2\lambda_2}\right)^{n_2}}{\sqrt{n_2!}} \frac{\left(\mathrm{i}\, J^*_{k_1\lambda_1}\right)^{n_1}}{\sqrt{n_1!}} < 0_+ \mid 0_- >, \tag{3.2.1}$$

with a $(+\mathrm{i})$ factor multiplying the complex conjugates $J^*_{k_j\lambda_j}$ for consistency with the minus sign in the exponent in Eqs. (3.1.25), and (3.1.6), as we shall see below. Equation (3.2.1) denotes the transition amplitude that all the created particles are re-absorbed by the current distribution ending by the vacuum state $|0_+ >$.

We consider the current $J^\mu(x)$, consisting of two parts:

$$J^\mu(x) = J_2^\mu(x) + J_1^\mu(x), \tag{3.2.2}$$

where $J_1^\mu(x)$, $J_2^\mu(x)$ are causally arranged such that the current $J_2^\mu(x)$ is switched on after $J_1^\mu(x)$ is switched off, i.e., removed. We may then write $< 0_+ \mid 0_- >$ explicitly as:

$$< 0_+ \mid 0_- >^J = < 0_+ \mid 0_- >^{J_2} \Big(\sum_n \sum_{n=n_1+n_2+\ldots} A(n_1, n_2, \ldots)\Big) < 0_+ \mid 0_- >^{J_1}, \tag{3.2.3}$$

$$A(n_1, n_2, \ldots) = < 0_+|n_1, n_2, \ldots >^{J_2} < n_1, n_2, \ldots|0_- >^{J_1}, \tag{3.2.4}$$

where the $n$ photons which have been created by $J_1$ were necessarily absorbed by $J_2$.

From the expressions in Eqs. (3.1.6), (3.2.1) and the covariant completeness relation in Eq. (3.1.13), we may explicitly carry out the expansion within the round brackets in Eq. (3.2.3), involving the transition amplitude $A$, giving

$$< 0_+ \mid 0_- >^{J_1+J_2} = < 0_+ \mid 0_- >^{J_2}\; T\; < 0_+ \mid 0_- >^{J_1} \equiv F[J_1 + J_2], \tag{3.2.5}$$

$$T = \exp\Big[ - \int (\mathrm{d}x) \int (\mathrm{d}x')\, J_2^\mu(x) J_{1\mu}(x') \int \frac{\mathrm{d}^3\mathbf{k}}{(2\pi)^3} \frac{1}{2|\mathbf{k}|}\, \mathrm{e}^{\mathrm{i}k(x-x')}\Big], \tag{3.2.6}$$

where $F[J_1 + J_2]$ is some functional of the *sum* of the two currents and must generate the $T$ term, involving the product $J_2 J_1$, and is readily determined. We note that the cross term $J_2 J_1$ occurs only in the middle exponential expression. The entire expression on the right-hand side of Eq. (3.2.5) will be a functional of the total current $J_1 + J_2$, if it has the general structure

$$F[J_1 + J_2] =$$
$$\exp\Big[\frac{i}{2}\int(dx)\int(dx')[J_1^{\mu}(x) + J_2^{\mu}(x)]D_+(x,x')[J_1^{\mu}(x') + J_2^{\mu}(x')], \qquad (3.2.7)$$

*such that* from Eqs. (3.2.5), (3.2.6),

$$\frac{i}{2}\int(dx)\int(dx')\big[J_2^{\mu}(x)D_+(x,x')J_1^{\mu}(x') + J_1^{\mu}(x)D_+(x,x')J_2^{\mu}(x')\big]$$
$$= -\int(dx)\int(dx')\,J_2^{\mu}(x)J_{1\mu}(x')\int\frac{d^3\mathbf{k}}{(2\pi)^3}\frac{1}{2|\mathbf{k}|}\,e^{ik(x-x')}, \qquad (3.2.8)$$

with $D_+(x,x')$ then worked out to be

$$\text{For } t > t' : D_+(x,x') = i\int\frac{d^3\mathbf{k}}{(2\pi)^3}\frac{1}{2|\mathbf{k}|}\,e^{i\mathbf{k}\cdot(\mathbf{x}-\mathbf{x}')}e^{-ik^0(t-t')}, \qquad (3.2.9)$$

$$\text{For } t < t' : D_+(x,x') = i\int\frac{d^3\mathbf{k}}{(2\pi)^3}\frac{1}{2|\mathbf{k}|}\,e^{-i\mathbf{k}\cdot(\mathbf{x}-\mathbf{x}')}e^{+ik^0(t-t')}. \qquad (3.2.10)$$

Clearly, we could have chosen the case that $J_2$ is switched off before $J_1$ is introduced, instead, and with the definition of $D(x,x') = D(x-x')$ in Eqs. (3.2.9)/(3.2.10), both cases may be covered. The general expression of the celebrated $< 0_+ \mid 0_- >$ then simply emerges as

$$< 0_+ \mid 0_- > = \exp\Big[\frac{i}{2}\int(dx)\int(dx')J^{\mu}(x)D_+(x-x')J_{\mu}(x')\Big]. \qquad (3.2.11)$$

In the next section, we investigate the physical meaning of $D_+(x-x')$.

## 3.3 Propagator and Energy Flow in Spacetime

In the present section, we provide the physical significance of the function $D_+(x-x')$ in Eqs. (3.2.9)/(3.2.10). This will, in turn, provide us two other equivalent forms for $< 0_+ \mid 0_- >$ given Eq. (3.2.11). To do this, we need to carry out an analysis involving integration in a complex plane.[7]

The function $D_+(x-x')$ may be written as a single integral in 4D combining the two equations for it mentioned above, in Eqs. (3.2.9), (3.2.10), in one equation, and most importantly its physical significance naturally emerges. We claim that it may be written as:

[7] Enough details of integration in a complex plane will be given here so that you wouldn't even need to consult books in complex variables.

$$D_+(x - x') = \int \frac{(\mathrm{d}k)}{(2\pi)^4} \frac{\mathrm{e}^{\mathrm{i}\,k(x-x')}}{(k^2 - \mathrm{i}\,\epsilon)}, \quad \text{where,} \tag{3.3.1}$$

$$(\mathrm{d}k) = \mathrm{d}^3\mathbf{k}\,\mathrm{d}k^0, \quad k(x - x') = \mathbf{k}\cdot(\mathbf{x} - \mathbf{x}') - k^0(t - t'), \quad k^2 = \mathbf{k}^2 - (k^0)^2, \tag{3.3.2}$$

$$\text{and the limit } \epsilon \to +0 \text{ (spelled out below) is to be taken.} \tag{3.3.3}$$

With $\epsilon$ small and positive, we may write in the denominator of

$$\frac{\mathrm{e}^{-\mathrm{i}\,k^0(t-t')}}{(k^2 - \mathrm{i}\epsilon)}, \quad \text{in Eq.(3.3.1) as,}$$

$$k^2 - \mathrm{i}\epsilon = \mathbf{k}^2 - k^{02} - \mathrm{i}\epsilon = -\Big(k^0 - (|\mathbf{k}| - \mathrm{i}\,\epsilon)\Big)\Big(k^0 - (-|\mathbf{k}| + \mathrm{i}\,\epsilon)\Big), \tag{3.3.4}$$

neglecting, in the process, a second order term involving $\epsilon^2$ on account of the fact that we eventually take the limit $\epsilon \to +0$, and $|\mathbf{k}| > 0$.

Equation (3.3.4), implies that $D_+(x - x')$ in Eq. (3.3.1) has two poles (singularities) in the complex $k^0$ energy plane, at

$$k^0 = |\mathbf{k}| - \mathrm{i}\,\epsilon, \quad k^0 = -|\mathbf{k}| + \mathrm{i}\,\epsilon, \quad \textit{with } \epsilon \to +0, \tag{3.3.5}$$

as shown in Fig. 3.1, and the approach of the above limit of $\epsilon$ is from positive values of the latter.

Consider the complex $k^0-$ plane. Write $k^0$ in terms of its real and imaginary parts: $k^0 = \mathcal{Re}(k^0) + \mathcal{Im}\,(k^0)$. For $t > t'$, $\exp[-\mathrm{i}\,k^0(t - t')] = \exp[-\mathrm{i}\,\mathcal{Re}(k^0)(t - t')]\ \times\exp[+\mathcal{Im}(k^0)(t - t')]$, which goes to zero for $\mathcal{Im}(k^0) \to -\infty$. That is, for $t > t'$, we may close the infinite semi circle contour of integration from *below*, picking up the pole at $k^0 = |\mathbf{k}|$. Similarly for $t < t'$, $\exp[-\mathrm{i}\,k^0(t - t')] = \exp[-\mathrm{i}\,\mathcal{Re}(k^0)(t - t')]\ \times\exp[+\mathcal{Im}(k^0)(t - t')]$, which now goes to zero for $\mathcal{Im}(k^0) \to +\infty$. That is, for $t < t'$, we may close the infinite semi circle contour of integration from *above*, picking up the pole at $k^0 = -|\mathbf{k}|$.

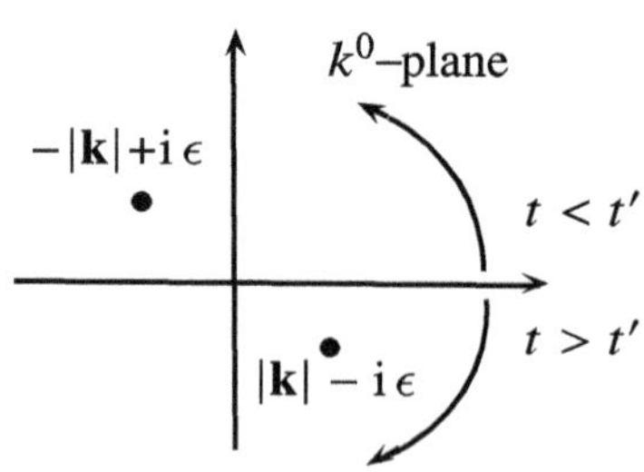


**Fig. 3.1** Figure showing the direction of the contour that is to be taken in the complex $k^0-$plane when applying the residue theorem, taken c.w. below the horizontal axis for $t > t'$, and c.c.w. above it for $t < t'$, generating, in turn, two semi-circles of infinite radii, one below and one above the horizontal axis, respectively, as described in the text

We are now ready to apply the **Residue Theorem** to the $k^0$-energy part of the integral for $D_+(x-x')$, for which the integral, in question, is equal to the sum of residues with a plus sign for a contour taken *counter*-clock-wise, and a negative sign for a contour taken clock-wise. For $t > t'$, the residue theorem, then gives (see Fig. 3.1)

$$\int \frac{\mathrm{d}k^0}{2\pi} \frac{\exp[-\mathrm{i}k^0(t-t']}{(k^0-|\mathbf{k}|)(k^0+|\mathbf{k}|)} = -\,\mathrm{i}\,\frac{\exp[-\mathrm{i}|\mathbf{k}|(t-t')]}{2|\mathbf{k}|}. \tag{3.3.6}$$

Similarly, for $t < t'$, we have

$$\int \frac{\mathrm{d}k^0}{2\pi} \frac{\exp[-\mathrm{i}k^0(t-t']}{(k^0-|\mathbf{k}|)(k^0+|\mathbf{k}|)} = +\,\mathrm{i}\,\frac{\exp[+\mathrm{i}|\mathbf{k}|(t-t')]}{-2|\mathbf{k}|}. \tag{3.3.7}$$

Hence

$$D_+(x-x') = \mathrm{i}\int \frac{\mathrm{d}^3\mathbf{k}}{(2\pi)^3\,2|\mathbf{k}|} \exp[\mathrm{i}\mathbf{k}\cdot(\mathbf{x}-\mathbf{x}')]\exp[-\mathrm{i}\,|\mathbf{k}||t-t'|], \tag{3.3.8}$$

which coincides with the corresponding expressions in Eqs. (3.2.9)/(3.2.10).

Accordingly, we may re-write the vacuum-to-vacuum transition amplitude, with $D_+$ as claimed in Eq. (3.3.1), in the following form

$$< 0_+ \mid 0_- > \; = \; \exp\Big[\frac{\mathrm{i}}{2}\int (\mathrm{d}x)\int (\mathrm{d}x')J^\mu(x)D_+(x-x')J_\mu(x')\Big]. \tag{3.3.9}$$

$$D_+(x-x') \; = \; \int \frac{(\mathrm{d}k)}{(2\pi)^4}\frac{\mathrm{e}^{\mathrm{i}\,k(x-x')}}{(k^2-\mathrm{i}\,\epsilon)}, \quad \epsilon \to +0. \tag{3.3.10}$$

It is easily verified by using, in the process the representation of the delta distribution in Eq. (3.6.24) in the Appendix at the end of the chapter, that $D_+(x-x')$ satisfies the partial differential equation

$$-\,\Box D_+(x-x') = \delta^{(4)}(x-x'), \quad \text{where } \Box = \boldsymbol{\nabla}^2 - \frac{\partial^2}{\partial t^2}, \tag{3.3.11}$$

and $\Box$ is referred to as the d'Alembertian..

Most importantly, $D_+(x-x')$ is referred to as the *photon propagator*, and as we have seen that *for* $t > t'$, $k^0 = +|\mathbf{k}|$ and energy flows from point $\mathbf{x}'$ in space to point $\mathbf{x}$, while for $t < t'$, the reverse situation occurs with energy loss $k^0 = -|\mathbf{k}|)$ from point $\mathbf{x}$ to point $\mathbf{x}'$. We will have ample opportunity to see how such an amount of energy is propagated, via the propagator, as an evolutionary process, in various physical situations as investigated in other chapters of the book.

We may also take advantage of the conservation law $\partial_i J^i(x) + \partial_0 J^0 = 0$, to rewrite $< 0_+ \mid 0_- >$ in another convenient form.

To the above end, we may write

$$\begin{aligned} J^{\mu}(x)J_{\mu}(x') &= J^{i}(x)J^{i}(x') - J^{0}(x)J^{0}(x') \\ &= J^{i}(x)J^{i}(x') - \Big(\frac{\partial^{i}}{\partial_0}J^{i}(x)\Big)\Big(\frac{\partial'^{j}}{\partial'_0}J^{j}(x')\Big), \end{aligned} \tag{3.3.12}$$

which allows us to write, $i, j = 1, 2, 3$ with a summation over repeated indices understood,

$$< 0_+ \mid 0_- > = \exp\Big[\frac{\mathrm{i}}{2}\int(\mathrm{d}x)\int(\mathrm{d}x' J_i(x) D_+^{ij}(x-x') J_j(x')\Big], \tag{3.3.13}$$

$$D_+^{ij}(x-x') = \int\frac{(\mathrm{d}k)}{(2\pi)^4}\Big[\delta^{ij} - \frac{k^i k^j}{k^{02}}\Big]\frac{\mathrm{e}^{\mathrm{i}k(x-x')}}{k^2 - \mathrm{i}\epsilon}, \quad \epsilon \to +0. \tag{3.3.14}$$

Students often ask me as to why we went to an integration in the complex plane from an integral originally defined over a real variable $k^0$ from $-\infty$ to $\infty$. The reason is that by going to the complex plane, we were able to introduce a *close* contour, apply the residue theorem, and most importantly show that part of the contour consisting of the infinite semi circle does not contribute, and thus ending up to an integration over the real axis.

It is easily verified that by using the delta distribution $\delta^{(4)}(x-x')$ in Eq. (3.6.24) in the Appendix, that $D_+^{ij}(x-x')$ satisfies the partial differential equation

$$\big[(-\nabla^2 + \partial^{02})\delta^{i\ell} + \partial^i\partial^\ell\big]D_+^{\ell j}(x-x') = \delta^{ij}\,\delta^{(4)}(x-x'), \tag{3.3.15}$$

where a sum over $\ell = 1, 2, 3$ is understood.

## 3.4 Propagation of a Photon with a Given Momentum and a Given polarization

One may use the expression of the vacuum-to-vacuum transition amplitude in Eqs. (3.3.13)/(3.3.14) to describe the propagation of a photon with a given momentum $\mathbf{k}$ and a given polarization $\lambda$.

As in Eq. (3.2.2), consider the current $J^{\mu}(x)$, consisting of two parts:

$$J^{\mu}(x) = J_2^{\mu}(x) + J_1^{\mu}(x), \tag{3.4.1}$$

where $J_1^{\mu}(x)$, $J_2^{\mu}(x)$ are causally arranged such that the current $J_2^{\mu}(x)$ is switched on after $J_1^{\mu}(x)$ is switched off, i.e., removed. We may then write $< 0_+ \mid 0_- >$ explicitly as:

$$< 0_+ \mid 0_- >^J = < 0_+ \mid 0_- >^{J_2} \; T \; < 0_+ \mid 0_- >^{J_1}, \tag{3.4.2}$$

$$T = \exp\Big[\int \frac{\mathrm{d}^3\mathbf{k}}{(2\pi)^3 2|\mathbf{k}|} \,(\mathrm{i}) J_2^{*i}(k)\Big[\delta^{ij} - \frac{\mathbf{k}^i}{|\mathbf{k}|}\frac{\mathbf{k}^j}{|\mathbf{k}|}\Big](\mathrm{i}) J_1^j(k)\Big], \tag{3.4.3}$$

where in writing the expression for $T$, we have used Eqs. (3.2.8)–(3.2.10), and (3.3.12). We may also use the completeness relation in (3.1.10) re-written as

$$[\delta^{ij} - \frac{\mathbf{k}^i}{|\mathbf{k}|}\frac{\mathbf{k}^j}{|\mathbf{k}|}] = \sum_{\lambda=1,2} \mathbf{e}^i_\lambda \mathbf{e}^j_\lambda, \quad i, j = 1, 2, 3, \tag{3.4.4}$$

to re-write $T$ in the convenient form:

$$T = \exp\Big[\int \frac{\mathrm{d}^3\mathbf{k}}{(2\pi)^3 2|\mathbf{k}|} \sum_{\lambda=1,2} \Big(\big(\mathrm{i}\, J_2^{*i}(k)\mathbf{e}^i_\lambda\big)\big(\mathrm{i}\,\mathbf{e}^j_\lambda\, J_1^j(k)\big)\Big)\Big] \tag{3.4.5}$$

$$\begin{aligned} &\equiv \exp\Big[\sum_{\lambda,\lambda'} \int \frac{\mathrm{d}^3\mathbf{k}}{(2\pi)^3 2|\mathbf{k}|} \int \frac{\mathrm{d}^3\mathbf{k}'}{(2\pi)^3 2|\mathbf{k}'|} \\ &\times \big(\mathrm{i} J_2^{*i}(k)\mathbf{e}^i_\lambda\big)\Big[(2\pi)^3 2|\mathbf{k}|\delta^{(3)}(\mathbf{k}-\mathbf{k}')\delta_{\lambda\lambda'}\Big]\big(\mathrm{i}\,\mathbf{e}^j_{\lambda'}\, J_1^j(k')\big)\Big)\Big]. \end{aligned} \tag{3.4.6}$$

From Eqs. (3.4.2) and (3.4.6), we may then describe the process that a photon, with momentum/polarization ($\mathbf{k}$, $\lambda$) is created by $J_1$ and then ends up at $J_2$ with momentum/polarization ($\mathbf{k}'$, $\lambda'$) by

$$< 0_+ \mid \mathbf{k}, \lambda >< \mathbf{k}, \lambda \mid \mathbf{k}', \lambda' >< \mathbf{k}', \lambda' \mid 0_- >, \tag{3.4.7}$$

where $< \mathbf{k}, \lambda \mid \mathbf{k}', \lambda' >$ denotes the amplitude that a photon of momentum/polarization ($\mathbf{k}'$, $\lambda'$) ends up with momentum/polarization ($\mathbf{k}$, $\lambda$),

$$< \mathbf{k}, \lambda \mid \mathbf{k}', \lambda' > = \sqrt{\frac{\mathrm{d}^3\mathbf{k}}{(2\pi)^3 2|\mathbf{k}|}}\Big[(2\pi)^3 2|\mathbf{k}|\delta^{(3)}(\mathbf{k}-\mathbf{k}')\delta_{\lambda\lambda'}\Big]\sqrt{\frac{\mathrm{d}^3\mathbf{k}'}{(2\pi)^3 2|\mathbf{k}'|}}, \tag{3.4.8}$$

$$< 0_+ \mid \mathbf{k}, \lambda >^{J_2} = < 0_+ \mid 0_- >^{J_2} \sqrt{\frac{\mathrm{d}^3\mathbf{k}}{(2\pi)^3 2|\mathbf{k}|}}\; \mathrm{i}\,\mathbf{J}_2^* \cdot \mathbf{e}_\lambda, \tag{3.4.9}$$

$$< \mathbf{k}', \lambda' \mid 0_- >^{J_1} = \mathbf{e}_{\lambda'} \cdot \mathrm{i}\,\mathbf{J}_1 \sqrt{\frac{\mathrm{d}^3\mathbf{k}'}{(2\pi)^3 2|\mathbf{k}'|}} < 0_+ \mid 0_- >^{J_1}, \tag{3.4.10}$$

where we have, in the process, used the definition in Eq. (3.13).

The moral of this simple derivation is that we may introduce obstacles on the way of the photon, generate the corresponding propagator $D^{ij}_+$, and describe all sorts of experimental situations in classical electrodynamics with photons of different energies of red, blue, ... colors, as we will experience later, in a physically clear way, by simply generalizing the expression $\big[(2\pi)^3 2|\mathbf{k}|\delta^{(3)}(\mathbf{k}-\mathbf{k}')\delta_{\lambda'\lambda}\big]$ of a free photon, accordingly.

## 3.5 The Phase Factor: Liénard-Wiechert, Ampère, and Coulomb Interactions

In the present section, we learn that the explicit current-current interactions energies of classical electrodynamics may be read directly from the phase of the vacuum-to-vacuum transition amplitude.

From Eqs. (3.3.9), (3.3.10), we recall the expression of the vacuum-to-vacuum transition amplitude given by

$$< 0_+ \mid 0_- > \;=\; \exp\Big[\frac{\mathrm{i}}{2}\int(\mathrm{d}x)\int(\mathrm{d}x')J^{\mu}(x)D_+(x-x')J_{\mu}(x')\Big], \tag{3.5.1}$$

$$D_+(x-x') \;=\; \int\frac{(\mathrm{d}k)}{(2\pi)^4}\frac{\mathrm{e}^{\mathrm{i}\,k(x-x')}}{(k^2-\mathrm{i}\,\epsilon)}, \quad \epsilon\to+0. \tag{3.5.2}$$

Quite generally, upon writing $J^{\mu}$ in terms of two currents,

$$J^{\mu}(x) \;=\; J_1^{\mu}(x)+J_2^{\mu}(x), \tag{3.5.3}$$

and using the symmetry property $D_+(x-x')=D_+(x'-x)$, the net contribution to the exponent

$$\frac{\mathrm{i}}{2}\int(\mathrm{d}x)\int(\mathrm{d}x')J^{\mu}(x)D_+(x-x')J_{\mu}(x'), \tag{3.5.4}$$

of $< 0_+ \mid 0_- >$, due to the product of the two currents $J_2^{\mu}J_{1\mu}$ is given by

$$\mathrm{i}\int(\mathrm{d}x)\int(\mathrm{d}x')J_2^{\mu}(x)D_+(x-x')J_{1\mu}(x') \tag{3.5.5}$$

On the other hand, the phase of $< 0_+ \mid 0_- >$ in Eq. (3.5.1) is, by definition,

$$\mathcal{E} \;=\; \frac{1}{2}\int(\mathrm{d}x)\int(\mathrm{d}x')J^{\mu}(x)J_{\mu}(x')\,\mathcal{R}e\big[D_+(x-x')\big]. \tag{3.5.6}$$

where $\mathcal{R}e$ denotes the real part.

Since the currents in Eq. (3.5.1) in our case are, in general, time-dependent, this phase must be of the form:[8]

$$-\int_{-\infty}^{\infty} \mathrm{d}T\, E(T) = \frac{1}{2}\int (\mathrm{d}x) \int (\mathrm{d}x') J^{\mu}(x) J_{\mu}(x')\, \mathcal{R}e\big[D_{+}(x-x')\big], \tag{3.5.7}$$

where $E(T)$ denotes the interaction energy of currents involved.

The contribution to the phase $\mathcal{E}$ due to the product of the two currents $J_2^{\mu} J_{\mu 1}$ emerges as

$$\mathcal{E}_{21} = \int (\mathrm{d}x) \int (\mathrm{d}x') J_2^{\mu}(x) \big[\mathcal{R}e[D_{+}(x-x')]\big] J_{1\mu}(x'). \tag{3.5.8}$$

Since $D_{+}(x-x') = D_{+}(x'-x)$, $\mathcal{R}e[D_{+}(x-x')]$ is simply,

$$\mathcal{R}e[D_{+}(x-x')] = \int \frac{(\mathrm{d}k)}{(2\pi)^4} \mathrm{e}^{\mathrm{i}\,k(x-x')} \mathcal{R}e\Big[\frac{1}{k^2 - \mathrm{i}\,\epsilon}\Big], \tag{3.5.9}$$

$$(\mathrm{d}k) = \mathrm{d}k^0 \mathrm{d}^3\mathbf{k},\ \ k(x-x') = \mathbf{k}\cdot(\mathbf{x}-\mathbf{x}') - k^0(t-t'). \tag{3.5.10}$$

In Eq. (3.6.16) of the following section, we will learn that for the imaginary part

$$\mathcal{I}m \frac{1}{k^2 - \mathrm{i}\,\epsilon} = +\pi\,\delta(k^2), \quad \epsilon \to +0. \tag{3.5.11}$$

On the other hand, for the real part,

$$\mathcal{R}e\Big[\frac{1}{k^2}\Big], \quad \text{with } k^2 = |\mathbf{k}|^2 - k^{02} = -(k^0 - |\mathbf{k}|)(k^0 + |\mathbf{k}|), \tag{3.5.12}$$

we have to understand how to handle the two poles $k^0 = \pm|\mathbf{k}|$, and examine, in turn, how the limits $k^0 \to \pm|\mathbf{k}|$ are to be taken. In a technical term, this is referred to as taking the principal part of the poles. This turns out to be easy, by working again in the complex $k^0-$ plane.

To the above end, referring to Eqs. (3.5.9)/(3.5.10) and (3.5.12), one has to consider the following integral

$$\int_{-\infty}^{\infty} \mathrm{d}k^0\, \frac{\mathrm{e}^{-\mathrm{i}k^0(t-t')}}{k^2} = \int_{-\infty}^{\infty} \frac{\mathrm{d}k^0}{2|\mathbf{k}|}\Big[\frac{\mathrm{e}^{-\mathrm{i}k^0(t-t')}}{k^0 + |\mathbf{k}|} - \frac{\mathrm{e}^{-\mathrm{i}k^0(t-t')}}{k^0 - |\mathbf{k}|}\Big]. \tag{3.5.13}$$

In particular for $t > t'$, one may integrate over $k^0$ from $-\infty$ to $\infty$, close the integral below the real axis, picking up the two poles as indicated here

---

[8] This simply generalizes a time dependent phase factor exp$[-E \times \text{time}]$, of phase $-E \times$ time, that one often encounters in quantum mechanics of a state of energy $E$.

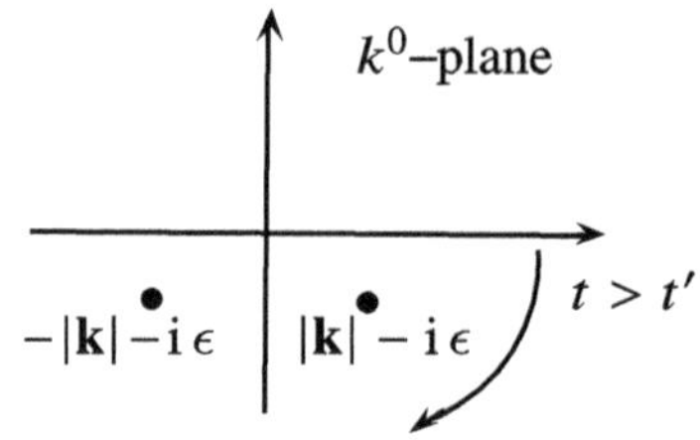


for which $\mathcal{I}mk^0 < 0$, and hence from

$$\mathrm{e}^{-\mathrm{i}k^0(t-t')} = \mathrm{e}^{(\mathcal{I}mk^0)(t-t')}\mathrm{e}^{-\mathrm{i}(\mathcal{R}ek^0)(t-t')}, \tag{3.5.14}$$

no contribution to the semi-circle of infinite radius arises. All told, the integral on the r.h.s. of Eq. (3.5.13) is, by the residue theorem, given by

$$-\frac{2\pi\mathrm{i}}{2|\mathbf{k}|}\big[\mathrm{e}^{\mathrm{i}|\mathbf{k}|(t-t')} - \mathrm{e}^{-\mathrm{i}|\mathbf{k}|(t-t')}\big], \tag{3.5.15}$$

where the minus sign arises because the contour of integration, below the axis, is taken clock-wise.

Upon substituting, the result of the integral in Eq. (3.5.15) into Eq. (3.5.9), we obtain

$$\mathcal{R}e[D_+(x-x')] = -\frac{\mathrm{i}}{16\pi^3}\int\frac{\mathrm{d}^3\mathbf{k}}{|\mathbf{k}|}\,\mathrm{e}^{\mathrm{i}\mathbf{k}\cdot(\mathbf{x}-\mathbf{x}')}\big[\mathrm{e}^{\mathrm{i}|\mathbf{k}|(t-t')} - \mathrm{e}^{-\mathrm{i}|\mathbf{k}|(t-t')}\big], \tag{3.5.16}$$

$$\mathrm{d}^3\mathbf{k} = 2\pi\,|\mathbf{k}|^2(\mathrm{d}\cos\theta), \quad \mathbf{k}\cdot(\mathbf{x}-\mathbf{x}') = |\mathbf{k}||\mathbf{x}-\mathbf{x}'|\cos\theta. \tag{3.5.17}$$

The angular integration is readily carried out giving,

$$\mathcal{R}e[D_+(x-x')] = -\frac{1}{8\pi^2|\mathbf{x}-\mathbf{x}'|}\int_0^\infty \mathrm{d}|\mathbf{k}|$$
$$\times\big[\mathrm{e}^{\mathrm{i}|\mathbf{k}|\,|\mathbf{x}-\mathbf{x}'|} - \mathrm{e}^{-\mathrm{i}|\mathbf{k}|\,|\mathbf{x}-\mathbf{x}'|}\big]\big[\mathrm{e}^{\mathrm{i}|\mathbf{k}|(t-t')} - \mathrm{e}^{-\mathrm{i}|\mathbf{k}|(t-t')}\big]. \tag{3.5.18}$$

Since the integrand is even in $|\mathbf{k}|$, we may integrate from $-\infty$ to $\infty$, and divide by 2, to obtain

$$\mathcal{R}e[D_+(x-x')] = -\frac{1}{16\pi^2|\mathbf{x}-\mathbf{x}'|}\int_{-\infty}^\infty \mathrm{d}k$$
$$\times\big[\mathrm{e}^{\mathrm{i}k[|\mathbf{x}-\mathbf{x}'|+(t-t')]} + \mathrm{e}^{-\mathrm{i}k\,|\mathbf{x}-\mathbf{x}'|+(t-t')]} - \mathrm{e}^{\mathrm{i}k[|\mathbf{x}-\mathbf{x}'|-(t-t')]} - \mathrm{e}^{-\mathrm{i}k\,|\mathbf{x}-\mathbf{x}'|-(t-t')]}\big]. \tag{3.5.19}$$

Using, by now, the familiar Dirac distribution

$$\frac{1}{2\pi}\int_{-\infty}^{\infty} \mathrm{d}k\, \mathrm{e}^{\pm \mathrm{i}k\,|a|} = \delta(|a|) \tag{3.5.20}$$

leads, by remembering that $t > t'$, to the expression

$$\mathcal{R}e[D_+(x-x')] = -\frac{1}{8\pi\,|\mathbf{x}-\mathbf{x}'|}\big[2\,\delta\big(t-t'+|\mathbf{x}-\mathbf{x}'|\big) - 2\,\delta\big(t-t'-|\mathbf{x}-\mathbf{x}'|\big)\big]. \tag{3.5.21}$$

Clearly, for non-vanishing $\big(t-t'+|\mathbf{x}-\mathbf{x}'|\big)$, the first Dirac delta is zero, and the following expression for the phase factor in Eq. (3.5.8) emerges as

$$\mathcal{E}_{21} = \int (\mathrm{d}x) \int (\mathrm{d}x')\, J_2^{\mu}(x) \frac{\delta\big(t-t'-|\mathbf{x}-\mathbf{x}'|\big)}{4\pi\,|\mathbf{x}-\mathbf{x}'|} J_{1\mu}(x'), \tag{3.5.22}$$

with relativity built in. By inserting, for convenience, the speed of light c in the argument of the delta distribution, it becomes

$$t - t' - \frac{|\mathbf{x}-\mathbf{x}'|}{\mathrm{c}}, \tag{3.5.23}$$

implying that for the causally arranged currents $J_2^{\mu}(x)$, $J_1^{\mu}(x')$, with $t > t'$, the disturbance from $\mathbf{x}'$ to $\mathbf{x}$ travels with speed of light.

For a given current $J^{\mu}(x)$, the expression

$$\int (\mathrm{d}x') \frac{\delta\big(t-t'-|\mathbf{x}-\mathbf{x}'|\big)}{4\pi\,|\mathbf{x}-\mathbf{x}'|} J_{\mu}(x'), \tag{3.5.24}$$

is the familiar Liénard-Wiechert potential.

To extract the interaction energy $E_{21}(T)$ between the two causally arranged currents $J_2^{\mu}(x)$, $J_1^{\mu}(x')$, we make a change of the integration variables $(t, t')$ to $(T, \tau)$ through

$$t = T + \frac{\tau}{2}, \qquad t' = T - \frac{\tau}{2}, \tag{3.5.25}$$

to rewrite Eq. (3.5.22) as

$$\mathcal{E}_{21} = -\int_{-\infty}^{\infty} \mathrm{d}T\, E_{21}(T) \tag{3.5.26}$$

$$= \int_{-\infty}^{\infty} \mathrm{d}T \int \mathrm{d}^3\mathbf{x}\,\mathrm{d}^3\mathbf{x}' \int_{-\infty}^{\infty} \mathrm{d}\tau\, J_2^{\mu}(x) \frac{\delta\big(\tau-|\mathbf{x}-\mathbf{x}'|\big)}{4\pi\,|\mathbf{x}-\mathbf{x}'|} J_{1\mu}(x'), \tag{3.5.27}$$

where $(t, t')$ are defined in Eq. (3.5.25), and we have extended the $\tau$ integration to $-\infty$ as, using the fact that the Dirac delta will not contribute for $\tau < 0$.

From the above equation, one may extract the interaction energy $E_{21}(T)$ of two causally arranged currents, with $t > t'$, to be

$$E_{21}(T) = -\int \mathrm{d}^3\mathbf{x}\,\mathrm{d}^3\mathbf{x}' \int_{-\infty}^{\infty} \mathrm{d}\tau\, J_2^{\mu}(T + \tau/2, \mathbf{x}) \frac{\delta\big(\tau - |\mathbf{x} - \mathbf{x}'|\big)}{4\pi\,|\mathbf{x} - \mathbf{x}'|} J_{1\mu}(T - \tau/2, \mathbf{x}'). \tag{3.5.28}$$

It is worth remembering that

$$J_2^{\mu}(x) J_{1\mu}(x') = \mathbf{J}_2(x) \cdot \mathbf{J}_1(x') - J_2^0(x) J_1^0(x'), \qquad J^0 = \rho. \tag{3.5.29}$$

In subsequent chapters, in particular, in the next one and Sect. 6.5, we will see how the above $T$ dependence is applied. On the other hand, if one may neglect the $\tau$ dependence in the argument of the currents, we may trivially integrate over $\tau$ in Eq. (3.5.28), to obtain,

$$E_{21}(T) \to -\int \mathrm{d}^3\mathbf{x}\,\mathrm{d}^3\mathbf{x}'\; J_2^{\mu}(T, \mathbf{x}) \frac{1}{4\pi\,|\mathbf{x} - \mathbf{x}'|} J_{1\mu}(T, \mathbf{x}'), \tag{3.5.30}$$

referred to as the Ampère law of current interactions, generalizing also the Coulomb interaction.

This chapter is based on several of the author's earlier publications.[9] The remarkable Schwinger's vacuum-to-vacuum transition appeared not only in his, relatively, more recent work,[10] of great importance, but it already appeared in most of his earlier monumental work on quantum electrodynamics.[11]

## 3.6 Appendix A: The Dirac Delta and the Levi-Civita Symbol

The Kronecker delta is defined by

$$\delta^{ij} = \delta_{ij} = \begin{cases} 1 & \text{for } i = j \\ 0 & \text{for } i \neq j \end{cases}. \tag{3.6.1}$$

[9] See, e.g., Manoukian (1986, 2015, 2016) [1–3], and references therein.

[10] See, Schwinger (1970) [6].

[11] See, e.g., Schwinger (1973) [7], for earlier references.

An immediate application of the Kronecker delta is in the following. If $\{\mathbf{v}_1, \mathbf{v}_2, \mathbf{v}_3\}$ is an orthonormal set of vectors in 3D, we may expand a vector $\mathbf{V}$ in terms of these unit vectors

$$\mathbf{V} = \big(\mathbf{V}\cdot\mathbf{v}_1\big)\mathbf{v}_1 + \big(\mathbf{V}\cdot\mathbf{v}_2\big)\mathbf{v}_2 + \big(\mathbf{V}\cdot\mathbf{v}_3\big)\mathbf{v}_3 = \mathbf{V}^i\delta_{ij}\mathbf{V}^j \tag{3.6.2}$$

$$\delta^{ij} = \mathbf{v}_1^i\mathbf{v}_1^j + \mathbf{v}_2^i\mathbf{v}_2^j + \mathbf{v}_3^i\mathbf{v}_3^j, \tag{3.6.3}$$

with the latter relation providing a so-called a completeness relation in 3D Euclidean space, and a summation over the repeated indices $i, j = 1, 2, 3$ in Eq. (3.6.3) understood, as discussed in the previous chapter.

A completeness relation in Minkowski spacetime was derived in Eqs. (3.1.13)–(3.1.16), and is given by

$$\eta^{\mu\nu} = \frac{k^\mu \underline{k}^\nu + \underline{k}^\mu k^\nu}{k\,\underline{k}} + \sum_{\lambda=1,2} \mathrm{e}_\lambda^\mu \mathrm{e}_\lambda^\nu, \tag{3.6.4}$$

where

$$\eta^{\mu\nu} = \mathrm{diag}[-1, 1, 1, 1], \quad \mathrm{e}_\lambda^\nu = (0, \mathbf{e}_\lambda), \quad \mathbf{e}_1, \mathbf{e}_2 \ \text{ unit vectors}, \tag{3.6.5}$$

$$\mathbf{e}_\lambda\cdot\mathbf{e}_{\lambda'} = \delta_{\lambda\lambda'}, \quad \mathbf{k}\cdot\mathbf{e}_\lambda = 0, \quad k = (k^0, \mathbf{k}), \ \underline{k} = (k^0, -\mathbf{k}), k^0 = |\mathbf{k}|, \tag{3.6.6}$$

$$k^2 = k^\mu k_\mu = \mathbf{k} - (k^0)^2, \quad k\underline{k} = k^\mu \underline{k}_\mu = -\mathbf{k}^2 - (k^0)^2. \tag{3.6.7}$$

The Levi-Civita symbol $\epsilon^{ijk}$ in 3D arises naturally when performing a cross product, e.g., between two vectors $\mathbf{A}$, $\mathbf{B}$:

$$(\mathbf{A}\times\mathbf{B})^i = \epsilon^{ijk} A^j B^k, \tag{3.6.8}$$

for the $i$th component of the cross product, and a sum over $j, k = 1, 2, 3$ is understood.

It is defined by

$$\epsilon^{ijk} = \begin{cases} +1 \ \text{ if } \ \{i, j, k\} \ \text{ is an even permutation of } \ \{1, 2, 3\} \\ -1 \ \text{ if } \ \{i, j, k\} \ \text{ is an odd permutation of } \ \{1, 2, 3\} \\ \ \ 0 \ \text{ if any two or three of } \ \{i, j, k\} \ \text{ are equal.} \end{cases} \tag{3.6.9}$$

It satisfies, in particular, the following key equation:

$$\epsilon^{ijk}\epsilon^{i\ell m} = \delta^{j\ell}\delta^{km} - \delta^{jm}\delta^{k\ell}. \tag{3.6.10}$$

The Levi-Civita symbol may be also easily defined in higher dimensions like the Minkowski space but no need of them will arise in this book.

The celebrated Dirac delta: $\delta-$distribution, involving real functions, is defined as follows

$$\int_{-\infty}^{\infty} \mathrm{d}x\, \delta(x-a)\, f(x) = f(a), \quad \int_{-\infty}^{\infty} \mathrm{d}x\, \delta(x) = 1, \quad \delta(-x) = \delta(x), \quad x\delta(x) = 0, \tag{3.6.11}$$

$$\delta(ax) = \frac{1}{|a|}\delta(x), \quad \delta^{(n)}(\mathbf{x}) = \delta(x_1)\ldots\delta(x_n), \quad \int \mathrm{d}^n\mathbf{x}\, \delta^{(n)}(\mathbf{x}-\mathbf{a})\, f(\mathbf{x}) = f(\mathbf{a}). \tag{3.6.12}$$

If $a_, a_2, \ldots a_k$ are roots a of a function $g(x) : g(a_i) = 0$, then

$$\delta\big(g(x)\big) = \sum_{i=1}^{k} \frac{\delta(x-a_i)}{|g'(a_i)|}, \quad g'(x) = \frac{\mathrm{d}}{\mathrm{d}x} g(x). \tag{3.6.13}$$

The following relations should be noted:

$$\delta(x^2 - a^2) = \frac{1}{2|a|}\big[\delta(x-a) + \delta(x+a)\big] \tag{3.6.14}$$

$$\delta\big((x-a)(x-b)\big) = \frac{1}{|a-b|}\big[\delta(x-a) + \delta(x-b)\big], \tag{3.6.15}$$

$$\mathcal{Im}\frac{1}{x \pm \mathrm{i}\epsilon} = \mp\pi\,\delta(x), \quad \epsilon \to +0, \tag{3.6.16}$$

where $\mathcal{Im}$ stands for the imaginary part.

In cylindrical coordinates $(r, \phi, z)$, $0 \leq r < \infty$, $0 \leq \phi < 2\pi$, $-\infty < z < \infty$,

$$\delta^3(\mathbf{r} - \mathbf{r}_\mathrm{o}) = \frac{\delta(r - r_\mathrm{o})}{r}\, \delta(\phi - \phi_\mathrm{o})\, \delta(z - z_\mathrm{o}), \tag{3.6.17}$$

$$\text{so that} \int_0^{\infty} r\mathrm{d}r \int_0^{2\pi} \mathrm{d}\phi \int_{-\infty}^{\infty} \mathrm{d}z\, \delta^3(\mathbf{r}) = 1, \tag{3.6.18}$$

while in spherical coordinates $(r, \theta, \phi)$, $0 \leq r < \infty$, $0 \leq \theta \leq \pi$, $0 \leq \phi < 2\pi$,

$$\delta^3(\mathbf{r} - \mathbf{r}_\mathrm{o}) = \frac{\delta(r - r_\mathrm{o})}{r^2}\, \frac{\delta(\theta - \theta_\mathrm{o})}{\sin\theta}\, \delta(\phi - \phi_\mathrm{o}), \tag{3.6.19}$$

$$\text{so that} \int_0^{\infty} r^2\mathrm{d}r \int_0^{\pi} \sin\theta\, \mathrm{d}\theta \int_0^{2\pi} \mathrm{d}\phi\, \delta^3(\mathbf{r}) = 1. \tag{3.6.20}$$

A mathematics enthusiasm, understands the Dirac delta distribution in the light of the *mean value theorem*, which states that for a function which is continues in an interval, e.g., $[-a, a]$ and is bounded,

$$\frac{1}{2a}\int_{-a}^{a} \mathrm{d}x\, f(x) = f(b), \quad \text{where } b \text{ is a number in the interval.} \tag{3.6.21}$$

Formally, one sees as $a$ is led to zero, $b$ goes to 0, and one is led to the value $f(0)$ for the integral.

The delta distribution is usually represented formally as a Fourier integral, e.g.,

$$\delta(x-a) = \frac{1}{2\pi}\int_{-\infty}^{\infty} \mathrm{d}k\, \mathrm{e}^{\mathrm{i}k(x-a)}, \tag{3.6.22}$$

$$\delta^{(3)}(\mathbf{x}-\mathbf{a}) = \frac{1}{(2\pi)^3}\int \mathrm{d}^3\mathbf{k}\, \mathrm{e}^{\mathrm{i}\mathbf{k}\cdot(\mathbf{x}-\mathbf{a})}, \tag{3.6.23}$$

$$\delta^{(4)}(x-a) = \frac{1}{(2\pi)^4}\int (\mathrm{d}k)\, \mathrm{e}^{\mathrm{i}\mathbf{k}\cdot(\mathbf{x}-\mathbf{a})}\, \mathrm{e}^{-\mathrm{i}k^0(x^0-a^0)}, \quad (\mathrm{d}k) = \mathrm{d}k^0\, \mathrm{d}^3\mathbf{k}. \tag{3.6.24}$$

The Dirac delta distribution may be represented by the following remarkable function.[12] It is continuous, differentiable, and unlike functions which have tails that go to infinity, it also vanishes rigorously at $z = \pm a$, as is easily check. It provides a representation of the Dirac delta distribution for $a \to 0$:

$$\delta_a(z) = \mathrm{e}^2 \begin{cases} 0, & z < -a, \\ \frac{a}{(z+a)^2}\exp\big[-\frac{2a}{z+a}\big], & -a \le z < 0, \\ \frac{a}{(a-z)^2}\exp\big[-\frac{2a}{a-z}\big], & 0 \le z < a, \\ 0, & a \le z. \end{cases} \tag{3.6.25}$$

Accordingly, we are led to consider the integral involving a function $f(z)$ of the variable $z$:

$$\int_{-\infty}^{\infty} \mathrm{d}z\, \delta_a(z) f(z) = \int_{-a}^{a} \mathrm{d}z\, \delta_a(z) f(z) \tag{3.6.26}$$

$$= \mathrm{e}^2\int_{-a}^{0} \mathrm{d}z\, \frac{a}{(z+a)^2}\exp\Big[-\frac{2a}{z+a}\Big] f(z)$$
$$+ \mathrm{e}^2\int_{0}^{a} \mathrm{d}z\, \frac{a}{(a-z)^2}\exp\Big[-\frac{2a}{a-z}\Big] f(z). \tag{3.6.27}$$

[12] This is based on a representation of a step function introduced in my book Manoukian (2016) [3], p. 129, p. 553.

Consider the first integral and make a change of variable $z = xa$ and then $x + 1 = 1/y$ to obtain

$$e^2 \int_{-1}^{0} dx \, \frac{1}{(x+1)^2} \exp\Big[-\frac{2}{x+1}\Big] f(ax) \tag{3.6.28}$$

$$= e^2 \int_{1}^{\infty} dy \, e^{-2y} \, f\Big(a\Big(1 - \frac{1}{y}\Big)\Big) \to f(0)\, e^2 \int_{1}^{\infty} dy \, e^{-2y} \tag{3.6.29}$$

$$= f(0) \frac{e^2}{(-2)} [0 - e^{-2}] = \frac{1}{2} f(0). \tag{3.6.30}$$

where note that for all $1 \leq y < \infty$, $a\big(1 - (1/y)\big) \to 0$ for $a \to 0$.

Similarly, consider the second integral in Eq. (3.6.27), change the variable $z = ax$, then $1 - x = 1/y$, to obtain

$$e^2 \int_{0}^{1} dx \, \frac{1}{(1-x)^2} \exp\Big[-\frac{2}{1-x}\Big] f(ax) \tag{3.6.31}$$

$$= e^2 \int_{1}^{\infty} dy \, e^{-2y} \, f\Big(a\Big(1 - \frac{1}{y}\Big)\Big) \to f(0)\, e^2 \int_{1}^{\infty} dy \, e^{-2y} \tag{3.6.32}$$

$$= f(0) \frac{e^2}{(-2)} [0 - e^{-2}] = \frac{1}{2} f(0). \tag{3.6.33}$$

again, note that for all $1 \leq y < \infty$, $a\big(1 - (1/y)\big) \to 0$ for $a \to 0$.

All told,

$$\lim_{a \to 0} \int_{-\infty}^{\infty} dz \, \delta_a(z) f(z) = f(0)\Big[\frac{1}{2} + \frac{1}{2}\Big] = f(0). \tag{3.6.34}$$

$$\int_{-\infty}^{\infty} dz \, \delta_a(z) = 1. \tag{3.6.35}$$

The latter equation is valid even if $a \neq 0$, and follows from the fact that by making a change $z = xa$ in the integral, we note, from the explicit expression of $\delta_a(z)$ given in Eq. (3.6.25), the dependence on $a$ completely disappears in the integral in Eq. (3.6.35). In the presence of $f(z)$ in the integrand as in Eq. (3.6.34), the $a$ dependence appears only in $f(z) \to f(ax)$ by such a change of variable, providing a more direct proof of the results.

The Dirac delta may be also generated from taking derivative of the Heaviside step function:[13]

[13] More often, it is denoted by $H(z)$.

$$\theta(z) = \begin{cases} 1, & \text{for } z > 0 \\ 0, & \text{for } z < 0 \end{cases}, \tag{3.6.36}$$

$$\theta(-z) = \begin{cases} 1, & \text{for } 0 < z \\ 0, & \text{for } z > 0 \end{cases}, \tag{3.6.37}$$

and

$$\frac{\mathrm{d}}{\mathrm{d}z}\,\theta(z) = \delta(z). \tag{3.6.38}$$

$$\frac{\mathrm{d}}{\mathrm{d}z}\,\theta(-z) = -\,\delta(z). \tag{3.6.39}$$

A representation of $\theta(z)$ may be obtained directly from $\delta_a(z)$ defined in Eq. (3.6.25) for $a \to 0$.[14]

When working in half-space with $z, z' > 0$, the Dirac delta takes the following form

$$\delta(z, z') = \frac{1}{\pi}\int_{-\infty}^{\infty} \mathrm{d}q\, \cos(qz)\, \cos(qz'). \tag{3.6.40}$$

On the other hand in half-space $z, z' > 0$, *and* $\delta(z, z') = 0$ *for* $z$ or $z' = 0$, it is expressed in terms of sine functions:

$$\delta(z, z') = \frac{1}{\pi}\int_{-\infty}^{\infty} \mathrm{d}q\, \sin(qz)\, \sin(qz') = \int_{-\infty}^{\infty} \frac{\mathrm{d}q}{2\pi}\big(\mathrm{e}^{\mathrm{i}q(z-z')} - \mathrm{e}^{\mathrm{i}q(z+z')}\big). \tag{3.6.41}$$

In a confined region with $z, z'$ in the range $(-a/2, a/2)$ and $\delta(z, z') = 0$ for $z$ or $z' = \pm a/2$, we have

$$\delta(z, z') = \frac{2}{a}\sum_{n=1}^{\infty} \sin\Big(n\pi\,\frac{z - a/2}{a}\Big)\, \sin\Big(n\pi\,\frac{z' - a/2}{a}\Big). \tag{3.6.42}$$

On the other hand, for $z, z'$ in the range $(0, d)$, and $\delta(z, z') = 0$ for $z,\ z' = 0$, we have

$$\delta(z, z') = \frac{1}{2d}\sum_{n=-\infty}^{\infty} \big[\mathrm{e}^{\mathrm{i}(z-z')\frac{n\pi}{d}} - \mathrm{e}^{-\mathrm{i}(z+z')\frac{n\pi}{d}}\big]. \tag{3.6.43}$$

[14] See Manoukian (2016) [3], p. 129, p. 553.

It is interesting to know that it may be also defined, formally, as the limit of a Gausian distribution of vanishing variance $\sigma^2 \to 0$:

$$\frac{1}{\sqrt{2\pi}\,\sigma}\exp\Big[-\frac{(z-z_0)^2}{2\sigma^2}\Big] \;\to\; \delta(z-z_0), \quad \text{for } \sigma^2 \to 0, \tag{3.6.44}$$

where

$$\frac{1}{\sqrt{2\pi}\,\sigma}\int_{-\infty}^{\infty} \mathrm{d}z \exp\Big[-\frac{(z-z_0)^2}{2\sigma^2}\Big] = 1. \tag{3.6.45}$$

One of the useful equations involving $\delta^{(3)}(\mathbf{x}-\mathbf{x}')$ in electrodynamics, is in the equation, referred to as a Poisson equation:

$$\nabla^2 \frac{1}{|\mathbf{x}-\mathbf{x}'|} = -4\pi\,\delta^{(3)}(\mathbf{x}-\mathbf{x}'), \tag{3.6.46}$$

and with $\mathbf{x} = (x^1, x^2, x^3)$, the Laplacian $\nabla^2$ is given by

$$\nabla^2 = \Big(\frac{\partial}{\partial x^1}\Big)^2 + \Big(\frac{\partial}{\partial x^2}\Big)^2 + \Big(\frac{\partial}{\partial x^3}\Big)^2. \tag{3.6.47}$$

To prove this, we may start from the equation

$$\frac{1}{\nabla^2}\delta^{(3)}(\boldsymbol{\rho}) = -\frac{1}{(2\pi)^3}\int \frac{\mathrm{d}^3\mathbf{k}}{|\mathbf{k}|^2}\,\mathrm{e}^{\mathrm{i}\mathbf{k}\cdot\boldsymbol{\rho}}, \qquad \boldsymbol{\rho} \equiv \mathbf{x}-\mathbf{x}'. \tag{3.6.48}$$

Upon using,

$$\mathrm{d}^3\mathbf{k} = |\mathbf{k}|^2\mathrm{d}|\mathbf{k}|\,\mathrm{d}\phi\,\sin\theta\,\mathrm{d}\sin\theta, \quad 0 \le |\mathbf{k}| < \infty,\ 0 \le \phi \le 2\pi,\ 0 \le \theta \le \pi, \tag{3.6.49}$$

$$\mathbf{k}\cdot\boldsymbol{\rho} = k|\boldsymbol{\rho}|\cos\theta, \tag{3.6.50}$$

we obtain

$$\frac{1}{\nabla^2}\delta^{(3)}(\boldsymbol{\rho}) = -\frac{1}{4\pi^2}\int_0^\infty \mathrm{d}|\mathbf{k}| \int_{-1}^{1} (\mathrm{d}\cos\theta)\,\mathrm{e}^{\mathrm{i}|\mathbf{k}||\boldsymbol{\rho}|\cos\theta} \tag{3.6.51}$$

$$= -\frac{1}{2\pi^2|\boldsymbol{\rho}|}\int_0^\infty \frac{\mathrm{d}|\mathbf{k}|}{|\mathbf{k}|}\sin(|\mathbf{k}||\boldsymbol{\rho}|). \tag{3.6.52}$$

Finally using the integral

$$\int_0^\infty \frac{\mathrm{d}|\mathbf{k}|}{|\mathbf{k}|}\sin(|\mathbf{k}||\boldsymbol{\rho}|) = \frac{\pi}{2}, \tag{3.6.53}$$

for $|\boldsymbol{\rho}| > 0$, and 0, otherwise, we obtain

$$\frac{1}{\nabla^2}\,\delta^{(3)}(\boldsymbol{\rho}) \;=\; -\frac{1}{4\pi\,|\boldsymbol{\rho}|}, \tag{3.6.54}$$

which is equivalent to Eq. (3.6.46).

Also note the following useful representation for the Dirac delta distribution

$$\mathcal{R}e\Big[\int_0^\infty \mathrm{d}t\; \mathrm{e}^{-\mathrm{i}\omega t}\Big] \;=\; \pi\,\delta(\omega). \tag{3.6.55}$$

The proof of this is simple. The l.h.s. of the equation is explicitly given by

$$\int_0^\infty \mathrm{d}t\,\cos\omega t = \frac{1}{2}\int_0^\infty \mathrm{d}t\big[\mathrm{e}^{-\mathrm{i}\omega t} + \mathrm{e}^{+\mathrm{i}\omega t}\big] = \frac{1}{2}\Big[\int_0^\infty \mathrm{d}t\;\mathrm{e}^{-\mathrm{i}\omega t} + \int_{-\infty}^0 \mathrm{d}t\;\mathrm{e}^{-\mathrm{i}\omega t}\Big], \tag{3.6.56}$$

where we have made a change of variable $t \to -t$ in the second integral, in the final step. Upon combining the two integrals, the above representation follows.

An important expression for the radial part of the delta distribution in Eq. (3.6.17) which is similar to the one dimensional delta distribution in Eq. (3.6.41) in the sense that it vanishes for $r$ taking a value, say, $r = a$, is given by

$$\frac{\delta(r,r')}{r} = \frac{2}{a^2}\sum_{n=1}^{\infty}\frac{J_0^{\mathrm{B}}(\gamma_n r)J_0^{\mathrm{B}}(\gamma_n r')}{|J_1^{\mathrm{B}}(\gamma_n a)|^2}, \qquad J_1^{\mathrm{B}}(x) \;=\; -\frac{\mathrm{d}}{\mathrm{d}x}J_0^{\mathrm{B}}(x), \tag{3.6.57}$$

where $J_0^{\mathrm{B}}(\gamma_\ell r)$ is the so-called Bessel function[15] of order zero, and where the summation is over positive roots $\gamma_n a$ of the Bessel function

$$J_0^{\mathrm{B}}(\gamma_n r)\Big|_{r=a} = 0, \quad n = 1, 2, \ldots, \tag{3.6.58}$$

The Bessel function $J_0^{\mathrm{B}}(\gamma_n r)$ satisfies the differential equation

$$\Big[\frac{\mathrm{d}^2}{\mathrm{d}r^2} + \frac{1}{r}\frac{\mathrm{d}}{\mathrm{d}r} + \gamma_n^2\Big]J_0^{\mathrm{B}}(\gamma_n r) \;=\; 0, \tag{3.6.59}$$

and

$$\int_0^a r\,\mathrm{d}r\,J_0^{\mathrm{B}}(\gamma_n r)J_0^{\mathrm{B}}(\gamma_{n'} r) \;=\; \delta_{n,n'}\frac{a^2}{2}|J_1^{\mathrm{B}}(\gamma_n a)|^2. \tag{3.6.60}$$

[15] We have conveniently introduced the letter B in defining the Bessel function in order not to confuse it with the current later on.

On the other hand, for the angular part in Eq. (3.6.17), with $\phi$ being restricted, one has,

$$\delta(\phi - \phi') = \frac{1}{2\pi} \sum_{N=-\infty}^{\infty} \mathrm{e}^{\mathrm{i}N(\phi-\phi')}. \tag{3.6.61}$$

Here note that

$$\int_0^{2\pi} \mathrm{d}\phi\, \delta(\phi - \phi') = \frac{1}{2\pi} \sum_{N=-\infty}^{\infty} \mathrm{e}^{-\mathrm{i}N\phi'}\, 2\pi\, \delta_{N0} = 1, \tag{3.6.62}$$

as expected, verifying self consistency.

## References


1. Manoukian, E. B. (1986). Exact solutions for stimulated enissions by external sources. *International Journal of Theoretical Physics, 25 (2)*, 147–158.
2. Manoukian, E. B. (2015). Vacuum-to-vacuum transition probability and the classic radiation theory. *Radiation Physics and Chemistry, 106*, 268–270. p. 1632.
3. Manoukian, E. B. (2016). *Quantum field theory I: Foundations and Abelian and Non-Abelian Gauge Theories*. Switzerland: Springer.
4. Navas, S., et al. PDG (2024). *Particle Data Group Collaboration. Physical Review D 110*, 030001.
5. Schwinger, J. (1969). *Particles and sources*. New York: Gordon and Breach.
6. Schwinger, J. (1970). *Particles, sources, and fields*. Reading, Massachusetts: Addison-Wesley.
7. Schwinger, J. (1973). A report on quantum electrodynamics. In. J. Mehra (Ed.), *The physicist's conception of nature*. Dordrecht/Holland: D. Reidel Publishing Company.

# Chapter 4
# $< 0_+ \mid 0_- >$: Theoretical Laboratory for Computation of Power of Radiation

*Power of the vacuum-to-vacuum transition amplitude in computing power of radiation*

To study the power of radiation emitted by a given current distribution $J^\mu(x)$, in this formalism, we need to derive the general expression of the total energy radiated by the latter. The power of radiation is then obtained from its very definition by deriving the expression of the energy of radiation emitted per unit time.[1]

Our starting point to derive the exact expression of the total energy of radiation due an arbitrary current distribution $J^\mu(x)$ from the vacuum-to-vacuum transition amplitude obtained in Sect. 3.3. An application is then worked out in detail to the classical problem of radiation emission by a relativistic accelerating charged particle.

## 4.1 $< 0_+ \mid 0_- >$ and the Power of Radiation

Our starting point is to derive the exact expression of the total energy of radiation due an arbitrary current distribution $J^\mu(x)$ from the expression of the vacuum-to-vacuum transition amplitude obtained in Eqs. (3.3.9)/(3.3.10), from which the exact expression of power of radiation is then readily extracted.

[1] No fields, no potentials, which were used for the relativistic charged particle, e.g., by Schwinger (1949) [4], Jackson (1998) [1], Zangwill [6], Pathak (2022) [3], are needed. Not even a quantum treatment of the charged particle is used, as was introduced by Schwinger (1973) [5]. The treatment is based on earlier work by the author, using the expression of the vacuum-to-vacuum transition amplitude: Manoukian (2015) [2].



E. B. Manoukian, *Radiation*,
https://doi.org/10.1007/978-3-032-21506-2_4

To the above end, recall the following expressions:

$$\langle 0_+ \mid 0_- \rangle = \exp\Big[\frac{\mathrm{i}}{2}\int (\mathrm{d}x)(\mathrm{d}x') J^\mu(x)\eta_{\mu\nu} D_+(x-x') J^\nu(x')\Big], \quad \partial_\mu J^\mu(x) = 0, \tag{4.1.1}$$

$$(\mathrm{d}x) = \mathrm{d}x^0\mathrm{d}x^1\mathrm{d}x^2\mathrm{d}x^3, \quad x^0 = t, \quad \eta^{\mu\nu} = \mathrm{diag}[-1, 1, 1, 1], \tag{4.1.2}$$

$$D_+(x-x') = \int \frac{(\mathrm{d}k)}{(2\pi)^4}\frac{\mathrm{e}^{\mathrm{i}k(x-x')}}{k^2 - \mathrm{i}\epsilon}, \quad \epsilon \to +0, \quad k^2 = \mathbf{k}^2 - k^{0^2}, \tag{4.1.3}$$

$$(\mathrm{d}k) = \mathrm{d}k^0\mathrm{d}k^1\mathrm{d}k^2\mathrm{d}k^3, \quad k(x'-x) = \mathbf{k}\cdot(\mathbf{x}'-\mathbf{x}) - k^0(t'-t). \tag{4.1.4}$$

In view of determining the average number of photons emitted by $J^\mu(x)$, described in Sect. 3.1, Eqs. (3.1.25), (3.1.26), note that

$$\mid \langle 0_+ \mid 0_- \rangle \mid^2 = \exp\Big[-\int (\mathrm{d}x)(\mathrm{d}x') J^\mu(x)\eta_{\mu\nu}(\mathcal{Im}\, D_+(x', x)) J^\nu(x')\Big], \tag{4.1.5}$$

where $\mathcal{Im}$ denotes the imaginary part. Due to the symmetry in the $x$, $x'$ variables of the above integrals, this amounts to take the imaginary part only of the denominator in the integrand,

$$\mathcal{Im}\; D_+(x-x') \;\to\; \int \frac{(\mathrm{d}k)}{(2\pi)^4}\mathrm{e}^{\mathrm{i}k(x-x')}\mathcal{Im}\Big(\frac{1}{k^2 - \mathrm{i}\epsilon}\Big) \tag{4.1.6}$$

$$= \pi \int \frac{(\mathrm{d}k)}{(2\pi)^4}\mathrm{e}^{\mathrm{i}k(x-x')}\delta(k^2), \tag{4.1.7}$$

where we have used Eq. (3.6.16) in Chap. 3, for the imaginary part of the expression in question. Accordingly,

$$\mid \langle 0_+ \mid 0_- \rangle \mid^2 = \exp\Big[-\pi \int (\mathrm{d}x)(\mathrm{d}x')\; J^\mu(x)\; J_\mu(x') \int \frac{(\mathrm{d}k)}{(2\pi)^4}\,\delta(k^2)\mathrm{e}^{\mathrm{i}k(x-x')}\Big], \tag{4.1.8}$$

where

$$\delta(k^2) = \frac{\delta(k^0 - \mid \mathbf{k} \mid) + \delta(k^0 + \mid \mathbf{k} \mid)}{2 \mid \mathbf{k} \mid}, \qquad \frac{(\mathrm{d}k)}{(2\pi)^4} \equiv \frac{\mathrm{d}^3\mathbf{k}}{(2\pi)^3}\frac{\mathrm{d}k^0}{2\pi}, \tag{4.1.9}$$

using, in the process, Eq. (3.6.14) of a property of a Dirac delta distribution.

Accordingly, from Eq. (3.1.25), the average number of photons emitted by $J^\mu(x)$ is given by

$$< N > = \pi \int (\mathrm{d}x)(\mathrm{d}x')\; J^\mu(x)\; J_\mu(x') \int \frac{(\mathrm{d}k)}{(2\pi)^4}\,\delta(k^2)\mathrm{e}^{\mathrm{i}k(x-x')}\Big]. \tag{4.1.10}$$

We are interested in the average number of photons, denoted by $N(\omega)$, emitted with a given energy $\hbar\omega \equiv \omega$ (with $\hbar$ set equal to one). To this end, we may, as a completeness over all possible energies, insert the identity

$$1 = \int_0^\infty \mathrm{d}\omega\, \delta(\omega - \mid \mathbf{k} \mid), \tag{4.1.11}$$

in the integrand of the expression of $< N >$ in Eq. (4.1.10), and write

$$\langle N \rangle = \int_0^\infty \mathrm{d}\omega\, N(\omega)\, \delta(\omega - \mid \mathbf{k} \mid). \tag{4.1.12}$$

Using, in the process, Eq. (4.1.9), we have

$$\delta(\omega - \mid \mathbf{k} \mid)\, \delta(k^2) = \frac{\delta(\omega - \mid \mathbf{k} \mid)}{2 \mid \mathbf{k} \mid}\big[\delta(k^0 - \omega) + \delta(k^0 + \omega)\big], \tag{4.1.13}$$

remembering that $k^0$ here takes both signs. After integrating Eq. (4.1.10) over $k^0$, the above gives

$$\begin{aligned} N(\omega) &= \frac{1}{2}\int (\mathrm{d}x)(\mathrm{d}x')J^\mu(x)J_\mu(x') \int \frac{\mathrm{d}^3\mathbf{k}}{(2\pi)^3 2|\mathbf{k}|}\delta(\omega - \mid \mathbf{k} \mid) \\ &\quad \times\, \mathrm{e}^{\mathrm{i}\mathbf{k}\cdot(\mathbf{x}-\mathbf{x}')}\big[\mathrm{e}^{-\mathrm{i}\omega(t-t')} + \mathrm{e}^{\mathrm{i}\omega(t-t')}\big]. \end{aligned} \tag{4.1.14}$$

Upon setting, $\mathbf{k}\cdot(\mathbf{x}-\mathbf{x}') = |\mathbf{k}||\mathbf{x}-\mathbf{x}'|\cos\theta$,

$$\mathrm{d}^3\mathbf{k} = \mathrm{d}\Omega \mid \mathbf{k} \mid^2 \mathrm{d} \mid \mathbf{k} \mid, \text{ with } \mathrm{d}\Omega = 2\pi\, \mathrm{d}(\cos\theta), \quad -1 \le \cos\theta \le 1, \tag{4.1.15}$$

with $\mathrm{d}\Omega$ corresponding to an integration over the solid angle, we obtain after integrating over $|\mathbf{k}|$

$$N(\omega) = \frac{\omega}{16\pi^3}\int (\mathrm{d}x)(\mathrm{d}x')J^\mu(x)J_\mu(x')\int \mathrm{d}\Omega\, \mathrm{e}^{\mathrm{i}\omega(\mathbf{x}-\mathbf{x}')\cdot\mathbf{n}}\frac{1}{2}\big[\mathrm{e}^{-\mathrm{i}\omega(t-t')} + \mathrm{e}^{\mathrm{i}\omega(t-t')}\big]. \tag{4.1.16}$$

We will see below in Eq. (4.1.21), that the $\Omega-$ integral is a function of $|\mathbf{x}-\mathbf{x}'|$, i.e., is even in the interchange $\mathbf{x} \leftrightarrow \mathbf{x}'$ in the *latter* integral. This allows us to combine the two terms in Eq. (4.1.16), by changing $x \leftrightarrow x'$ in the second term and combine it with the first term, and re-write Eq. (4.1.16) in the more convenient form consisting one term, by involving a factor of 2 in the integral. That is,

$$< N > = \int_0^\infty \mathrm{d}\omega\, N(\omega), \tag{4.1.17}$$

$$\begin{aligned} N(\omega) = \frac{\omega}{16\pi^3} \int (\mathrm{d}x) \int (\mathrm{d}x') \Big[\mathbf{J}(x)\cdot\mathbf{J}(x') - J^0(x)J^0(x')\Big] \\ \times \int \mathrm{d}\Omega\, \mathrm{e}^{\mathrm{i}\omega\,\mathbf{n}\cdot(\mathbf{x}-\mathbf{x}')}\mathrm{e}^{-\,\mathrm{i}\omega(t-t')}. \end{aligned} \tag{4.1.18}$$

Since each photon has an energy $\hbar\omega \equiv \omega$, the energy of radiation of (angular) frequency $\omega$ emitted is given by

$$\begin{aligned} E(\omega) = \frac{\omega^2}{16\pi^3} \int (\mathrm{d}x) \int (\mathrm{d}x') \Big[\mathbf{J}(x)\cdot\mathbf{J}(x') - J^0(x)J^0(x')\Big] \\ \times \int \mathrm{d}\Omega\, \mathrm{e}^{\mathrm{i}\omega\,\mathbf{n}\cdot(\mathbf{x}-\mathbf{x}')}\mathrm{e}^{-\,\mathrm{i}\omega(x^0-x'^0)}. \end{aligned} \tag{4.1.19}$$

We note, in passing, that the integration over the solid angle may be explicitly carried out by using

$$\mathrm{d}\Omega = 2\pi d(\cos\theta), \quad -1 \leq \cos\theta \leq 1, \quad \mathbf{n}\cdot(\mathbf{x}-\mathbf{x}') = |\mathbf{x}-\mathbf{x}'|\cos\theta, \tag{4.1.20}$$

$$\Rightarrow \int \mathrm{d}\Omega\, \mathrm{e}^{\mathrm{i}\omega\,\mathbf{n}\cdot(\mathbf{x}-\mathbf{x}')} = \frac{-\,2\pi\,\mathrm{i}}{\omega|\mathbf{x}-\mathbf{x}'|}\Big[\mathrm{e}^{\mathrm{i}\omega|\mathbf{x}-\mathbf{x}'|} - \mathrm{e}^{-\mathrm{i}\omega|\mathbf{x}-\mathbf{x}'|}\Big], \tag{4.1.21}$$

which, being a function of $|\mathbf{x}-\mathbf{x}'|$, is even in the interchange $\mathbf{x} \leftrightarrow \mathbf{x}'$.

Using the conservation law $\partial_\mu J^\mu = 0$, we may equivalently write

$$\Big[\mathbf{J}(x)\cdot\mathbf{J}(x') - J^0(x)J^0(x')\Big] = \Big[\delta_{ij} - \frac{\partial_i\,\partial_j{}'}{\partial_0\,\partial_0'}\Big]J^i(x)J^j(x'), \tag{4.1.22}$$

and integrate by parts over $x$, $x'$ in Eq. (4.1.19), to obtain

$$\begin{aligned} E(\omega) = \frac{\omega^2}{16\pi^3} \int (\mathrm{d}x) \int (\mathrm{d}x') J^i(x)J^j(x') \\ \times \int \mathrm{d}\Omega\,[\delta^{ij} - \mathbf{n}^i\mathbf{n}^j]\mathrm{e}^{\mathrm{i}\omega\,\mathbf{n}\cdot(\mathbf{x}-\mathbf{x}')}\mathrm{e}^{-\,\mathrm{i}\omega(t-t')}. \end{aligned} \tag{4.1.23}$$

Upon introducing integration variables $T$, $\tau$ as follows

$$t = T + \frac{\tau}{2}, \quad t' = T - \frac{\tau}{2}, \quad t - t' = \tau, \tag{4.1.24}$$

the power of radiation per unit frequency about $\omega$ emerges as

$$\begin{aligned} \mathrm{P}(\omega, T) &= \frac{\omega^2}{16\pi^3} \int \mathrm{d}^3\mathbf{x} \int \mathrm{d}^3\mathbf{x}' \int_{-\infty}^{\infty} \mathrm{d}\tau \, J^i(x) J^j(x') \\ &\times \int \mathrm{d}\Omega \left[\delta^{ij} - \mathbf{n}^i \mathbf{n}^j\right] \mathrm{e}^{\mathrm{i}\,\omega \mathbf{n}\cdot(\mathbf{x}-\mathbf{x}')} \mathrm{e}^{-\,\mathrm{i}\omega\tau}, \end{aligned} \tag{4.1.25}$$

$x = (t, \mathbf{x})$, $x' = (t', \mathbf{x}')$, with $t, t'$ defined in Eq. (4.1.24).

Needless to say, we may equivalently re-write the above expression in the form,

$$\begin{aligned} \mathrm{P}(\omega, T) &= \frac{\omega^2}{16\pi^3} \int \mathrm{d}^3\mathbf{x} \int \mathrm{d}^3\mathbf{x}' \int_{-\infty}^{\infty} \mathrm{d}\tau \left[\mathbf{J}(x) \cdot \mathbf{J}(x') \; - \; J^0(x) J^0(x')\right] \\ &\times \int \mathrm{d}\Omega \, \mathrm{e}^{\mathrm{i}\omega\,\mathbf{n}\cdot(\mathbf{x}-\mathbf{x}')} \mathrm{e}^{-\,\mathrm{i}\omega\tau}. \end{aligned} \tag{4.1.26}$$

Equation (4.1.25), for example, also leads to the expression of the power of radiation per unit frequency, per solid unit angle, with radiation emitted along the unit vector $\mathbf{n}$ given by:

$$\begin{aligned} \mathrm{P}_{\mathbf{n}}(\omega, T) &= \frac{\omega^2}{16\pi^3} \int \mathrm{d}^3\mathbf{x} \int \mathrm{d}^3\mathbf{x}' \int_{-\infty}^{\infty} \mathrm{d}\tau \, J^i(x) J^j(x') \\ &\times \left[\delta^{ij} - \mathbf{n}^i \mathbf{n}^j\right] \mathrm{e}^{\mathrm{i}\omega\,\mathbf{n}\cdot(\mathbf{x}-\mathbf{x}')} \mathrm{e}^{-\,\mathrm{i}\omega\tau}. \end{aligned} \tag{4.1.27}$$

We note from Eq. (4.1.23), that $E(\omega)$ is invariant under the simultaneous interchange $(x, x', \omega) \to (x', x, -\omega)$. Accordingly to obtain the total energy of radiation $E$, we may instead of integrating $E(\omega)$ in Eq. (4.1.23) from 0 to $\infty$, we may integrate over $\omega$ from $-\infty$ to $\infty$, and divide the resulting integral by 2, to obtain

$$\begin{aligned} E &= \frac{1}{32\pi^3} \int_{-\infty}^{\infty} \mathrm{d}\omega \, \omega^2 \int (\mathrm{d}x) \int (\mathrm{d}x') J^i(x) J^j(x') \\ &\times \int \mathrm{d}\Omega \left[\delta^{ij} - \mathbf{n}^i \mathbf{n}^j\right] \mathrm{e}^{\mathrm{i}\omega\,\mathbf{n}\cdot(\mathbf{x}-\mathbf{x}')} \mathrm{e}^{-\,\mathrm{i}\omega(t-t')}. \end{aligned} \tag{4.1.28}$$

Upon introducing the following Fourier transforms:

$$J^{i*}(\omega, \mathbf{x}) = \int_{-\infty}^{\infty} \mathrm{d}t \, \mathrm{e}^{-\mathrm{i}\omega t} \, J^i(t, \mathbf{x}), \tag{4.1.29}$$

$$J^j(\omega, \mathbf{x}') = \int_{-\infty}^{\infty} \mathrm{d}t' \, \mathrm{e}^{\mathrm{i}\omega t'} \, J^j(t', \mathbf{x}'), \tag{4.1.30}$$

as well as the Dirac distribution

$$\delta\big(\omega - \omega'\big) = \frac{1}{2\pi} \int_{-\infty}^{\infty} \mathrm{d}T \, \mathrm{e}^{\mathrm{i}\,(\omega-\omega')T}, \tag{4.1.31}$$

the energy of radiation emitted per unit time, i.e., the power of radiation $\mathrm{P}(T)$, becomes

$$\mathrm{P}(T) = \frac{1}{64\pi^4}\int \mathrm{d}\Omega \left| \int \mathrm{d}^3\mathbf{x}\, \mathrm{e}^{-\mathrm{i}\,\omega \mathbf{n}\cdot\mathbf{x}} \int_{-\infty}^{\infty} \mathrm{d}\omega\, \omega\, \mathrm{e}^{-\mathrm{i}\omega T}\, \hat{\mathbf{J}}(\omega, \mathbf{x}) \right|^2, \tag{4.1.32}$$

where

$$\hat{J}^j(\omega, \mathbf{x}) = J^i(\omega, \mathbf{x})\big[\delta^{ij} - \mathbf{n}^i\mathbf{n}^j\big], \quad \text{and we have used that} \tag{4.1.33}$$

$$\big[\delta^{ik} - \mathbf{n}^i\mathbf{n}^k\big]\big[\delta^{kj} - \mathbf{n}^k\mathbf{n}^j\big] = \big[\delta^{ij} - \mathbf{n}^i\mathbf{n}^j\big]. \tag{4.1.34}$$

In particular, the expression of the power of radiation emitted within a solid angle in the direction of a unit vector $\mathbf{n}$, emerges as

$$\mathrm{P}_{\mathbf{n}}(T) = \frac{1}{64\pi^4}\left| \int \mathrm{d}^3\mathbf{x}\, \mathrm{e}^{-\mathrm{i}\,\omega \mathbf{n}\cdot\mathbf{x}} \int_{-\infty}^{\infty} \mathrm{d}\omega\, \omega\, \mathrm{e}^{-\mathrm{i}\omega T}\, \hat{\mathbf{J}}(\omega, \mathbf{x}) \right|^2. \tag{4.1.35}$$

Finally, the average number $< N >$ of photons emitted per unit time may be also expressed from Eqs. (4.1.17), (4.1.18), by integrating over the solid angle as given in Eq. (4.1.21), to be explicitly given by

$$\begin{aligned} <N> = {} & \frac{1}{4\pi^2}\int_0^{\infty} \mathrm{d}\omega \int_{-\infty}^{\infty} \mathrm{d}\tau\, \mathrm{e}^{-\mathrm{i}\omega\tau} \int \mathrm{d}^3\mathbf{x}\, \mathrm{d}^3\mathbf{x}' \\ & \times \frac{\sin\omega|\mathbf{x}-\mathbf{x}'|}{|\mathbf{x}-\mathbf{x}'|}\Big[\mathbf{J}(x)\cdot\mathbf{J}(x') - J^0(x)J^0(x')\Big], \end{aligned} \tag{4.1.36}$$

where $t, t', \tau = t - t'$ are given in Eq. (4.1.24).

Now we come to the important point of bringing polarization aspects of radiation in evidence. Referring to $\mathrm{P}(\omega, T)$ in Eq. (4.1.25), and to the completeness relation in 3D space in Eq. (3.4.4),

$$[\delta^{ij} - \mathbf{n}^i\mathbf{n}^j] = \sum_{\lambda=1,2} \mathbf{e}^i_\lambda \mathbf{e}^j_\lambda, \tag{4.1.37}$$

express the power of radiation per unit frequency about $\omega$ as follows

$$\begin{aligned} \mathrm{P}(\omega, T) = {} & \frac{\omega^2}{16\pi^3}\int \mathrm{d}^3\mathbf{x} \int \mathrm{d}^3\mathbf{x}' \int_{-\infty}^{\infty} \mathrm{d}\tau \\ & \times \int \mathrm{d}\Omega \sum_{\lambda=1,2} \big(J^i(x)\,\mathbf{e}^i_\lambda\big)\big(J^j(x')\,\mathbf{e}^j_\lambda\big)\, \mathrm{e}^{\mathrm{i}\omega\,\mathbf{n}\cdot(\mathbf{x}-\mathbf{x}')}\mathrm{e}^{-\mathrm{i}\omega\tau}, \end{aligned} \tag{4.1.38}$$

consisting a sum of two terms, providing information of radiation emitted within a solid angle in the direction of the unit vector $\mathbf{n}$, with polarizations $\mathbf{e}_1$, $\mathbf{e}_2$. An explicit application, for example, of this, singling out polarizations of radiation will be given in Sect. 6.5 of Chap. 6 on synchrotron radiation.

Likewise, we may express the average number $< N >$ of photon emitted per unit time, from Eqs. (4.1.17), (4.1.18), (4.1.38), as

$$< N > = \int_0^\infty \mathrm{d}\omega\, N(\omega), \tag{4.1.39}$$

$$\begin{aligned} N(\omega) &= \frac{\omega}{16\pi^3} \int (\mathrm{d}^3\mathbf{x}) \int (\mathrm{d}^3\mathbf{x}') \int_{-\infty}^{\infty} \mathrm{d}\tau \int \mathrm{d}\Omega \sum_{\lambda=1,2} \\ &\times \big(J^i(x)\,\mathbf{e}^i_\lambda\big)\big(J^j(x')\,\mathbf{e}^j_\lambda\big)\,\mathrm{e}^{\mathrm{i}\omega\,\mathbf{n}\cdot(\mathbf{x}-\mathbf{x}')}\,\mathrm{e}^{-\,\mathrm{i}\omega\tau}, \end{aligned} \tag{4.1.40}$$

where in both Eqs. (4.1.38) and (4.1.40), $t = T + \tau/2,\ t' = T - \tau/2,\ t - t' = \tau$.

## 4.2 Power of Radiation: Accelerating Relativistic Charged Particle

The components of the current $J^\mu(t, \mathbf{x})$ associated with a charged particle of charge e, may be spelled out as:

$$\mathbf{J}(x) = \mathrm{e}\,\dot{\mathbf{R}}(t)\,\delta^{(3)}\big(\mathbf{x} - \mathbf{R}(t)\big), \quad \dot{\mathbf{R}} = \frac{\mathrm{d}}{\mathrm{d}t}\mathbf{R(t)}, \quad J^0(x) = \mathrm{e}\,\delta^{(3)}\big(\mathbf{x} - \mathbf{R}(t)\big). \tag{4.2.1}$$

We will work out the power of radiation for the problem at hand from the expression of the total energy $E$ of radiation obtained by integrating $E(\omega)$ in Eq. (4.1.19) over $\omega$ from $-\infty$ to $\infty$ and dividing by 2. From Eq. (4.2.1), $E$ now takes the form

$$\begin{aligned} E &= \frac{\mathrm{e}^2}{32\pi^3} \int_{-\infty}^{\infty} \mathrm{d}\omega\,\omega^2 \int \mathrm{d}\Omega \int_{-\infty}^{\infty} \mathrm{d}t \int_{-\infty}^{\infty} \mathrm{d}t' \times \\ &\Big[\dot{\mathbf{R}}(t') \cdot \dot{\mathbf{R}}(t) - 1\Big]\mathrm{e}^{\mathrm{i}\omega[(\mathbf{R}(t')-\mathbf{R}(t))\cdot\mathbf{n}]}\mathrm{e}^{-\mathrm{i}\omega(t'-t)}, \end{aligned} \tag{4.2.2}$$

where, using symmetry, we have found it convenient to interchange $t \leftrightarrow t'$.

We set[2]

$$t' - t = \tau, \quad \tau\left(1 - \frac{[\mathbf{R}(t') - \mathbf{R}(t)]}{\tau} \cdot \mathbf{n}\right) = \gamma, \tag{4.2.3}$$

$$\text{and note that } \frac{|\,\mathbf{R}(t') - \mathbf{R}(t)\,|}{\tau} < 1, \text{ and for } \tau \to \pm\infty, \Rightarrow \gamma \to \pm\infty, \tag{4.2.4}$$

[2] We make use of the ingenious approach of change of variables by Schwinger (1949) [4] in his classic paper.

where, regarding the inequality in Eq. (4.2.4), remember we have set c equal to one, which would otherwise replace the 1 on the r.h.s. of the inequality. Thus we obtain,

$$E = \frac{e^2}{32\pi^3}\int_{-\infty}^{\infty} d\omega\,\omega^2 \int d\Omega \int_{-\infty}^{\infty} dt \int_{-\infty}^{\infty} d\gamma\, e^{-i\omega\gamma}\left(\frac{d\tau}{d\gamma}\right) \mid \dot{\mathbf{R}}(t')\cdot\dot{\mathbf{R}}(t) - 1 \mid . \tag{4.2.5}$$

Using the elementary integral

$$\int_{-\infty}^{\infty} d\omega\,\omega^2\, e^{-i\omega\gamma} = -(2\pi)\left(\frac{d}{d\gamma}\right)^2 \delta(\gamma), \tag{4.2.6}$$

gives, as a consequence of the delta function, the expression

$$E = -\frac{e^2}{16\pi^2 c}\int d\Omega \int_{-\infty}^{\infty} dt \int_{-\infty}^{\infty} d\gamma\,\delta(\gamma)\left(\frac{d}{d\gamma}\right)^2\left[\frac{d\tau}{d\gamma}\left(\mathbf{v}(t+\tau)\cdot\mathbf{v}(t) - 1\right)\right], \tag{4.2.7}$$

where $\mathbf{v}(t) = \dot{\mathbf{R}}$, $t' = t + \tau$.

Consider the chain rule,

$$\frac{d}{d\gamma} = \frac{d\tau}{d\gamma}\frac{d}{d\tau}, \quad \frac{d\tau}{d\gamma} = \frac{1}{[1 - \mathbf{v}(t+\tau)\cdot\mathbf{n}]} \equiv f(t+\tau), \tag{4.2.8}$$

where we have used Eq. (4.2.3), defining the expression for $\tau$, and $t' = t + \tau$, to solve for $d\tau/d\gamma$. Note that, in considering the $\gamma$-integral in Eq. (4.2.7), that $\gamma = 0$, gives $\tau = 0$, and we may write

$$\left(\frac{d}{d\gamma}\right)^2 = \frac{d\tau}{d\gamma}\frac{d}{d\tau}\frac{d\tau}{d\gamma}\frac{d}{d\tau}, \tag{4.2.9}$$

$$\begin{aligned}
&\int_{-\infty}^{\infty} d\gamma\,\delta(\gamma)\,\frac{d\tau}{d\gamma}\frac{d}{d\tau}\frac{d\tau}{d\gamma}\frac{d}{d\tau}\left[\frac{d\tau}{d\gamma}[\mathbf{v}(t+\tau)\cdot\mathbf{v}(t) - 1]\right] = \\
&\quad \mathbf{v}(t)\cdot f(t+\tau)\frac{d}{d\tau} f(t+\tau)\frac{d}{d\tau} f(t+\tau)\,\mathbf{v}(t+\tau)\Big|_{\tau=0} \\
&\quad - f(t+\tau)\frac{d}{d\tau} f(t+\tau)\frac{d}{d\tau} f(t+\tau)\Big|_{\tau=0} \\
&= \mathbf{v}(t)\cdot f(t)\frac{d}{dt} f(t)\frac{d}{dt} f(t)\,\mathbf{v}(t) - f(t)\frac{d}{dt} f(t)\frac{d}{dt} f(t),
\end{aligned} \tag{4.2.10}$$

where in the last step, we used the property of $\delta(\gamma)$ implying to set $\tau = 0$. The expression on the r.h.s of the equality in Eq. (4.2.10) may be explicitly re-written as

$$- (\dot{\mathbf{v}})^2 f^3 \;-\; 2[(\mathbf{v}\cdot\dot{\mathbf{v}})\,\dot{\mathbf{v}}\cdot\mathbf{n}]f^4 \;+\; (1-\mathbf{v}^2)(\dot{\mathbf{v}}.\mathbf{n})^2 f^5 \;+\; \frac{\mathrm{d}}{\mathrm{d}t}F(t), \qquad (4.2.11)$$

$$\text{where } F(t) = f^2(t)\Big[\mathbf{v}\cdot\frac{\mathrm{d}}{\mathrm{d}t}(f(t)\mathbf{v}) - \frac{\mathrm{d}}{\mathrm{d}t}f(t)\Big],\quad f(t) = \frac{1}{[1-\mathbf{v}(t)\cdot\mathbf{n}]} \qquad (4.2.12)$$

The last term $\mathrm{d}F(t)/\mathrm{d}t$ in Eq. (4.2.11) involves a total derivative in $t$ and will not contribute to the $t$-integral in (4.2.7) and may be omitted.

All told, the total energy emerges in the form

$$E = \frac{\mathrm{e}^2}{16\pi^2}\int \mathrm{d}\Omega \int_{-\infty}^{\infty}\mathrm{d}t\Big[\dot{\mathbf{v}}^2 f^3 \;+\; 2[(\mathbf{v}\cdot\dot{\mathbf{v}})(\dot{\mathbf{v}}\cdot\mathbf{n})]f^4 \;-\; (1-\mathbf{v}^2)(\dot{\mathbf{v}}.\mathbf{n})^2 f^5\Big]. \qquad (4.2.13)$$

By omitting the $t$-integration, the expression of the power of radiation $P$, by definition, becomes

$$P(t) = \frac{\mathrm{e}^2}{16\pi^2}\int \mathrm{d}\Omega\Big[\dot{\mathbf{v}}^2 f^3 \;+\; 2[(\mathbf{v}\cdot\dot{\mathbf{v}})(\dot{\mathbf{v}}\cdot\mathbf{n})]f^4 \;-\; (1-\mathbf{v}^2)(\dot{\mathbf{v}}.\mathbf{n})^2 f^5\Big], \qquad (4.2.14)$$

$$f(t) = \frac{1}{\big(1-\mathbf{v}(t)\cdot\mathbf{n}\big)}. \qquad (4.2.15)$$

We may express the power $P(t)$ in terms of the power of radiation $\mathrm{P}(\mathbf{n}, t)$ in a direction specified by the unit vector $\mathbf{n}$ emitted by the accelerating charged particle as:

$$P(t) \;=\; \int \mathrm{d}\Omega\,\mathrm{P}(\mathbf{n}, t), \qquad (4.2.16)$$

$$\mathrm{P}(\mathbf{n}, t) \;=\; \frac{\mathrm{e}^2}{16\pi^2}\Big[\frac{\dot{\mathbf{v}}^2}{(1-\mathbf{v}\cdot\mathbf{n})^3} + 2\,\frac{(\mathbf{v}\cdot\dot{\mathbf{v}})\,(\dot{\mathbf{v}}\cdot\mathbf{n})}{(1-\mathbf{v}\cdot\mathbf{n})^4} - \frac{(1-\mathbf{v}^2)(\dot{\mathbf{v}}.\mathbf{n})^2}{(1-\mathbf{v}\cdot\mathbf{n})^5}\Big]. \qquad (4.2.17)$$

To obtain the total power of radiation $\mathrm{P}(t)$, we use, in the process, the following useful integrals:

$$\int \mathrm{d}\Omega\,\frac{1}{(1-\mathbf{v}\cdot\mathbf{n})^3} \;=\; \frac{4\pi}{(1-\mathbf{v}^2)^2}, \qquad (4.2.18)$$

$$\int \mathrm{d}\Omega\,\frac{n^i}{(1-\mathbf{v}\cdot\mathbf{n})^4} \;=\; \frac{16\pi}{3}\,\frac{\mathrm{v}^i}{(1-\mathbf{v}^2)^3},\qquad i = 1, 2, 3. \qquad (4.2.19)$$

$$\int \mathrm{d}\Omega\,\frac{n^i\,n^j}{(1-\mathbf{v}\cdot\mathbf{n})^5} \;=\; \frac{4\pi}{(1-\mathbf{v}^2)^3}\Big[\frac{\delta^{ij}}{3} + 2\,\frac{\mathrm{v}^i\,\mathrm{v}^j}{(1-\mathbf{v}^2)}\Big],\qquad i, j = 1, 2, 3, \qquad (4.2.20)$$

with the latter two integrals obtained, for example, from the first elementary one by taking derivatives with respect $\mathrm{v}^i$ in successions. The third one may be also obtained by writing the result as $[A\delta^{ij} + B\mathrm{v}^i\mathrm{v}^j]$, and readily determine $A$ an $B$. $\delta^{ij}$ is the

Kronecker delta. Accordingly, the power of radiation is, from Eq. (4.2.17), given by

$$\mathrm{P}(t) = \int \mathrm{d}\Omega\, \mathrm{P}(\mathbf{n}, t) = \frac{2}{3}\frac{\mathrm{e}^2}{4\pi}\frac{1}{(1-\mathbf{v}^2)^2}\left[\dot{\mathbf{v}}^2 + \frac{(\dot{\mathbf{v}}\cdot\mathbf{v})^2}{(1-\mathbf{v}^2)}\right], \qquad \mathbf{v} = \dot{\mathbf{R}}. \quad (4.2.21)$$

The r.h.s of the above equality may be re-written in several equivalent forms. For example, upon writing

$$(\dot{\mathbf{v}}\cdot\mathbf{v})^2 = (|\dot{\mathbf{v}}|\,|\mathbf{v}|\cos\theta)^2 = (|\dot{\mathbf{v}}|\,|\mathbf{v}|)^2 - (\mathbf{v}\times\dot{\mathbf{v}})^2, \quad (4.2.22)$$

with $\times$ denoting the cross product notation, the expression, within the square brackets in Eq. (4.2.21) simply becomes

$$\dot{\mathbf{v}}^2 + \frac{(\dot{\mathbf{v}}\cdot\mathbf{v})^2}{(1-\mathbf{v}^2)} = \frac{\left[\dot{\mathbf{v}}^2 - \left(\mathbf{v}\times\dot{\mathbf{v}}\right)^2\right)}{(1-\mathbf{v}^2)}, \quad (4.2.23)$$

leading to the well known expression[3] to $P(t)$ in Eq. (4.2.21), given by

$$P(t) = \frac{2}{3}\frac{\mathrm{e}^2}{4\pi}\frac{1}{(1-\mathbf{v}^2)^3}\left[\dot{\mathbf{v}}^2 - \left(\mathbf{v}\times\dot{\mathbf{v}}\right)^2\right], \quad (4.2.24)$$

in generalised CGS units, involving the $1/4\pi$ factor.

This chapter is based on the author's paper cited below.[4]

This classical power of radiation of an accelerating charged particle has been treated by several authors by completely different methods and at different levels of rigor. In particular, it has been studied by the traditional approach by introducing, in the process, electromagnetic fields/potentials,[5] none of which were introduced here. It has been also considered by Schwinger more from a quantum view point in which he introduces a propagator for the charged particle itself.[6]

## References

1. Jackson, J. D. (1998). *Classical Electrodynamics, Third Edition*. New York: Wiley.
2. Manoukian, E. B. (2015). Vacuum-to-vacuum transition probability and the classic radiation theory. *Radiation Physics and Chemistry, 106*, 268–270.
3. Pathak, P.H. & Burkholder, R. J. (2022). *Electromagnetic radiation, scattering, and diffraction*. New Jersey: Wiley.
4. Schwinger, J. (1949). Classical radiation of accelerated electrons. *Physical Review, 75*, 1912–1925.

[3] Schwinger (1949) [4], Eq. (I.4).

[4] Manoukian (2015) [2].

[5] See e.g., Schwinger (1949) [4], Jackson (1998) [1], Zangwill [6], Pathak (2022) [3].

[6] Schwinger (1973) [5].

5. Schwinger, J. (1973). Classical radiation of accelerated electrons, II. Quantum viewpoint. *Physical Review, D7*, 1696–1701.
6. Zangwill, A. (2012). *Modern Electrodynamics*. Cambridge: Cambridge.

# Chapter 5
# The Field Concept, The Gauge Problem and Derivation of Maxwell's Equations

*The field concept, and as a consequence of gauge invariance, what one calls electric and magnetic fields emergence from the formalism, and, in turn, their field equations simply pop up, famously called Maxwell's equations*

In the present chapter, we make contact with the traditional theory by rather starting from the vacuum-to-vacuum amplitude which was expressed in terms of a conserved current $J^{\mu}(x)$.

The field concept is introduced as the effect on an infinitesimal conserved probe current $\delta J^{\mu}(x)$, as a test source, due to a pre-existing current $J^{\mu}(x)$. As a consequence of the conservation of the currents, this naturally, leads to the necessity of specifying a so-called gauge in order to define the field. The physical requirement that the underlying theory be a gauge invariant one, gives rise to two 3-vector fields, one calls the electric and magnetic fields. The field equations emerge naturally in a unified manner and branch out into the familiar four celebrated equations one refers to as Maxwell's equations. The corresponding equations involving phenomenological parameters in matter are then considered and the relevant boundary conditions of the theory are derived by simply using Dirac delta distributions encountered earlier. Moreover, the propagation of the electromagnetic field in vacuum and in matter are described, as well as basic equations of the subject of electrodynamics are derived.

## 5.1 The Field Concept and the Gauge Problem: Emergence of Two 3-Vector Fields One calls Electric and Magnetic Fields

A field, denoted by $A^{\mu}(x)$ at a spacetime point $x$, is considered as a measure of the strength of excitation that the effect a current $J^{\mu}(x')$ may have on a test infinitesimal conserved current $\delta J^{\mu}(x)$ at the spacetime point $x$ in question. This is obtained by

 
E. B. Manoukian, *Radiation*,
https://doi.org/10.1007/978-3-032-21506-2_5

adding an infinitesimal current $\delta J^\mu(x)$ to a pre-existing current $J^\mu$ in the vacuum-to-vacuum transition amplitude $< 0_+ \mid 0_- >$ in Eq. (3.2.11):

$$< 0_+ \mid 0_- >_J = \mathrm{e}^{\mathrm{i}W[J]}, \tag{5.1.1}$$

$$W[J] = \frac{1}{2}\int(\mathrm{d}x)\int(\mathrm{d}x')J^\mu(x)D_+(x-x')J_\mu(x'), \tag{5.1.2}$$

leading to the change

$$W[J] \to W[J+\delta J] = W[J] + \int(\mathrm{d}x)\,\delta J_\mu(x)A^\mu(x), \tag{5.1.3}$$

where the additional term

$$\int(\mathrm{d}x)\,\delta J_\mu x A^\mu(x), \tag{5.1.4}$$

gives a measure of the effect, or a measure of the strength of the excitation at $x$, where the test current is located, due to the pre-existing current $J^\mu(x)$.

For an infinitesimal test current $\delta J^\mu(x)$,

$$W[J+\delta J] = W[J] +$$
$$\frac{1}{2}\int(\mathrm{d}x)\int(\mathrm{d}x')\big[\delta J^\mu(x)D_+(x-x')J_\mu(x') + J^\mu(x)D_+(x-x')\delta J_\mu(x')\big]. \tag{5.1.5}$$

Or, using the fact that $D_+(x-x') = D_+(x'-x)$, we have

$$W[J+\delta J] = W[J] + \int(\mathrm{d}x)\,\delta J_\mu(x)\big[\int(\mathrm{d}x')D_+(x-x')J^\mu(x')\big]. \tag{5.1.6}$$

From Eqs. (5.1.3), (5.1.6), the expression of the field thus emerges naturally from the equation

$$\int(\mathrm{d}x)\,\delta J_\mu(x)\big[\int(\mathrm{d}x')D_+(x-x')J^\mu(x')\big] = \int(\mathrm{d}x)\,\delta J_\mu(x)A^\mu(x). \tag{5.1.7}$$

The physical significance of this equation is clear. The l.h.s. describes, via the propagator $D_+(x-x')$, the effect that the current $J^\mu(x')$ may have, from spacetime point $x'$ to spacetime point $x$, on the test current $\delta J_\mu(x)$. While the r.h.s. gives a measure of this effect, or a measure of the strength of this excitation, at $x$, where the test current is located.

The above equation may be elegantly re-written in the equivalent form

$$\int (\mathrm{d}x)\, \delta J_\mu(x) A^\mu(x) = (-\mathrm{i})\frac{\delta < 0_+ \mid 0_- >}{< 0_+ \mid 0_- >}, \tag{5.1.8}$$

as is easily verified.

We are almost ready to extract the expression of the field $A^\mu(x)$ in question. To this end note that if we add any divergence term $\partial^\mu \Lambda(x)$, to $A^\mu(x)$ on the r.h.s. of Eq. (5.1.7), with $\Lambda(x)$ an arbitrary function, then an integration by parts shows that the integral in question does not change, due to the conservation law $\partial^\mu \delta J_\mu(x) = 0$.

That is, $A^\mu(x)$ *may be defined up to an arbitrary additive divergence term* $\partial^\mu \Lambda$. This is ***the gauge problem of electrodynamics***. It is not sufficiently emphasized in the classical literature. The importance of this will be obvious when one demands, on physical grounds, that the resulting theory must be gauge invariant.

From Eq. (5.1.7), we may infer that

$$A^\mu(x) = \int (\mathrm{d}x') D_+(x - x') J^\mu(x') + \partial^\mu \Lambda(x), \tag{5.1.9}$$

up to an *arbitrary* $\partial^\mu \Lambda$ term shown, as mentioned above.

In particular, in considering the expression $\partial_\mu A^\mu(x)$, the above equation gives

$$\partial_\mu A^\mu(x) = \int (\mathrm{d}x')\, \partial_\mu D_+(x - x') J^\mu(x') + \partial_\mu \partial^\mu \Lambda(x), \tag{5.1.10}$$

$$= \int (\mathrm{d}x')\, D_+(x - x') \partial'_\mu J^\mu(x') + \partial_\mu \partial^\mu \Lambda(x), \tag{5.1.11}$$

where in the last step we used $\partial_\mu D_+(x - x') = -\partial'_\mu D_+(x - x')$, and integrated by parts with respect to $x'$. Hence by using the conservation law $\partial'_\mu J_\mu(x') = 0$, it follows that

$$\partial_\mu A^\mu(x) = \partial_\mu \partial^\mu \Lambda(x) \equiv \Box \Lambda(x). \tag{5.1.12}$$

Since $\partial_\mu \partial^\mu \Lambda$ is arbitrary, one choice would be to set it equal to zero, giving

$$\partial_\mu A^\mu(x) = 0. \tag{5.1.13}$$

This choice of gauge is referred to as the ***Lorentz gauge***.

Another popular gauge, used especially in radiation theory, is to choose

$$\partial^0 \Lambda(x) = -\int (\mathrm{d}x') D_+(x - x') J^0(x'). \tag{5.1.14}$$

In this gauge, referred to as the ***temporal gauge***, Eq. (5.1.9) gives

$$A^0(x) = 0. \tag{5.1.15}$$

Quite generally, Eq. (5.1.9) leads to the following two equations,

$$\partial^\nu \partial_\mu A^\mu(x) = \partial^\nu \Box \Lambda(x), \tag{5.1.16}$$

$$\partial_\mu \partial^\mu A^\nu(x) = -J^\nu(x) + \partial^\nu \Box \Lambda, \tag{5.1.17}$$

where we have used Eq. (3.3.11)

$$-\Box D_+(x-x') = \delta^{(4)}(x-x'), \quad \text{and } \Box = \partial^\mu \partial_\mu = \nabla^2 - \frac{\partial^2}{\partial t^2}, \quad \nabla^2 \equiv \boldsymbol{\partial}^2. \tag{5.1.18}$$

and the property of the Dirac delta distribution in Eq. (3.6.12).

By considering the combination of Eq. (5.1.17) minus Eq. (5.1.16), we finally obtain the ***gauge invariant field equations of electrodynamics***:

$$\partial_\mu F^{\mu\nu}(x) = -J^\nu(x), \tag{5.1.19}$$

$$\textbf{where} \quad F^{\mu\nu}(x) = \partial^\mu A^\nu(x) - \partial^\nu A^\mu(x). \tag{5.1.20}$$

Clearly, $F^{\mu\nu}(x)$ is gauge invariant, i.e., it is independent of arbitrary functions like $\Lambda(x)$ in Eq.(5.1.9). This is easily verified from the cancellation of the gauge term in the transformation embodied in the following equation

$$F^{\mu\nu}(x) \rightarrow F^{\mu\nu}(x) + \big[\partial^\mu \partial^\nu \Lambda(x) - \partial^\nu \partial^\mu \Lambda(x)\big] = F^{\mu\nu}(x), \tag{5.1.21}$$

by using, in the process, the gauge transformation $A^\sigma \rightarrow A^\sigma + \partial^\sigma \Lambda$, for $\sigma = \nu$, and $\sigma = \mu$ in Eq. (5.1.20).

We note that upon applying $\Box$ to $A^\mu(x)$ in Eq. (5.1.9) and using Eq. (5.1.18), give

$$\Box A^\mu(x) = -J^\mu(x) + \partial^\mu \Box \Lambda(x). \tag{5.1.22}$$

On the other hand, in a convenient notation, Eq. (5.1.16) gives

$$\partial^\mu \partial^\nu A_\nu(x) = \partial^\mu \Box \Lambda(x). \tag{5.1.23}$$

Upon combining Eqs. (5.1.17)/(5.1.23), we obtain

$$\big[-\eta^{\mu\nu}\Box + \partial^\mu \partial^\nu\big] A_\nu(x) = J^\mu(x). \tag{5.1.24}$$

This equation, however, does not give rise to a unique solution due to the inherited arbitrariness of $\Lambda(x)$.[1]

---

[1] For the mathematical keen reader, this, in a technical language, means that the operator that is applied to $A_\nu(x)$ is not invertible.

It is thus important to work out the equations of the field $A^{\mu}(x)$, e.g., in the two gauges discussed above.

In the Lorentz gauge, specified in Eq. (5.1.13), the field equations Eq. (5.1.19) give,

$$\Box A^{\nu}(x) = -J^{\nu}(x), \Rightarrow A^{\nu}(x) = \int (\mathrm{d}x')\, D_{+}(x-x')J^{\nu}(x'). \tag{5.1.25}$$

While in the temporal gauge, specified in Eq. (5.1.15), the field equations in Eq. (5.1.19) give,

$$\Box A^{i} - \partial^{i}\partial^{j}A^{j}(x) = -J^{i}(x), \quad \text{or} \tag{5.1.26}$$
$$\big[(-\boldsymbol{\partial}^{2} + \partial^{02})\delta^{ij} + \partial^{i}\partial^{j}\big]A^{j}(x) = J^{i}(x). \tag{5.1.27}$$

Quite generally, upon writing,

$$A^{i}(x) = \int (\mathrm{d}x') D_{+}^{ij}(x,x')\, J^{j}(x'), \tag{5.1.28}$$

we immediately have from Eq. (5.1.27), the equation satisfied by the propagator $D^{ij}(x,x')$ to be given by

$$\big[(-\,\boldsymbol{\partial}^{2} + \partial^{02})\delta^{i\ell} + \partial^{i}\partial^{\ell}\big]D_{+}^{\ell j}(x,x') = \delta^{ij}\,\delta^{(4)}(x-x'), \tag{5.1.29}$$

coinciding with Eq. (3.3.15), using the equivalent notations $\boldsymbol{\nabla}^{2} = \boldsymbol{\partial}^{2}$. This is easily verified by applying the partial differential operator preceding $D^{ij}(x,x')$ in Eq. (5.1.29) to $A^{i}(x)$ in Eq. (5.1.28), using finally Eq. (5.1.27), and referring to the property of the Dirac delta distribution in Eqs. (3.6.12), (3.6.24).

Another useful relation satisfied by the propagator $D_{+}^{ij}(x,x')$ is the following. We may rewrite the right-hand side of Eq. (5.1.28) as

$$\int (\mathrm{d}x'') \int (\mathrm{d}x') D_{+}^{ik}(x,x'')\delta^{kj}\delta^{(4)}(x''-x')J^{j}(x'), \quad \text{but} \tag{5.1.30}$$
$$\big[(-\,\boldsymbol{\partial}''^{2} + \partial''^{02})\delta^{k\ell} + \partial''^{k}\partial''^{\ell}\big]D_{+}^{\ell j}(x'',x') = \delta^{kj}\,\delta^{(4)}(x''-x'). \tag{5.1.31}$$

Upon replacing $\delta^{kj}\delta^{(4)}(x''-x')$ in Eq. (5.1.30), by the expression on the left-hand side of Eq. (5.1.31), and integrating by parts with respect to $x''$, Eq. (5.1.30) becomes

$$\int (\mathrm{d}x'') \int (\mathrm{d}x') \Big(\big[(-\,\boldsymbol{\partial}''^{2} + \partial''^{02})\delta^{k\ell} + \partial''^{k}\partial''^{\ell}\big]D_{+}^{ik}(x,x'')\Big) \times D_{+}^{\ell j}(x'',x')J^{j}(x'). \tag{5.1.32}$$

This will coincide with $A^{i}(x)$ itself in Eq. (5.1.28), if

$$\big[(-\,\partial''^2 + \partial''^{02})\delta^{k\ell} + \partial''^k\partial''^\ell\big]D_+^{ik}(x, x'') = \delta^{\ell i}\delta^{(4)}(x - x''). \tag{5.1.33}$$

Using the notation in Eq. (5.1.29), thus we obtain another equation satisfied by the propagator $D_+^{\ell j}(x, x')$, now with differentiations carried out with respect to $x'$ rather than $x$, i.e., unlike what we have in Eq. (5.1.29),

$$\big[(-\,\partial'^2 + \partial'^{02})\delta^{ji} + \partial'^j\partial'^i\big]D_+^{\ell j}(x, x') = \delta^{i\ell}\,\delta^{(4)}(x - x'), \tag{5.1.34}$$

Here, note that the derivatives are on the second variable $x'$ of $D^{\ell j}(x, x')$, also note positions of the indices $\ell, i, j$.

Equations (5.1.29) and (5.1.34) are the basic equations satisfied by the propagator in free space in the subject of ***classical electrodynamics*** in the temporal gauge. They will readily be generalized in a medium in Eqs. (5.4.52) and (5.4.53) of Sect. 5.4.

Equations (5.1.27)–(5.1.29) and (5.1.34) constitute the basic equations of classical electrodynamics in free space, in the temporal gauge.

Equation (5.1.8) is historically of great importance as it is a special case of the so-called *Schwinger's action principle*, or also called, more appropriately, as *Schwinger's dynamical principle*, well known to field theorists. It was introduced by Schwinger in his monumental work early in his career.[2] It is interesting to point out that the *entire* subject of quantum field theories of the fundamental particles may be developed from his dynamical principle.[3]

It is remarkable that *gauge invariance* led to the introduction of an anti-symmetric second rank tensor field with components $F^{\mu\nu}(x)$, i.e., which may have only 6 independent components, due to its anti-symmetric property. These may be explicitly spelled out as

$$\big(F^{01}(x), F^{02}(x), F^{03}\big), \quad \big(F^{23}(x), F^{31}(x), F^{12}(x)\big), \tag{5.1.35}$$

representing two 3-vector fields and may be denoted as $\mathbf{E}(x), \mathbf{B}(x)$ which have precise transformations rules under coordinate transformations which will be discussed at the end of Sect. 5.3.

We have, in particular, learnt that, *from gauge invariance, two 3-vector fields emerge from the theory, which one calls the electric and magnetic fields* $\mathbf{E}(x), \mathbf{B}(x)$. Details on these field components will be given in Sects. 5.3–5.6.

---

[2] See, e.g., Schwinger (1951), (1953) [7, 8], also used later in his career, e.g., Schwinger (1970) [10] as used here.

[3] For a complete text-book treatment of this, including the proof of the principle itself in field theory, see one of my earlier books: Manoukian (2016) [6]; see also Manoukian (1986) [5], for the interested reader.

## 5.2 Action Integral of Electrodynamics

Although, Dirac hasn't called the vacuum-to-vacuum transition amplitude as such, he postulated[4] several years ago, that it is essentially given by the exponential of the action-integral (times $\mathrm{i}/\hbar$). Feynman was inspired by this expression and it re-appeared later in his fundamental work on path integrals.[5] With field equations of electrodynamics derived in Eq. (5.1.19), we may directly use them to recast the vacuum-to-vacuum transition amplitude in the form stated by Dirac.

To the above end, we use the field equations given through Eqs. (5.1.19), (5.1.20), (5.1.24), and the equation satisfied by the propagator in Eq. (5.1.18) spelled out as

$$-\,\partial_\nu F^{\nu\mu}(x) = J^\mu(x), \qquad F^{\mu\nu}(x) = \partial^\mu A^\nu(x) - \partial^\nu A^\mu(x), \quad \text{or in detail} \tag{5.2.1}$$

$$\big[-\eta^{\mu\nu}\Box + \partial^\mu\partial^\nu\big]A_\nu(x) = J^\mu(x). \tag{5.2.2}$$

$$-\,\Box\, D_+(x-x') = \delta^{(4)}(x-x'), \tag{5.2.3}$$

up to gauge fixing.

We re-write the expression of the vacuum-to-vacuum transition amplitude in Eq. (3.3.9) in the form

$$< 0_+ \mid 0_- > = \mathrm{e}^{\mathrm{i}\mathcal{A}}, \tag{5.2.4}$$

where

$$\mathcal{A} = \frac{1}{2}\int(\mathrm{d}x)\int(\mathrm{d}x')J^\mu(x)D_+(x-x')J_\mu(x'), \tag{5.2.5}$$

and replace $J^\mu(x)$ appearing to the left of $D_+$, of the above equation, by its equivalent given in Eq. (5.2.2), to obtain

$$\begin{aligned}\mathcal{A} &= \frac{1}{2}\int(\mathrm{d}x)(\mathrm{d}x')\Big(\big[-\eta^{\mu\nu}\Box + \partial^\mu\partial^\nu\big]A_\nu(x)\Big)D_+(x-x')J_\mu(x')\\ &= \frac{1}{2}\int(\mathrm{d}x)(\mathrm{d}x')A^\mu(x)\delta^{(4)}(x-x')J_\mu(x') = \frac{1}{2}\int(\mathrm{d}x)A^\mu(x)J_\mu(x),\end{aligned} \tag{5.2.6}$$

where note that the $\partial^\mu\partial^\nu$ term in Eq. (5.2.2) does not contribute by integrating by parts and then using the conservation law $\partial'_\mu J^\mu(x') = 0$. We have also used Eq. (5.2.3), again, after integrating by parts generating the Dirac delta $\delta^{(4)}(x-x')$ term.

---

[4] Dirac (1933) [1].

[5] See, e.g., Feynman (1948), (1965) [2], [3].

On the other hand,

$$\begin{aligned}\int (\mathrm{d}x) A^\mu(x) J_\mu(x) &= \int (\mathrm{d}x) A^\mu(x)[-\eta_{\mu\nu}\Box + \partial_\mu\partial_\nu]A^\nu(x) \\ &= \int (\mathrm{d}x)\, [\partial^\mu A^\nu \partial_\mu A_\nu - \partial^\mu A^\nu \partial_\nu A_\mu] \\ &= \frac{1}{2}\int (\mathrm{d}x) F_{\mu\nu}(x) F^{\mu\nu}(x),\end{aligned} \tag{5.2.7}$$

where we have integrated by parts while going from the first equation to the second one above, and recall that $\Box = \partial^\lambda \partial_\lambda$.

From Eqs. (5.2.6), (5.2.7), we, respectively, have

$$\mathcal{A} = \frac{1}{2}\int (\mathrm{d}x) A^\mu(x) J_\mu(x) \tag{5.2.8}$$

$$\equiv -\frac{1}{2}\int (\mathrm{d}x) A^\mu(x) J_\mu(x) + \int (\mathrm{d}x) A^\mu(x) J_\mu(x) \tag{5.2.9}$$

$$= \int (\mathrm{d}x)\Big[ -\frac{1}{4} F_{\mu\nu}(x) F^{\mu\nu}(x) + A^\mu(x) J_\mu(x)\Big]. \tag{5.2.10}$$

Therefore, from Eqs. (5.2.5), (5.2.10), the following equations emerge

$$< 0_+ \mid 0_- > = \mathrm{e}^{\mathrm{i}\mathcal{A}}, \tag{5.2.11}$$

$$\mathcal{A} = \int (\mathrm{d}x)\, \mathcal{L}(x), \tag{5.2.12}$$

$$\mathcal{L}(x) = -\frac{1}{4} F_{\mu\nu}(x) F^{\mu\nu}(x) + A^\mu(x) J_\mu(x), \tag{5.2.13}$$

where the expression in Eq. (5.2.11) is in the Dirac form mentioned above, with $\mathcal{A}$, now, given in Eq. (5.2.12) denoting the *action*, and $\mathcal{L}(x)$ in Eq. (5.2.13) denoting the *Lagrangian density of electrodynamics*, necessarily, up to gauge fixing.

## 5.3 Derivation of Maxwell's Equations

We recall the field equations of electrodynamics derived in Eqs. (5.1.19), (5.1.20), and the definition of the 4-current and its conservation law given in Eqs. (3.8), and (3.9) in Chap. 3, respectively,

$$\partial_\mu F^{\mu\nu}(x) = -\, J^\nu(x), \qquad F^{\mu\nu}(x) = \partial^\mu A^\nu(x) - \partial^\nu A^\mu(x), \tag{5.3.1}$$

$$J^\mu = (\rho, \mathbf{J}), \qquad \nabla\cdot\mathbf{J}(x) + \frac{\partial}{\partial t}\rho(x) = 0. \tag{5.3.2}$$

Also note that the tensor field $F^{\mu\nu}(x)$ is anti-symmetric in its indices $(\mu, \nu)$: $F^{\mu\nu}(x) = -F^{\nu\mu}(x)$, and the following equation readily follows,

$$\partial^{\mu} F^{\sigma\nu} + \partial^{\nu} F^{\mu\sigma} + \partial^{\sigma} F^{\nu\mu} = 0, \tag{5.3.3}$$

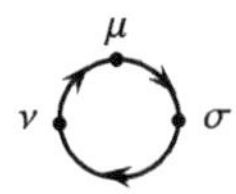


Moreover, this anti-symmetry property implies $\partial_\nu \partial_\mu F^{\mu\nu} = 0$, which, from the first equation in Eq. (5.3.1), is, of course, the statement of the conservation of the current $\partial_\mu J^\mu = 0$.

We recall that in making a transition from the vector field $A^\mu(x)$ to the tensor field $F^{\mu\nu}(x)$ arose to ensure that the ensuing theory be *gauge invariant*. As a consequence of the anti-symmetry property of $F^{\mu\nu}(x)$ in its indices, its diagonal elements $\{F^{00}, F^{11}, F^{22}, F^{33}\}$ are zero, and only *six* of its components may be independent. This is crying out loud to introduce two 3-vectors, i.e., with three components, as we have touched upon at the end of Sect. 5.1. Accordingly, we may simply set:

$$F^{0i} \equiv E^i, \quad i = 1, 2, 3, \tag{5.3.4}$$

$$F^{12} \equiv B^3, \; F^{31} \equiv B^2, \; F^{23} \equiv B^1, \quad \text{or} \quad \frac{1}{2}\epsilon^{ijk} F^{jk} \equiv B^i, \; i, j, k, = 1, 2, 3, \tag{5.3.5}$$

and a summation over the repeated indices is understood, where from Eq. (3.6.9) of the Appendix of Chap. 3,

$$\epsilon^{ijk} = \begin{cases} +1 & \text{if } \{i, j, k\} \text{ is an even permutation of } \{1, 2, 3\} \\ -1 & \text{if } \{i, j, k\} \text{ is an oddpermutation of } \{1, 2, 3\} \\ 0 & \text{if two or three of the indices in } \{i, j, k\} \text{ are equal} \end{cases}, \tag{5.3.6}$$

referred to as the Levi-Civita symbol, and from Eq. (3.6.10), it satisfies, in particular, the following key equation:

$$\epsilon^{ijk}\epsilon^{i\ell m} = \delta^{j\ell}\delta^{km} - \delta^{jm}\delta^{k\ell}. \tag{5.3.7}$$

As a result, one may express $F^{ij}$ in terms of $B^i$ as follows

$$F^{ij} = \epsilon^{ijk} B^k, \; B^i = \frac{1}{2}\epsilon^{ijk}\big[\partial^j A^k - \partial^k A^j\big] = \epsilon^{ijk}\partial^j A^k, \; \text{i.e,} \; \mathbf{B} = \nabla \times \mathbf{A}, \tag{5.3.8}$$

$$\text{also note that} \quad (\nabla \times \mathbf{B})^i = \epsilon^{ijk}\partial_j B^k, \tag{5.3.9}$$

with the latter following from the definition of the curl of a vector.

We consider, in turn, the 4 components of the field equations $\partial_\mu F^{\mu\nu} = -J^\mu$. For $\nu = 0$,

$$\partial_\mu F^{\mu 0} = -J^0, \;\Rightarrow\; \partial_i F^{i0} = -J^0 \;\Rightarrow\; -\partial_i F^{0i} = -J^0, \tag{5.3.10}$$

Therefore from Eq. (5.3.4)

$$\nabla \cdot \mathbf{E} = \rho, \qquad J^0 = \rho. \tag{5.3.11}$$

On the other hand, for $\nu = j = 1, 2, 3$,

$$\begin{aligned} \partial_\mu F^{\mu j} &= -J^j, \;\Rightarrow\; \partial_0 F^{0j} + \partial_i F^{ij} = -J^j \;\Rightarrow \\ \partial_0 F^{0j} + \partial_i \epsilon^{ijk} B^k &= -J^j, \;\Rightarrow\; \partial_0 E^j - \epsilon^{jik} \partial_i B^k = -J^j, \end{aligned} \tag{5.3.12}$$

where on the second line, we have used the first equation in Eqs. (5.3.8) and (5.3.4). From Eq. (5.3.9), we may re-write the last entry of Eq. (5.3.12) as

$$\frac{\partial}{\partial t}\mathbf{E} - \nabla \times \mathbf{B} = -\mathbf{J}. \tag{5.3.13}$$

Referring to Eq. (5.3.3), choose: $\mu = k, \sigma = 0, \nu = j$, to get,

$$\begin{aligned} \partial^\mu F^{\sigma\nu} + \partial^\nu F^{\mu\sigma} + \partial^\sigma F^{\nu\mu} &= 0, \;\Rightarrow \\ \partial^k F^{0j} + \partial^j F^{k0} + \partial^0 F^{jk} &= 0. \end{aligned} \tag{5.3.14}$$

Multiply the last equation by $\epsilon^{ijk}$, and sum over repeated indices. This gives

$$\big( - \epsilon^{ikj} \partial^k F^{0j} - \epsilon^{ijk} \partial^j F^{0k} \big) - \partial_0 \epsilon^{ijk} F^{jk} = 0, \;\Rightarrow \tag{5.3.15}$$

$$- 2\, \nabla \times \mathbf{E} - 2 \frac{\partial}{\partial t}\mathbf{B} = 0, \tag{5.3.16}$$

using the anti-symmetry property of $F^{\mu\nu}$ and Eqs. (5.3.5), (5.3.8).

Finally note from Eq. (5.3.8)

$$\nabla \cdot \mathbf{B} = \epsilon^{ijk} \partial^i \partial^j A^k, \tag{5.3.17}$$

$$= \frac{1}{2}\big[\epsilon^{ijk} + \epsilon^{jik}\big]\, \partial^i \partial^j A^k = 0, \tag{5.3.18}$$

using $\epsilon^{ijk} = -\epsilon^{jik}$.

From Eqs. (5.3.11), (5.3.16), (5.3.13), and (5.3.18), the celebrated Maxwell equations emerge:

$$\nabla \cdot \mathbf{E} = \rho. \tag{5.3.19}$$

$$\nabla \times \mathbf{E} = -\frac{\partial}{\partial t}\mathbf{B}, \tag{5.3.20}$$

$$\nabla \times \mathbf{B} = \frac{\partial}{\partial t}\mathbf{E} + \mathbf{J}. \tag{5.3.21}$$

$$\nabla \cdot \mathbf{B} = 0, \tag{5.3.22}$$

in rationalized CGS units.[6]

In making a transition from the vector field $A^\mu(x)$ to the tensor field $F^{\mu\nu}(x)$ arose to ensure that the ensuing theory be *gauge invariant*. That is, *as a consequence of gauge invariance, the electric and magnetic fields popped out of the theory.*

The interest in having a gauge invariant theory, is that we may choose any gauge we want to make computations easier.

For example, in the *temporal* gauge $A^0 = 0$, defined in Eq. (5.1.15), we have from Eq. (5.3.4)

$$E^i = F^{0i} = \partial^0 A^i - \partial^i A^0 = \partial^0 A^i, \quad \Rightarrow \quad \mathbf{E} = -\frac{\partial}{\partial t}\mathbf{A}, \tag{5.3.23}$$

while $\mathbf{B}$ does not involve $A^0$, as seen in Eq. (5.3.5), and is given by $\mathbf{B} = \nabla \times \mathbf{A}$ in Eq. (5.3.8).

In the *Lorentz* gauge, defined in Eq. (5.1.13),

$$\partial_\mu A^\mu = 0, \quad \partial_0 A^0 = -\partial_i A^i, \quad E^i = \partial^0 A^i - \partial^i A^0,$$
$$\Rightarrow \quad E^i = \partial^0\Big[\delta^{ij} - \frac{\partial^i \partial^j}{\partial^{02}}\Big]A^j, \qquad \mathbf{B} = \nabla \times \mathbf{A}. \tag{5.3.24}$$

Don't be scared of an expression like $1/\partial_0 = 1/(\partial/\partial t)$. It may be interpreted by considering a Fourier transform of a given function of $f(t)$:

$$f(t) = \int_{-\infty}^{\infty} \frac{\mathrm{d}\omega}{2\pi}\, \mathrm{e}^{-\mathrm{i}\omega t} f(\omega), \quad \Rightarrow \quad \frac{1}{\partial/\partial t} f(t) = \mathrm{i}\int_{-\infty}^{\infty} \frac{\mathrm{d}\omega}{2\pi}\, \mathrm{e}^{-\mathrm{i}\omega t}\frac{f(\omega)}{\omega}. \tag{5.3.25}$$

You don't even have to use a different notation for the Fourier transform of $f$. The dependence on the variable, in question, would make it evident if you are using $f$ or its Fourier transform, unless otherwise stated.

Quite generally, we may express the electric and magnetic fields in terms of the fields $(A^0, \mathbf{A})$ as follows:

$$\mathbf{E} = -\nabla A^0 - \frac{\partial \mathbf{A}}{\partial t}, \qquad \mathbf{B} = \nabla \times \mathbf{A}. \tag{5.3.26}$$

[6] Equation (5.3.20) is referred to as Faraday's Law, while Eq. (5.3.21) as Ampère's Law.

$F^{\mu\nu}(x)$ being the components of a second rank tensor, involving two indices, has, according to Eq. (2.4.2), the following transformation rule from one Lorentz frame F to another one F′,

$$F'^{\sigma\lambda}(x') = \Lambda^{\sigma}{}_{\mu}\,\Lambda^{\lambda}{}_{\nu}\,F^{\mu\nu}(x), \tag{5.3.27}$$

where the matrix elements $\Lambda^{\sigma}{}_{\mu}$ of the transformation matrix $\Lambda$ have been spelled out in Eq. (2.1.16). In particular, if we have just an electric field in frame F, with components $E^i = F^{0i}$, as defined in Eq. (5.3.4), then the above equation gives

$$F'^{\sigma\lambda}(x') = \Lambda^{\sigma}{}_{0}\,\Lambda^{\lambda}{}_{i}\,F^{0i}(x) + \Lambda^{\sigma}{}_{i}\,\Lambda^{\lambda}{}_{0}\,F^{i0}(x) \tag{5.3.28}$$

$$= \big(\Lambda^{\sigma}{}_{0}\,\Lambda^{\lambda}{}_{i} - \Lambda^{\sigma}{}_{i}\,\Lambda^{\lambda}{}_{0}\big)\,E^i(x), \qquad E^i = F^{0i} = -F^{i0}, \tag{5.3.29}$$

where frame F′ is moving with constant velocity $\mathbf{v}$ relative to frame F, as described in Sect. 2.1. The above equation leads to

$$E'^{j}(x') = \big(\Lambda^{0}{}_{0}\Lambda^{j}{}_{i} - \Lambda^{0}{}_{i}\,\Lambda^{j}{}_{0}\big)E^i(x), \quad F'^{jk}(x') = \big(\Lambda^{j}{}_{0}\Lambda^{k}{}_{i} - \Lambda^{j}{}_{i}\,\Lambda^{k}{}_{0}\big)E^i(x). \tag{5.3.30}$$

in frame F′.

From the explicit expressions of $\Lambda^{i}{}_{j}$, $\Lambda^{0}{}_{0}$, and $\Lambda^{i}{}_{0}$ in Eq. (2.1.16),

$$\Lambda^{i}{}_{j} = \delta^{i}{}_{j} + (\gamma - 1)\frac{\mathrm{v}^i \mathrm{v}_j}{v^2}, \quad \Lambda^{0}{}_{0} = \gamma, \quad \Lambda^{i}{}_{0} = -\gamma\,\mathrm{v}^i = \Lambda^{0}{}_{i}, \quad \gamma = \frac{1}{\sqrt{1-\mathrm{v}^2}}, \tag{5.3.31}$$

it readily follows, by using, in the process, $B'^i = \epsilon^{ijk}F'^{jk}/2$ given in Eq. (5.3.5), that

$$\mathbf{E}'_{\perp} = \gamma\,\mathbf{E}_{\perp}, \qquad \mathbf{E}'_{\parallel} = \mathbf{E}_{\parallel}, \qquad \mathbf{B}' = -\gamma\,\mathbf{v}\times\mathbf{E}, \tag{5.3.32}$$

where $\perp$, and $\parallel$ correspond, respectively, to components perpendicular and parallel to $\mathbf{v}$.

The above particular case, out of many other similar cases, shows that the components of the electromagnetic field, one calls electric and magnetic fields, depend very much on the frame in consideration. In the above case only an electric field component $\mathbf{E}$ was observed in one frame, while in the other, moving with velocity $\mathbf{v}$ relative to the former, not only an electric field was observed but a magnetic field as well for $\mathbf{v}\times\mathbf{E}\neq\mathbf{0}$, i.e., for $\mathbf{v}$ not parallel to the electric field $\mathbf{E}$. This frame dependence of these two field components become identified by going to a particular frame. Moreover, Maxwell's equations Eqs. (5.3.20) and (5.3.21), show that a changing magnetic field can create an electric field and vice versa, respectively.

Later in Chap. 8, when we develop propagator theory in more detail to describe scattering theory of electromagnetic phenomena, including radiation emission, we will learn that propagators describe, in a time evolutionary process, the time at which photons are emitted to the time they are detected in a physical process. This is, perhaps, more in the spirit of recent developments of fundamental theoretical physics in describing nature, and this together with applications of electric and magnetic fields, a much richer language for studying classical electrodynamics certainly emerges.

## 5.4 Physics in Matter and Boundary Conditions: Phenomenological Parameters

A dielectric (or dielectric medium) is an electrical insulator that can be polarised by an applied electric field. When a dielectric material is placed in an electric field, electric charges do not flow through the material as they do in an electrical conductor, because they have no loosely bound, or free, electrons that may drift through the material, but instead they shift, only slightly, from their average equilibrium positions, causing dielectric polarisation.

Because of dielectric polarisation, positive charges are displaced in the direction of the field and negative charges shift in the direction opposite to the field. This creates an internal electric field that reduces the overall field within the dielectric itself. If a dielectric is composed of weakly bonded molecules, those molecules not only become polarised, but also reorient so that their symmetry axes align to the field. As a result of polarization and change of the electric field within the medium, a phenomenological parameter, referred to as the dielectric constant $\varepsilon$, also called permitivity of the material, is introduced to take this into account as follows.

In our rationalized CGS system of units, the permitivity of free space $\varepsilon_0$ is taken to be one like $\hbar$ and the speed of light c. If $E_0$ denotes the electric field in the absence of a dielectric, then the permitivity of the medium is defined by $\varepsilon = E_0/E$, where $E$ is the electric field in the dielectric. Here, the value of $E_0$, not being less than $E$, the value of the permitivity is also not less than 1. Thus, in turn, one may introduce a measure of degree of polarisability of a dielectric material, referred to as electric susceptibility $\chi_e$ defined by $\chi_e = \varepsilon - 1$.

Finally, regarding a dielectric material, we define the refractive index of a dieletric by $n = \sqrt{\varepsilon}$. We will generalize this further below, when we consider the magnetic property of a medium.

Similar to the concept of electric susceptibility, one may define the magnetic susceptibility $\chi_m$, indicating the degree of magnetization of a material in response to an applied magnetic field, by introducing, in the process, a parameter $\mu$, referred to as the permeability, which, as discussed below, may not only be greater than one but may be also less than one, and is defined by $\chi_m = \mu - 1$. The permeability of free space $\mu_0$ is taken to be one in our rationalized CGS system of units. The index of refraction of a medium will, in general, be taken as $n = \sqrt{\varepsilon\mu}$.

The value of the permeability $\mu$ depends on the special type of magnetization a material acquires in response to an applied magnetic field. In particular, diamagnetism describes the property of a material which gets repelled by an external applied magnetic field resulting to a permeability $\mu$ less than 1, i.e., to a negative magnetic susceptibility. It is quite a weak effect of magnetisation in most materials and occurs only in the presence of an externally applied magnetic field.

On the other hand, unlike diamagnetism, paramagnetism describes the property of a material which gets attracted by an applied magnetic field, resulting to a magnetic permeability $\mu$ greater than one, i.e., to a positive magnetic susceptibility.

Some materials, such as iron, have a property referred to as ferromagnetism that give rise to a large magnetic permeability, allowing such a material to form a permanent magnet, and are thus well attracted to a magnet.

Generally, when the wavelength of light in a given medium is too big to unravel its underlying detailed structure then, as a working hypothesis, phenomenological parameters may be introduced to, hopefully, describe the interaction of light with matter, in a satisfactory way and not dwell on a detailed description of such a structure. On the other hand, when one deals with X-rays, as treated in Chap. 15, the small wavelength of such rays reveal such a structure and the rays are considered to interact directly with the electrons in atoms making up the medium. In such a problem, phenomenological parameters are not needed to do physics, and, regarding this, Feynman considered this improvement in this theoretical description as[7] "Chemistry becoming Physics". On the other hand, when such parameters are needed to develop a theory, Einstein would be expected to say,[8] they must be "freely invented" and inserted.

With analyses based on observations, with the physics just described as how these phenomenological parameters $\varepsilon$ and $\mu$, summarizing the property of matter, arise in considering the electric and magnetic field, Maxwell's equations, now including them, take on the following forms:

$$\nabla \cdot \varepsilon \mathbf{E} = \rho, \tag{5.4.1}$$

$$\nabla \times \mathbf{E} = -\frac{\partial \mathbf{B}}{\partial t}, \tag{5.4.2}$$

$$\nabla \times \frac{\mathbf{B}}{\mu} = \frac{\partial \varepsilon \mathbf{E}}{\partial t} + \mathbf{J}, \tag{5.4.3}$$

$$\nabla \cdot \mathbf{B} = 0. \tag{5.4.4}$$

in rationalized CGS units. By comparing these equations with those in Eqs. (5.3.19)–(5.3.22) in free space with parameters of permitivity and permeability ($\varepsilon_0$, $\mu_0$), the rationalized CGS units are then spelled out to be defined by setting

$$\hbar = 1, \quad \mathrm{c} = 1, \quad \varepsilon_0 = 1, \quad \mu_0 = 1. \tag{5.4.5}$$

Accordingly, in the rationalized CGS system of units, ($\varepsilon$, $\mu$), automatically, denote relative permitivity and relative permeability, respectively.

---

[7] Feynman (1967) [4].

[8] Schwinger (1957) [9].

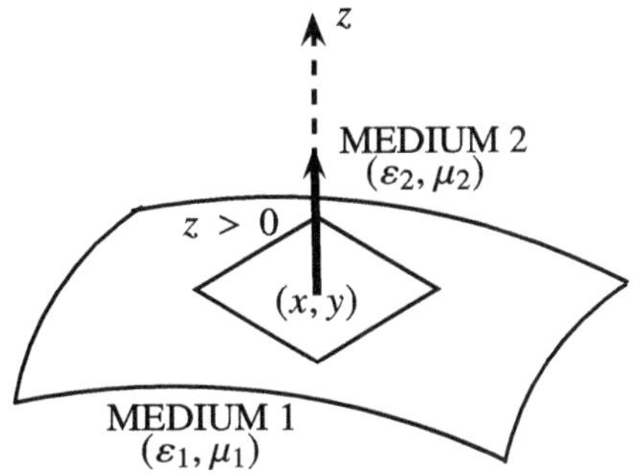


**Fig. 5.1** Introduce a unit vector $\mathbf{n}$ normal to a tangent plane at a point $(x, y)$ on the interface of two different media, with parameters $(\varepsilon_2, \mu_2)$, and $(\varepsilon_1, \mu_1)$. Set up an axis along $\mathbf{n}$, call it the $z$ axis, with $z > 0$ above the interface, and $z < 0$ below it, at the point $(x, y)$

Since different materials would generally have different phenomenological parameters $\varepsilon$ and $\mu$, it is important to investigate the inherited boundary conditions that would arrive from a transition from one medium to the other. Within specified regions of space, these parameters will be considered constants.

To the above end, recall the Heaviside Side Step Function introduced in Eqs. (3.6.36)–(3.6.39) in Sect. exrefsec3.6 of Chap. 3:

$$\theta(z) = \begin{Bmatrix} 1, & \text{for}\, z > 0, \\ 0, & \text{for}\, z < 0. \end{Bmatrix}, \quad \theta(-z) = \begin{Bmatrix} 1, & \text{for}\, 0 > z, \\ 0, & \text{for}\, z > 0. \end{Bmatrix}. \tag{5.4.6}$$

$$\frac{\partial}{\partial z}\theta(z) = \delta(z), \quad \frac{\partial}{\partial z}\theta(-z) = -\delta(z). \tag{5.4.7}$$

Consider first Eq. (5.4.1). Here we have to consider the following equations: Referring to Fig. 5.1, all considered at a point $(x, y)$, on the surface separating the two media in the figure, for arbitrary $z$:

$$\nabla \cdot \varepsilon\,\mathbf{E} = \rho, \tag{5.4.8}$$

$$\nabla \cdot \varepsilon_2\,\mathbf{E}_2 = \rho_2, \quad \nabla \cdot \varepsilon_1\,\mathbf{E}_1 = \rho_1, \tag{5.4.9}$$

$$\varepsilon\mathbf{E} = \theta(z)\varepsilon_2\mathbf{E}_2 + \theta(-z)\varepsilon_1\mathbf{E}_1, \tag{5.4.10}$$

$$\rho = \theta(z)\rho_2 + \theta(-z)\rho_1 + \delta(z)\Sigma_s, \tag{5.4.11}$$

where $\Sigma_s$ denotes a surface charge density that may reside on the surface separating the two media at a point $(x, y)$ on the interface shown in Fig. 5.1.

We use the fact that $\partial^j\theta(z) = \delta^{j3}\delta(z)$, and that $E^3 = \mathbf{n}\cdot\mathbf{E}$, for the 3$^{\text{rd}}$-component, where $\mathbf{n}$ is a unit vector perpendicular to the surface of separation at point $(x, y)$ in the direction specified by $z > 0$, to infer from Eqs. (5.4.8)–(5.4.11) that

$$\theta(z)\rho_2 + \theta(-z)\rho_1 + \Sigma_s\delta(z) = \rho = \nabla \cdot \varepsilon\,\mathbf{E} = \tag{5.4.12}$$

$$\theta(z)\nabla \cdot \varepsilon_2\,\mathbf{E}_2 + \theta(-z)\nabla \cdot \varepsilon_1\,\mathbf{E}_1 + \delta(z)\mathbf{n}\cdot\big(\varepsilon_2\,\mathbf{E}_2 - \varepsilon_1\,\mathbf{E}_1\big)$$

$$= \theta(z)\rho_2 + \theta(-z)\rho_1 + \delta(z)\mathbf{n}\cdot\big(\varepsilon_2\,\mathbf{E}_2 - \varepsilon_1\,\mathbf{E}_1\big). \tag{5.4.13}$$

Upon cancelling out $\big(\theta(z)\rho_2 + \theta(-z)\rho_1\big)$ common in the extreme l.h.s. and extreme r.h.s. of the above equalities, and using, in the process Eqs. (5.4.9), (5.4.10), one is led to the equation

$$\big(\varepsilon_2 \mathbf{n}\cdot\mathbf{E}_2 \;-\; \varepsilon_1 \mathbf{n}\cdot\mathbf{E}_1\big)\,\delta(z) \;=\; \Sigma_s\,\delta(z), \tag{5.4.14}$$

which upon integration over $z$ gives, at an arbitrary point $(x, y)$ on the surface separating the two media, i.e., corresponding to $z = 0$,

$$\varepsilon_2 \mathbf{n}\cdot\mathbf{E}_2 \;-\; \varepsilon_1 \mathbf{n}\cdot\mathbf{E}_1 \;=\; \Sigma_s, \tag{5.4.15}$$

for the normal components of the electric fields.

An identical analysis in reference to Eq. (5.4.4), leads to

$$\mathbf{n}\cdot\mathbf{B}_2 \;-\; \mathbf{n}\cdot\mathbf{B}_1 \;=\; 0, \tag{5.4.16}$$

for the normal components of the magnetic fields, establishing the continuity of the normal component of the magnetic field.

Regarding Eq. (5.4.3), consider first the following relations by using, in the process, $\partial^j\theta(z) = \delta^{j3}\delta(z)$,

$$\big(\nabla\times\mathbf{B}_2\theta(z)\big)^i = \epsilon^{ijk}\partial^j\big(\mathbf{B}_2^k\theta(z)\big) \;=\; \theta(z)\epsilon^{ijk}\partial^j\mathbf{B}_2^k \;+\; \epsilon^{i3k}\mathbf{B}_2^k\delta(z) \tag{5.4.17}$$

$$= \theta(z)(\nabla\times\mathbf{B}_2)^i \;+\; \big(\delta^{i1}\epsilon^{132}\mathbf{B}_2^2 \;+\; \delta^{i2}\epsilon^{231}\mathbf{B}_2^1\big)\delta(z) \tag{5.4.18}$$

$$= \theta(z)(\nabla\times\mathbf{B}_2)^i \;+\; \big(\delta^{i2}\mathbf{B}_2^1 \;-\; \delta^{i1}\mathbf{B}_2^2\big)\delta(z). \tag{5.4.19}$$

Moreover, with the upper indices 1, 2 in the following first equation below denoting, as above, component ones, the above equation leads, from

$$\big(\mathbf{n}\times\mathbf{B}_2\big)^2 \;=\; \mathbf{B}_2^1, \quad \big(\mathbf{n}\times\mathbf{B}_2\big)^1 \;=-\,\mathbf{B}_2^2, \tag{5.4.20}$$

to

$$\nabla\times(\mathbf{B}_2\theta(z)) \;=\; \theta(z)\nabla\times\mathbf{B}_2 \;+\; \mathbf{n}\times\mathbf{B}_2\,\delta(z), \tag{5.4.21}$$

where in Eq. (5.4.20), we used the facts that $\big(\mathbf{n}\times\mathbf{B}_2\big)^i = \epsilon^{i3k}B_2^k$, and $\mathbf{n}^j = \delta^{j3}$.

Accordingly, with

$$\mathbf{J}_1 \;=\; \nabla\times\frac{\mathbf{B}_1}{\mu_1} \;-\; \frac{\partial\varepsilon_1\,\mathbf{E}_1}{\partial t}, \tag{5.4.22}$$

$$\mathbf{J}_2 \;=\; \nabla\times\frac{\mathbf{B}_2}{\mu_2} \;-\; \frac{\partial\varepsilon_2\,\mathbf{E}_2}{\partial t}, \tag{5.4.23}$$

and all the following equations evaluated at $(x, y)$ and arbitrary $z$, we may similarly write

$$\theta(z)\mathbf{J}_2 + \theta(-z)\mathbf{J}_1 + \mathbf{K}_s\delta(z) = \boldsymbol{\nabla} \times \frac{\mathbf{B}}{\mu} - \frac{\partial \varepsilon\, \mathbf{E}}{\partial t} = \theta(z)\Big(\boldsymbol{\nabla} \times \frac{\mathbf{B}_2}{\mu_2} - \frac{\partial \varepsilon_2\, \mathbf{E}_2}{\partial t}\Big)$$
$$+ \theta(-z)\Big(\boldsymbol{\nabla} \times \frac{\mathbf{B}_1}{\mu_1} - \frac{\partial \varepsilon_1\, \mathbf{E}_1}{\partial t}\Big) + \Big(\mathbf{n} \times \frac{\mathbf{B}_2}{\mu_2} - \mathbf{n} \times \frac{\mathbf{B}_1}{\mu_1}\Big)\delta(z)$$
$$= \theta(z)\mathbf{J}_2 + \theta(-z)\mathbf{J}_1 + \Big(\mathbf{n} \times \frac{\mathbf{B}_2}{\mu_2} - \mathbf{n} \times \frac{\mathbf{B}_1}{\mu_1}\Big)\delta(z). \tag{5.4.24}$$

where $\mathbf{K}_s$ is a surface current density that may reside on the interface.

Upon comparing the extreme r.h.s. of the above equation with its extreme l.h.s., and integrate over $z$, we get

$$\mathbf{n} \times \frac{\mathbf{B}_2}{\mu_2} - \mathbf{n} \times \frac{\mathbf{B}_1}{\mu_1} = \mathbf{K}_s. \tag{5.4.25}$$

In the exact manner, Eq. (5.4.2) leads to the following boundary condition

$$\mathbf{n} \times \mathbf{E}_2 - \mathbf{n} \times \mathbf{E}_1 = 0. \tag{5.4.26}$$

Hence vis-à-vis, Eqs. (5.4.1)–(5.4.4), the boundary conditions emerge as

$$\varepsilon_2\mathbf{n} \cdot \mathbf{E}_2 - \varepsilon_1\mathbf{n} \cdot \mathbf{E}_1 = \Sigma_s, \tag{5.4.27}$$
$$\mathbf{n} \times \mathbf{E}_2 - \mathbf{n} \times \mathbf{E}_1 = 0. \tag{5.4.28}$$
$$\mathbf{n} \times \frac{\mathbf{B}_2}{\mu_2} - \mathbf{n} \times \frac{\mathbf{B}_1}{\mu_1} = \mathbf{K}_s. \tag{5.4.29}$$
$$\mathbf{n} \cdot \mathbf{B}_2 - \mathbf{n} \cdot \mathbf{B}_1 = 0. \tag{5.4.30}$$

In a conductor, electrons in the outer layers of atoms are nearly free and when an external field is applied negative charges move in a direction opposite to it, while, the positive ones along the same direction. In turn an electric field arises due to the separated charges in an opposite direction to the applied field going from the positive to the negative ones thus tending to cancel out the applied field. In a so-called perfect conductor a complete cancellation arises giving a net zero electric field. One may infer, directly from Maxwell's equations that the magnetic field also becomes zero in this case.

From the boundary conditions derived above and recorded in Eqs. (5.4.27)–(5.4.30), we may infer that for a *perfect* conductor at the interface of a medium with parameters $(\epsilon, \mu)$ and the conductor itself, the corresponding boundary conditions at point $(x, y)$, $z = 0$, at which point the conductor itself corresponds to region 1, read

$$\varepsilon \mathbf{n} \cdot \mathbf{E} = \Sigma_{\mathrm{s}}, \tag{5.4.31}$$

$$\mathbf{n} \times \mathbf{E} = 0. \tag{5.4.32}$$

$$\mathbf{n} \times \frac{\mathbf{B}}{\mu} = \mathbf{K}_{\mathrm{s}}. \tag{5.4.33}$$

$$\mathbf{n} \cdot \mathbf{B} = 0. \tag{5.4.34}$$

In general, in considering a conductor, not a perfect one, one invokes Ohm's law to an applied electric field to the conductor, as it generates an electric *current*, and the current density takes the form

$$\mathbf{J} = \sigma \, \mathbf{E}, \tag{5.4.35}$$

where $\sigma$ is a constant, referred to as the *conductivity* of the material, as another parameter pertinent to the material. Concerning Ohm's law, it is more appropriate to define $1/\sigma$ as the *resistivity*. The magnitude of $\sigma$ tells us how the nearly free electrons would respond to the electric field.

Conductors include metals. Silver, which is a very good conductor, will be studied in detail in Chap. 11, and investigate how radiation may seep through a thin film of such a metal, as it is applied to red and blue light.

Before providing a careful description of the underlying theory of the electrodynamics and radiation involving a non-perfect conductor in Chap. 11, as mentioned above, a rough estimate may be given to show the attenuation of the electric and magnetic field in a conductor. To this end, consider a "Fourier" component of the electric field of the form

$$\mathbf{E}(x) = \mathbf{E}_0 \, \mathrm{e}^{-\gamma \mathbf{n} \cdot \mathbf{x}} \, \mathrm{e}^{\mathrm{i}[|\mathbf{k}| \mathbf{n} \cdot \mathbf{x} - \omega t]} = \mathbf{E}_0 \exp\big(\mathrm{i}\,[\,(|\mathbf{k}| + \mathrm{i}\gamma)\,\mathbf{n} \cdot \mathbf{x} - \omega t\,]\big), \tag{5.4.36}$$

for propagation in a direction specified by the unit vector $\mathbf{n}$, with $\mathbf{n} \cdot \mathbf{x} > 0$, where the parameter $\gamma$ is to be determined. By taking the curl of $\nabla \times \mathbf{E}$ in Eq. (5.4.2) and use Eq. (5.4.3) with $\mathbf{J}$ given in Eq. (5.4.35) and zero charge density $\rho$, one has

$$\nabla \times \big(\nabla \times \mathbf{E}\big) = \nabla\big(\nabla \cdot \mathbf{E}\big) - \nabla^2 \mathbf{E} = -\nabla^2 \mathbf{E} = -\frac{\partial\big(\nabla \times \mathbf{B}\big)}{\partial t} \tag{5.4.37}$$

$$= -\frac{\partial}{\partial t}\Big(\mu_{\mathrm{c}} \varepsilon \frac{\partial \mathbf{E}}{\partial t} + \mu_{\mathrm{c}} \sigma \, \mathbf{E}\Big) = \big(\mu_{\mathrm{c}} \varepsilon \omega^2 + \mathrm{i}\,\mu_{\mathrm{c}} \sigma \omega\big) \mathbf{E}, \tag{5.4.38}$$

or, from $-\nabla^2 \mathbf{E} = (\,|\mathbf{k}| + \mathrm{i}\gamma\,)^2 \mathbf{E}$,

$$\big(\,|\mathbf{k}| + \mathrm{i}\gamma\,\big)^2 = \mu_{\mathrm{c}} \varepsilon \omega^2 + \mathrm{i}\,\mu_{\mathrm{c}} \sigma \omega, \tag{5.4.39}$$

assuming constant phenomenological parameters. This leads to two equations

$$|\mathbf{k}|^2 - \gamma^2 = \mu_c \varepsilon \omega^2, \qquad 2\gamma |\mathbf{k}| = \mu_c \sigma \omega, \tag{5.4.40}$$

$$\Rightarrow \gamma = \frac{\mu_c \sigma \omega}{2|\mathbf{k}|}, \qquad |\mathbf{k}|^4 - \mu_c \varepsilon \omega^2 |\mathbf{k}|^2 - \frac{(\mu_c \sigma \omega)^2}{4} = 0. \tag{5.4.41}$$

For a good conductor $\sigma \gg \varepsilon\omega$, from which,

$$|\mathbf{k}| \simeq \sqrt{\frac{\mu_c \sigma \omega}{2}}, \qquad \gamma = \sqrt{\frac{\mu_c \sigma \omega}{2}}. \tag{5.4.42}$$

Upon introducing the *skin depth* $\delta$, defined by

$$\delta = \sqrt{\frac{2}{\mu_c \sigma \omega}}, \tag{5.4.43}$$

the following expression emerges for $\mathbf{E}$

$$\mathbf{E}(x) = \mathbf{E}_0 \, \mathrm{e}^{-\mathbf{n}\cdot\mathbf{x}/\delta} \, \mathrm{e}^{\mathrm{i}[\frac{\mathbf{n}\cdot\mathbf{x}}{\delta} - \omega t]}. \tag{5.4.44}$$

On the other hand, for the magnetic field, with a time dependence $\exp(-\mathrm{i}\omega t)$, again, we have from Eq. (5.4.2), for the corresponding Fourier component,

$$\mathrm{i}\omega \mathbf{B} = \nabla \times \mathbf{E} = \mathbf{n} \times \mathbf{E} \frac{\mathrm{i}}{\delta}\big(1 + \mathrm{i}\big) = \mathbf{n} \times \mathbf{E}\, \mathrm{i}\sqrt{\mu_c \sigma \omega}\, \frac{(1+\mathrm{i})}{\sqrt{2}}, \tag{5.4.45}$$

$$\Rightarrow \mathbf{B} = \mathbf{B}_0 \, \mathrm{e}^{-\mathbf{n}\cdot\mathbf{x}/\delta} \, \mathrm{e}^{\mathrm{i}[\frac{\mathbf{n}\cdot\mathbf{x}}{\delta} - \omega t + \frac{\pi}{4}]}, \qquad \mathbf{B}_0 = \mathbf{n} \times \mathbf{E}_0 \sqrt{\frac{\mu_c \sigma}{\omega}}, \tag{5.4.46}$$

where we have used that $(1 + \mathrm{i})/\sqrt{2} = \exp(\mathrm{i}\pi/4)$.

From the definition of the skin depth in Eq. (5.4.43), $\delta \to 0$ as $\sigma \to \infty$, hence the electric and magnetic fields given in Eqs. (5.4.44), (5.4.46), respectively, vanish within a perfect conductor.

This is, however, a very rough treatment, and an explicit derivation of radiation seeping through a thin metal will be given later, as mentioned above. For silver, for example, $\mu_c$ is nearly one.

Using the fact that $\mathbf{B} = \nabla \times \mathbf{A}$, and by working in the temporal gauge $A^0 = 0$, for which $\mathbf{E} = -\partial_0 \mathbf{A}$, for constant phenomenological parameters corresponding to a given medium, we may refer to Eq. (5.4.3), to write

$$\nabla \times \mathbf{B} = \varepsilon\mu \, \partial_0 \, \mathbf{E} + \mu \, \mathbf{J}, \tag{5.4.47}$$

$$\nabla \times (\nabla \times \mathbf{A}) = \varepsilon\mu \, \partial_0 \, \partial^0 \mathbf{A} + \mu \, \mathbf{J}, \tag{5.4.48}$$

and using the vector identity

$$\nabla \times (\nabla \times \mathbf{A}) = -\nabla^2 \mathbf{A} + \nabla(\nabla \cdot \mathbf{A}), \tag{5.4.49}$$

as before, and $\partial^0 = -\partial_0$, we obtain the basic equations of classical electrodynamics in a medium

$$\Big[\big(-\boldsymbol{\partial}^2 + \varepsilon\mu\partial_0^2\big)\delta^{ij} + \partial^i\partial^j\Big]A^j(x) = \mu\, J^i(x). \tag{5.4.50}$$

Upon writing

$$A^i(x) = \int (\mathrm{d}x')\, D_+^{ij}(x, x')J^j(x'), \tag{5.4.51}$$

we may readily extract from Eqs. (5.4.50), (5.4.51), the two basic equations satisfied by the propagator $D_+^{ij}(x, x')$ in a medium in the subject of ***classical electrodynamics*** in the temporal gauge, $(\boldsymbol{\nabla}^2 = \boldsymbol{\partial}^2)$

$$\big[(-\boldsymbol{\partial}^2 + \varepsilon\mu\,\partial^{02})\delta^{i\ell} + \partial^i\partial^\ell\big]D_+^{\ell j}(x, x') = \mu\,\delta^{ij}\,\delta^{(4)}(x - x'), \tag{5.4.52}$$

$$\big[(-\boldsymbol{\partial}'^2 + \varepsilon\mu\,\partial'^{02})\delta^{\ell i} + \partial'^\ell\partial'^i\big]D_+^{j\ell}(x, x') = \mu\,\delta^{ij}\,\delta^{(4)}(x - x'), \tag{5.4.53}$$

exactly as we have done in Eqs. (5.1.29), (5.1.34) for free space.

Equations (5.4.50)–(5.4.53) constitute the *basic equations of classical electrodynamics in a medium* in the temporal gauge.

Obviously, if we do not impose a specific gauge, then Eq. (5.4.50) simply becomes

$$\big(-\boldsymbol{\partial}^2 + \varepsilon\mu\,\partial_0^2\big)\,A^\alpha(x) + \partial^\alpha\partial_\beta A^\beta(x) = \mu\,J^\alpha(x). \tag{5.4.54}$$

For a conductor of finite conductivity $\sigma$, we may use Ohm's law in Eq. (5.4.35), to infer that the propagator $D_+(x, x')$, with $x' = (t', x'^1, x'^2, z')$, and $z'$ outside of a conductor, while $z$ within a conductor, and infer from Eqs. (5.4.50) and (5.4.52) that for $t > t'$,

$$\big[\big(-\boldsymbol{\partial}^2 + \varepsilon\mu\partial^{02} - \mu\sigma\partial^0\big)\,\delta^{i\ell} + \partial^i\partial^\ell\big]\,D_+^{\ell j}(x, x') = 0, \tag{5.4.55}$$

where we have used $E^i = \partial^0 A^i$ in temporal gauge. This is a very useful equation regarding radiation moving into a conductor, with the phenomenological indicated, in a time evolutionary process, as we will see in Chap. 11.

## 5.5 Electromagnetic Waves

Away from charge distributions and current densities, i.e. in regions in which these densities are zero, Maxwell's equations, from Eqs. (5.4.1)–(5.4.4), read as

$$\nabla \cdot \varepsilon \boldsymbol{E} = 0, \tag{5.5.1}$$
$$\nabla \times \mathbf{E} = -\frac{\partial \mathbf{B}}{\partial t}, \tag{5.5.2}$$
$$\nabla \times \mathbf{B} = \mu\varepsilon \frac{\partial \mathbf{E}}{\partial t}, \tag{5.5.3}$$
$$\nabla \cdot \boldsymbol{B} = 0. \tag{5.5.4}$$

We take the time derivative of Eq. (5.5.3), and use Eqs. (5.5.2), (5.5.1) to obtain

$$\mu\epsilon \frac{\partial^2 \mathbf{E}}{\partial t^2} = \frac{\partial(\nabla \times \mathbf{B})}{\partial t} = -\nabla \times (\nabla \times \mathbf{E}) = -\nabla(\nabla \cdot \mathbf{E}) + \nabla^2 \mathbf{E} = \nabla^2 \mathbf{E}, \tag{5.5.5}$$

leading to

$$\nabla^2 \mathbf{E} = \frac{\mu\epsilon}{\mathrm{c}^2} \frac{\partial^2 \mathbf{E}}{\partial t^2}, \quad \text{and similarly to} \quad \nabla^2 \mathbf{B} = \frac{\mu\epsilon}{\mathrm{c}^2} \frac{\partial^2 \mathbf{B}}{\partial t^2}, \tag{5.5.6}$$

where we have, momentarily, re-inserted the speed of light c for additional clarity. These equations, as we will see in detail, are wave equations of fields propagating with the *speed of light* in the medium given by $\mathrm{c}/\sqrt{\mu\epsilon}$, with c denoting the speed of light in vacuum.

We carry a double Fourier transform and use the first equation for **E** above, with c set equal to 1, to write

$$\mathbf{E}(t, \mathbf{x}) = \int \frac{\mathrm{d}^3\mathbf{k}}{(2\pi)^3} \mathrm{e}^{\mathrm{i}\mathbf{k}\cdot\mathbf{x}} \int_{-\infty}^{\infty} \frac{\mathrm{d}\omega}{(2\pi)} \mathrm{e}^{-\mathrm{i}\omega t} \mathbf{E}(\omega, \mathbf{k}), \tag{5.5.7}$$
$$\Rightarrow \Big(\mathbf{k}^2 - \mu\epsilon\omega^2\Big)\mathbf{E}(\mathbf{k}, \omega) = 0. \tag{5.5.8}$$

Applying equations in Eqs. (3.6.11) and (3.6.14) of the Appendix in Chap. 3:

$$x\delta(x) = 0, \quad \delta(x^2 - a^2) = \frac{1}{2|a|}\big[\delta(x - a) + \delta(x + a)\big] \tag{5.5.9}$$

give

$$\mathbf{E}(\omega, \mathbf{k}) = \Big[\delta\Big(\,|\,\mathbf{k}\,| - \sqrt{\mu\epsilon}\omega\Big) + \delta\Big(\,|\,\mathbf{k}\,| + \sqrt{\mu\epsilon}\omega\Big)\Big]\mathbf{F}(\omega, \mathbf{k}), \tag{5.5.10}$$

where $\mathbf{F}(\omega, \mathbf{k})$ is arbitrary depending, in general, on $(\omega, \mathbf{k})$ for both signs of $\omega$, and we have absorbed the $1/(2|\sqrt{\mu\epsilon}\omega|)$ factor in **F**. This provides two solutions, which when substituted in Eq. (5.5.7), using $\mathrm{d}^3\mathbf{k} = |\mathbf{k}|^2\mathrm{d}|\mathbf{k}|\mathrm{d}\Omega$, and integrating over $|\mathbf{k}|$, gives the following general expression corresponding to the two signs of $\omega$:

$$\mathbf{E}(t,\mathbf{x}) = \int \mathrm{d}\Omega \left[ \boldsymbol{\mathcal{E}}^{+}_{\mathbf{n}}\left(\mathbf{n}\cdot\mathbf{x} - \frac{t}{\sqrt{\mu\epsilon}}\right) + \boldsymbol{\mathcal{E}}^{-}_{\mathbf{n}}\left(\mathbf{n}\cdot\mathbf{x} + \frac{t}{\sqrt{\mu\epsilon}}\right)\right], \tag{5.5.11}$$

where,

$$\boldsymbol{\mathcal{E}}^{+}_{\mathbf{n}}\left(\mathbf{n}\cdot\mathbf{x} - \frac{t}{\sqrt{\mu\epsilon}}\right) = \int_0^\infty \frac{\mu\epsilon\,\omega^2}{(2\pi)^3}\frac{\mathrm{d}\omega}{(2\pi)}\,\mathbf{F}(\omega,\sqrt{\mu\epsilon}\omega\mathbf{n})\exp\left[\mathrm{i}\left(\mathbf{n}\cdot\mathbf{x} - \frac{t}{\sqrt{\mu\epsilon}}\right)\sqrt{\mu\epsilon}\,\omega\right], \tag{5.5.12}$$

and

$$\begin{aligned}\boldsymbol{\mathcal{E}}^{-}_{\mathbf{n}}\left(\mathbf{n}\cdot\mathbf{x} + \frac{t}{\sqrt{\mu\epsilon}}\right) &= \int_{-\infty}^0 \frac{\mu\epsilon\,\omega^2}{(2\pi)^3}\frac{\mathrm{d}\omega}{(2\pi)}\,\mathbf{F}(\omega,-\sqrt{\mu\epsilon}\omega\mathbf{n})\exp\left[-\mathrm{i}\left(\mathbf{n}\cdot\mathbf{x} + \frac{t}{\sqrt{\mu\epsilon}}\right)\sqrt{\mu\epsilon}\,\omega\right],\\ &= \int_0^\infty \frac{\mu\epsilon\,\omega^2}{(2\pi)^3}\frac{\mathrm{d}\omega}{(2\pi)}\,\mathbf{F}(\omega,\sqrt{\mu\epsilon}\omega\mathbf{n})\exp\left[\mathrm{i}\left(\mathbf{n}\cdot\mathbf{x} + \frac{t}{\sqrt{\mu\epsilon}}\right)\sqrt{\mu\epsilon}\,\omega\right]. \end{aligned} \tag{5.5.13}$$

Consider first $\boldsymbol{\mathcal{E}}^{+}_{\mathbf{n}}(\mathbf{n}\cdot\mathbf{x} - t/\sqrt{\mu\epsilon})$ in Eq. (5.5.12). Equation (5.5.1) implies that

$$\mathbf{n}\cdot\boldsymbol{\mathcal{E}}^{+}_{\mathbf{n}}\left(\mathbf{n}\cdot\mathbf{x} - \frac{t}{\sqrt{\mu\epsilon}}\right) = 0, \quad \text{i.e.} \perp \quad \mathbf{n}. \tag{5.5.14}$$

For $\mathbf{n}\cdot\mathbf{x} - t/\sqrt{\mu\epsilon} =$Const., and a given fixed $t$, it defines a plane perpendicular to $\mathbf{n}$:

$$\mathbf{n}\cdot\mathbf{x}_1 - \frac{t}{\sqrt{\mu\epsilon}} = \text{Const.},\quad \mathbf{n}\cdot\mathbf{x}_2 - \frac{t}{\sqrt{\mu\epsilon}} = \text{Const.}, \tag{5.5.15}$$

$$\mathbf{n}\cdot(\mathbf{x}_2 - \mathbf{x}_1) = 0, \quad t \text{ fixed}, \tag{5.5.16}$$

with $\mathbf{x}_1$ and $\mathbf{x}_2$ denoting two points on the plane. See Fig. 5.2.

The incremental change, $\Delta\mathbf{x}/\Delta t = \mathbf{n}/\sqrt{\mu\epsilon}$, corresponding to the propagation speed of light in the medium, and the equation $\mathbf{n}\cdot\mathbf{x} - t/\sqrt{\mu\epsilon} =$ Const. remains invariant:

$$\mathbf{n}\cdot(\mathbf{x}+\Delta\mathbf{x}) - \frac{1}{\sqrt{\mu\epsilon}}(t+\Delta t) = \text{Const.} \Rightarrow \mathbf{n}\cdot\mathbf{x} - \frac{t}{\sqrt{\mu\epsilon}} = \text{Const.}, \tag{5.5.17}$$

implying propagation with speed $1/\sqrt{\mu\epsilon}$ along the unit vector $\mathbf{n}$.

From Eqs. (5.5.2)/(5.5.7) we also have, noting, in the process, the minus sign on the r.h.s. of Eq. (5.5.2), that

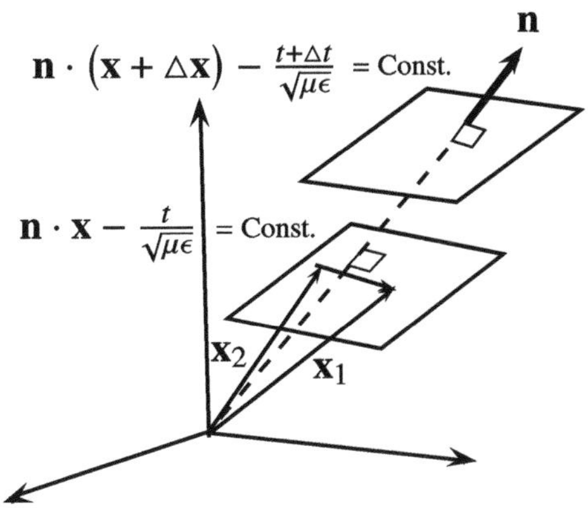


**Fig. 5.2** A figure showing the direction of the line of wave propagation, specified by the unit vector **n**, piercing the so-called wavefront planes, perpendicular to the latter. An incremental change $\Delta\mathbf{X}/\Delta t = \mathbf{n}c/\sqrt{\varepsilon\mu}$ corresponding to the speed of light in the medium, the equation $\mathbf{n}\cdot\mathbf{x} - \frac{c}{\sqrt{\mu\epsilon}}t =$ Const., remains invariant. We work in the convenient rationalized CGS units, in which, in particular, c is set equal to one

$$\mathbf{B}(t,\mathbf{x}) = \int \mathrm{d}\Omega\left[\mathcal{B}_{\mathbf{n}}^{+}\left(\mathbf{n}\cdot\mathbf{x} - \frac{t}{\sqrt{\mu\epsilon}}\right) - \mathcal{B}_{\mathbf{n}}^{-}\left(\mathbf{n}\cdot\mathbf{x} + \frac{t}{\sqrt{\mu\epsilon}}\right)\right], \tag{5.5.18}$$

$$\mathcal{B}_{\mathbf{n}}^{\pm}\left(\mathbf{n}\cdot\mathbf{x} \mp \frac{t}{\sqrt{\mu\epsilon}}\right) = \int_0^\infty \left(\mu\epsilon\right)^{3/2}\frac{\omega^2}{(2\pi)^3}\frac{\mathrm{d}\omega}{(2\pi)}\times$$
$$\mathbf{n}\times\mathbf{F}\left(\omega, \sqrt{\mu\epsilon}\omega\mathbf{n}\right)\exp\left[\mathrm{i}\left(\mathbf{n}\cdot\mathbf{x} \mp \frac{t}{\sqrt{\mu\epsilon}}\right)\sqrt{\mu\epsilon}\,\omega\right], \tag{5.5.19}$$

where, for any given $t$, $\mathcal{B}_{\mathbf{n}}^{+}\left(\mathbf{n}\cdot\mathbf{x} - \frac{t}{\sqrt{\mu\epsilon}}t\right)$, is perpendicular to $\mathcal{E}_{\mathbf{n}}^{+}\left(\mathbf{n}\cdot\mathbf{x} - \frac{t}{\sqrt{\mu\epsilon}}\right)$, lies in the same plane and, in an electromagnetic wave, they are both in phase, reaching their maximum and minimum simultaneously.

A similar analysis may be carried out for the $\mathcal{E}_{\mathbf{n}}^{-}$, $\mathcal{B}_{\mathbf{n}}^{-}$ vectors with propagation described in the opposite direction to **n**.

From Eqs. (5.5.2), and (5.5.19) also note that

$$\mathbf{B}(\omega,\mathbf{k}) = \mathbf{n}\times\mathbf{E}(\omega,\mathbf{k}), \tag{5.5.20}$$

where $\mathbf{B}(\omega,\mathbf{k})$ is similarly defined as $\mathbf{E}(\omega,\mathbf{k})$ in Eq. (5.5.7). **E**, **B**, and $\mathbf{k} = |\mathbf{k}|\mathbf{n}$ form a right-handed system, with **n** along the $z-$ axis, **E** along the $x-$ one, and **B** along the $y-$ one.

## 5.6 Charged Particle Dynamics in an External Electromagnetic Field

To find an equation of motion of a charged particle in an external electromagnetic field $F^{\mu\nu}$ consistent with relativity, we first recall, from Chap. 2, Eqs. (2.3.3)–(2.3.6), the variables relevant to a massive particle with 3 velocity $\mathbf{v}$ (with c, now, set equal to one in our system of units):

$$\mathbf{p} = m\gamma\,\mathbf{v}, \quad p^0 = m\gamma\,, \quad p^\mu p_\mu = -m^2 \tag{5.6.1}$$

$$\gamma = 1/\sqrt{1-\mathbf{v}^2}, \quad \mathrm{d}t = \gamma\,\mathrm{d}\tau. \tag{5.6.2}$$

We also recall from Eq. (2.3.8) in Chap. 2, that

$$E = m\gamma = \frac{m}{\sqrt{(1-\mathbf{v}^2)}} \simeq m + \frac{1}{2}m\,\mathbf{v}^2 + O(\mathbf{v}^4), \tag{5.6.3}$$

denotes the total energy of a particle, with $m$ denoting its rest energy.

We recall that in non-relativistic physics, the equation of motion of a particle is given by the time derivative of the 3-momentum set equal to the force on it. Accordingly, a covariant equation may be now defined by taking the derivative of the four momentum $p^\mu$ with respect to the proper time $\tau$ and set

$$\frac{\mathrm{d}}{\mathrm{d}\tau}p^\mu = f^\mu, \tag{5.6.4}$$

and, in turn, determine the 4-vector with components $f^\mu$, self consistently.

We first take the $\tau$ derivative of $p^\mu p_\mu = -m^2$ and infer the following

$$\frac{\mathrm{d}}{\mathrm{d}\tau}(p^\mu p_\mu) = 0, \Rightarrow p_\mu\frac{\mathrm{d}}{\mathrm{d}\tau}p^\mu = 0, \quad p_\mu f^\mu = 0, \quad \text{from which} \quad f^0 = \frac{\mathbf{p}\cdot\mathbf{f}}{p^0}. \tag{5.6.5}$$

On the other hand, using $\mathrm{d}t/\mathrm{d}\tau = \gamma$ given in Eq. (5.6.2), we have from Eq. (5.6.4),

$$\frac{\mathrm{d}}{\mathrm{d}\tau}\,\mathbf{p} = \mathbf{f}, \quad \text{and} \quad \frac{\mathrm{d}t}{\mathrm{d}\tau}\frac{\mathrm{d}}{\mathrm{d}t}(\mathbf{p}) = \gamma\frac{\mathrm{d}}{\mathrm{d}t}(\mathbf{p}) = \mathbf{f}, \tag{5.6.6}$$

while the force $\mathbf{F}$ is defined, with a derivative with respect to $t$, by

$$\mathbf{F} = \frac{\mathrm{d}}{\mathrm{d}t}\mathbf{p}, \qquad \text{hence from the last equality in Eq.(5.6.6)} \quad \mathbf{f} = \gamma\,\mathbf{F}. \tag{5.6.7}$$

From Eq. (5.6.1), using the fact that $\mathbf{p}/p^0 = \mathbf{v}$, we may infer from the last equality above, and the last equality in Eq. (5.6.5) that

$$f^{0} = \frac{\mathbf{p} \cdot \mathbf{f}}{p^0} = \mathbf{v} \cdot \mathbf{f} = \gamma \mathbf{v} \cdot \mathbf{F}. \tag{5.6.8}$$

Therefore, Eqs. (5.6.7), (5.6.8) allow us to re-write Eq. (5.6.4) as

$$\frac{\mathrm{d}}{\mathrm{d}\tau} p^{\mu} = f^{\mu}, \quad f^{\mu} = \big(\gamma\, \mathbf{v} \cdot \mathbf{F}, \gamma\, \mathbf{F}\big). \tag{5.6.9}$$

where $\mathbf{v} \cdot \mathbf{F} = \mathrm{d}E/\mathrm{d}t$ is the expression for power.

Let $q$ be the charge of a test particle moving in external electromagnetic field $F^{\mu\nu}$ produced by a *given charge* $Q$. Consider the rest frame F$'$ in which the charge $Q$ is instantaneously at rest (a co-moving frame). Let $\mathbf{v}$ denote the instantaneous velocity of charge $q$ in frame F, and let $\mathbf{v}'$ denote its velocity in frame F$'$ at the instant when charge $Q$ is at rest in it. It is an empirical fact that the force on a *test* charge $q$ due to a static charge $Q$ is given by

$$\mathbf{F}' = q\mathbf{E}', \tag{5.6.10}$$

where $\mathbf{E}'$ is the electric field produced by charge $Q$ instantaneously at rest in frame F$'$. Hence

$$\mathbf{f}' = q\gamma'\mathbf{E}', \quad f'^{0} = q\,\gamma'\mathbf{v}' \cdot \mathbf{E}', \quad \gamma' = \frac{1}{\sqrt{(1-\mathbf{v}'^2)}}. \tag{5.6.11}$$

Recall, the expression of the 4-velocity $U^{\mu} = (\gamma, \gamma\mathbf{v})$, $U_0 = -U^0$ in Eq. (2.3.5) of Chap. 2, and $E^i = F^{0i}$ in Eq. (5.3.4). Accordingly, the components of $f'^{\mu}$ in Eq. (5.6.11), remembering that $F'^{ij} = 0$ and $F'^{i0} = -F'^{0i}$, may be expressed as

$$f'^{\mu} = q\, F'^{\mu\nu} U'_{\nu} \quad \text{in F}'. \tag{5.6.12}$$

We may invoke covariance, meaning that the form of the above equation should hold true in frame F as well, witnessing the power of relativity at work, to conclude, from Eqs. (5.6.9) and (5.6.12), that

$$\frac{\mathrm{d}p^{\mu}}{\mathrm{d}\tau} = f^{\mu} = q\, F^{\mu\nu} U_{\nu}, \tag{5.6.13}$$

as the Lorentz covariant equation of motion of a test charged particle in an electromagnetic field $F^{\mu\nu}$.

Since

$$F^{ij} = \varepsilon^{ijk}\mathbf{B}^k, \qquad F^{i0} = -\,\mathbf{E}^i, \tag{5.6.14}$$

$$\varepsilon^{ijk}\mathbf{v}^j\mathbf{B}^k = \big(\mathbf{v} \times \mathbf{B}\big)^i, \tag{5.6.15}$$

and $U^\mu = (\gamma, \gamma \mathbf{v}), \mathbf{f} = \gamma \mathbf{F}$ in Eq. (5.6.7), the explicit expression for the force emerges from Eq. (5.6.13) to be given by

$$\mathbf{F} = q\big(\mathbf{E} + \mathbf{v} \times \mathbf{B}\big), \qquad (5.6.16)$$

referred to as the **Lorentz Force**.

## References

1. Dirac, P. A. M. (1933). The Lagrangian in quantum mechanics. *Physikalische Zeitschrift der Sowjetunion, 3*, 64–72.
2. Feynman, R. P. (1948). Space-time approach to non-relativistic quantum mechanics. *Reviews of Modern Physics, 20*, 367–387.
3. Feynman, R. P., & Hibbs, A. R. (1965). *Quantum mechanics and path integrals*. New York: McGraw-Hill.
4. Feynman, R. (1967). *The character of physical law*. Cambridge, MA: MIT Press.
5. Manoukian, E. B. (1986). Action principle and quantization of gauge fields. *Physical Review D, 34*, 3739–3749.
6. Manoukian, E. B. (2016). *Quantum field theory I: Foundations and abelian and non-abelian gauge theories*. Switzerland: Springer.
7. Schwinger, J. (1951). On Green's function of quantized fields. I. *Proceedings of the National Academy of Sciences USA, 37*, 452–455.
8. Schwinger, J. (1953). The theory of quantized fields. I, II. *Physical Review, 91*, 713–728, 728–740.
9. Schwinger, J. (1957). A theory of the fundamental interactions. *Annals of Physics (NY), 2, 407*, 407–434.
10. Schwinger, J. (1970). *Particles, sources, and fields*. Reading, Massachusetts: Addison-Wesley.

# Chapter 6
# Synchrotron Radiation

*Power of the vacuum-to-vacuum transition amplitude in describing a form of high intensity radiation, referred to as synchrotron radiation, practically used in all the sciences, in technology, medicine and, in general, in studying the structure of matter.*

In its simplest possible description, synchrotron radiation is radiation emitted by an ultra-relativistic charged particle in motion in a magnetic field. Although such radiation had been observed for years and years in nature, it was not understood in physical terms to be due to the motion of charged particles emitting radiation while experiencing a centripetal acceleration until, in 1898, Alfred-Marie Liénard[1] and Oliver Heaviside[2] in 1902, explained and developed the basics of the underlying physical theory with further more analysis carried out by George Adolphus Schott[3] in 1907. Notable theoretical work was already done by several physicists such as Dmitri Dmitrievich Ivanenko, Isaak Yakovlevich Pomeranchuk,[4] and Julian Seymour Schwinger,[5] in the mid forties, before it was visually observed experimentally.

It was observed as such in 1947 by a staff member named Floyd Haber, of the laboratory of Herbert C. Pollock and Robert Langmuir in the General Electric Research Laboratory in Schenectady, NY. It was such an excitement for the research team, that the radiation was termed as "Ivanenko-Pomeranchuk-radiation" as well as,[6] "Schwinger-radiation".[7]

---

[1] Liénard 1898 [8].

[2] Heaviside 1902 [5].

[3] Schott (1907, 1912) [15, 16].

[4] Ivanenko and Pomeranchuk (1944) [6].

[5] Schwinger (1946) [17], see also Schwinger (1949) [18].

[6] Schwinger (1946) [17] see also Schwinger (1949) [18].

[7] For excellent historical accounts see, in particular, Pollock 1983 [13], and Bordovitsyn (1999) [2].



E. B. Manoukian, *Radiation*,
https://doi.org/10.1007/978-3-032-21506-2_6

Such ultra-relativistic electrons produce radiation of such a high-energy corresponding to wavelengths so small of the order of an Angstrom Å$= 10^{-10}$ m or so, and is referred to as X-ray radiation. With such small wavelengths, one is able to unravel the structure of matter by bombarding it with X-rays. This will be applied in Chap. 15 to study the structure of some elements, e.g., of copper, silver, chromium and others. Other applications of X-ray scattering will be mentioned in Chap. 15 as well.

In the present chapter, we develop the theoretical intricacies of the theory. We first take advantage of the equation of motion of a relativistic charged particle in an external electromagnetic field, developed in Sect. 5.6 of the last chapter, to describe its motion in a uniform magnetic field.

We consider an electron of charge e and mass m in an external constant magnetic field $\mathbf{B} = B(0, 0, 1)$, with motion described by the relativistic expression derived in Eq. (5.6.13), given by

$$\frac{\mathrm{d}p^\mu}{\mathrm{d}\tau} = \mathrm{e}\, F^{\mu\nu} U_\nu, \qquad F^{ij} = \epsilon^{ijk} B^k, \quad F^{0i} = 0, \quad i, j, k \in \{1, 2, 3\}, \tag{6.1}$$

$$U^\nu = (\gamma, \gamma \mathbf{v}(t)), \ p^0 = \mathrm{m}\, U^0 = \mathrm{m}\gamma, \ \mathbf{p} = \mathrm{m}\, \mathbf{U} = \mathrm{m}\gamma \mathbf{v}(t), \ \gamma = \frac{1}{\sqrt{1 - \mathbf{v}^2(t)}}. \tag{6.2}$$

In particular, using $\mathrm{d}\tau = \mathrm{d}t/\gamma$, we have

$$\frac{\mathrm{d}p^0}{\mathrm{d}\tau} = \mathrm{e} F^{0i} U_i = 0, \ \text{ i.e. } \gamma\, \mathrm{m} \frac{\mathrm{d}\gamma}{\mathrm{d}t} = 0, \tag{6.3}$$

$$\Rightarrow \ \frac{\mathrm{d}}{\mathrm{d}t} \mathbf{v}^2(t) = 0, \ \ |\mathbf{v}| \equiv \mathrm{v} \ \text{ is a constant, and, } \ \mathbf{v}(t) \cdot \dot{\mathbf{v}}(t) = 0. \tag{6.4}$$

Also[8]

$$\frac{\mathrm{d}\mathbf{p}^i}{\mathrm{d}\tau} = \mathrm{e} F^{ij} U_j = \mathrm{e}\, \gamma \epsilon^{ijk} \mathbf{v}^j \mathbf{B}^k, \ \text{ i.e., } \frac{\mathrm{d}\mathbf{p}}{\mathrm{d}\tau} = \mathrm{e}\, \gamma \mathbf{v} \times \mathbf{B}, \tag{6.5}$$

$$\text{using } \ \mathbf{p} = \mathbf{m}\gamma \mathbf{v}, \ \mathrm{d}\tau = \frac{\mathrm{d}t}{\gamma}, \quad \text{lead to} \quad \frac{\mathrm{d}\mathbf{v}}{\mathrm{d}t} = \frac{\mathrm{e}}{\gamma \mathrm{m}} \mathbf{v} \times \mathbf{B}. \tag{6.6}$$

We explicitly have

$$\mathbf{v} \times \mathbf{B} = B(\mathbf{v}^2, -\mathbf{v}^1, 0), \qquad \frac{\mathrm{d}(\mathbf{v}^1, \mathbf{v}^2, \mathbf{v}^3)}{\mathrm{d}t} = -\, \omega_\mathrm{o}\, (\mathbf{v}^2, -\mathbf{v}^1, 0) \tag{6.7}$$

$$\text{where } \omega_\mathrm{o} = \frac{|\mathrm{e}|\, B}{\gamma \mathrm{m}}, \ \Rightarrow \ \frac{\mathrm{d}\mathbf{v}^1}{\mathrm{d}t} = -\, \omega_\mathrm{o}\, \mathbf{v}^2, \quad \frac{\mathrm{d}\mathbf{v}^2}{\mathrm{d}t} = \omega_\mathrm{o}\, \mathbf{v}^1, \quad \frac{\mathrm{d}\mathbf{v}^3}{\mathrm{d}t} = 0. \tag{6.8}$$

[8] Needless to say, the second equation in Eq. (6.5) denotes the Lorentz Force.

Equation of motion of the electron now readily follows to be given by

$$\frac{\mathrm{d}}{\mathrm{d}t}\big[\mathbf{v}^1 + \mathrm{i}\mathbf{v}^2\big] = \mathrm{i}\,\omega_\mathrm{o}\,\big[\mathbf{v}^1 + \mathrm{i}\mathbf{v}^2\big], \quad \text{from which,} \tag{6.9}$$

$$\big[\mathbf{v}^1 + \mathrm{i}\mathbf{v}^2\big] = \big[\mathbf{v}^1 + \mathrm{i}\mathbf{v}^2\big]\Big|_{t=0}\,\mathrm{e}^{\mathrm{i}\omega_\mathrm{o}\,t}, \quad \frac{\mathrm{d}\mathbf{v}^3}{\mathrm{d}t} = 0. \tag{6.10}$$

We define convenient variables and we set up initial conditions:

$$\mathbf{v}(t) \equiv \dot{\mathbf{R}}(t), \qquad \mathbf{R}(t)\Big|_{t=0} = (R, 0, 0), \qquad \dot{\mathbf{R}}(t)\Big|_{t=0} = (0, \mathrm{v}, 0). \tag{6.11}$$

From Eq. (6.10), we explicitly have

$$\mathbf{R}(t) = R\,\big(\cos\omega_\mathrm{o}\,t, \sin\omega_\mathrm{o}\,t, 0\big), \quad \dot{\mathbf{R}}(t) = \mathrm{v}\,\big(-\sin\omega_\mathrm{o}\,t, \cos\omega_\mathrm{o}\,t, 0\big), \tag{6.12}$$

$$R = \frac{\mathrm{v}}{\omega_\mathrm{o}}, \qquad \omega_\mathrm{o} = \frac{|\mathrm{e}|\,B}{\gamma\mathrm{m}}, \qquad \gamma = \frac{1}{\sqrt{1-\mathrm{v}^2}}. \tag{6.13}$$

This corresponds to circular motion in a plane perpendicular to the magnetic field **B**, along a circle of radius $R$ as shown in Fig. 6.1. The parameter $\omega_\mathrm{o}$ is usually referred to as the cyclotron frequency.

The so-called Bessel function $J_\ell(z)$, in particular, plays a special role in synchrotron physics studies. They satisfy well established mathematical properties. A brief account of these properties are collected and spelled out in the Appendix, Sect. 6.7. They will be used in Sects. 6.4–6.7.

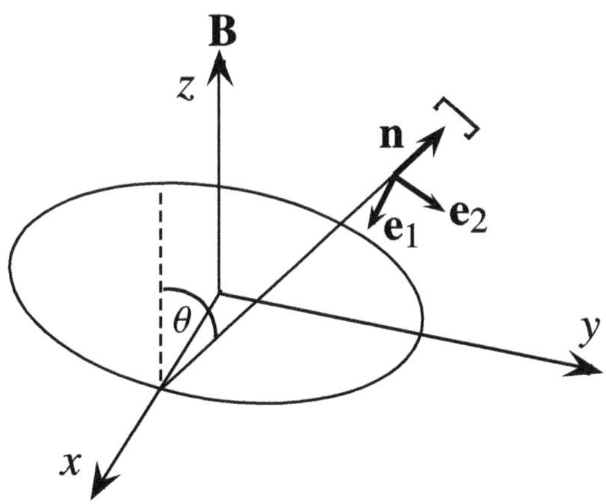


**Fig. 6.1** Figure shows a circular orbit of an electron in a plane perpendicular to the magnetic field **B**. It also depicts emission of radiation by the electron in its orbit, for example, along a given unit vector **n** with corresponding polarization vectors

## 6.1 Power of Radiation in Synchrotron Radiation

Using the notation in Eq. (4.2.1), the four-current density, in question, is given by

$$\mathbf{J}(x) = \mathrm{e}\,\dot{\mathbf{R}}(t)\,\delta^{(3)}\big(\mathbf{x} - \mathbf{R}(t)\big), \quad \dot{\mathbf{R}} = \frac{\mathrm{d}}{\mathrm{d}t}\mathbf{R(t)}, \quad J^0(x) = \mathrm{e}\,\delta^{(3)}\big(\mathbf{x} - \mathbf{R}(t)\big), \tag{6.1.1}$$

for which the following properties follow from Eq. (6.12) are satisfied

$$\dot{\mathbf{R}}(t)^2 = \mathrm{v}^2, \quad \ddot{\mathbf{R}}(t)^2 = \mathrm{v}^2\,\omega_\mathrm{o}^2, \quad \dot{\mathbf{R}}(t)\cdot\ddot{\mathbf{R}}(t) = 0, \quad \dot{\mathbf{R}}(t) = \mathbf{v}(t), \quad \mathbf{v}^2(t) = \text{Const.} \tag{6.1.2}$$

The general expression for the power of radiation corresponding the four-current density in Eq. (6.1.1) was derived in Eq. (4.2.21), and is given by

$$\mathrm{P}(t) = \int \mathrm{d}\Omega\,\mathrm{P}(\mathbf{n}, t) = \frac{2}{3}\frac{\mathrm{e}^2}{4\pi}\frac{1}{(1-\mathrm{v}^2)^2}\left[\dot{\mathbf{v}}^2 + \frac{(\dot{\mathbf{v}}\cdot\mathbf{v})^2}{(1-\mathrm{v}^2)}\right], \tag{6.1.3}$$

from which and Eq. (6.1.2), we readily obtain

$$\mathrm{P}(t) = \frac{2}{3}\frac{\mathrm{e}^2}{4\pi}\frac{1}{(1-\mathrm{v}^2)^2}\big[\mathrm{v}^2\omega_\mathrm{o}^2 + 0\big], \qquad \mathbf{v} = \dot{\mathbf{R}}, \tag{6.1.4}$$

and the exact power of radiation for synchrotron radiation, becomes

$$P = \frac{2}{3}\frac{\mathrm{e}^2\omega_\mathrm{o}^2}{4\pi}\frac{\mathrm{v}^2}{(1-\mathrm{v}^2)^2} = \frac{2}{3}\frac{\omega_\mathrm{o}}{R}\left(\frac{\mathrm{e}^2}{4\pi}\right)\frac{\mathrm{v}^3}{(1-\mathrm{v}^2)^2}, \tag{6.1.5}$$

where $\omega_\mathrm{o} = |\mathrm{e}|B/\gamma\mathrm{m}$, $R = v/\omega_\mathrm{o}$ (see Eq. (6.13)).

The power of radiation per revolution, takes the form

$$P_{\mathrm{rev}} = P\,\frac{2\pi}{\omega_\mathrm{o}} = \frac{4\pi}{3R}\frac{\mathrm{e}^2}{4\pi}\frac{\mathrm{v}^3}{(1-\mathrm{v}^2)^2}. \tag{6.1.6}$$

Upon using the expression for the energy of the electron

$$E = \frac{\mathrm{m}}{\sqrt{1-\mathrm{v}^2}}. \tag{6.1.7}$$

the following expression for the power, per revolution, in rationalized CGS units, emerges as,

$$P_{\mathrm{rev}} = \frac{4\pi}{3}\left(\frac{\mathrm{e}^2}{4\pi}\right)\frac{\mathrm{v}^3}{R}\left(\frac{E}{\mathrm{m}}\right)^4. \tag{6.1.8}$$

The simplicity of the derivation should be noted.

Of particular interest is the power of radiation emitted per unit angular frequency about any given value $\omega$. This may be obtained directly from Eq. (4.1.26) expressed as

$$\mathrm{P}(\omega, T) = \frac{\omega^2}{16\pi^3}\int \mathrm{d}^3\mathbf{x}\int \mathrm{d}^3\mathbf{x}'\int_{-\infty}^{\infty}\mathrm{d}\tau\; J^{\mu}(x)J_{\mu}(x')\int \mathrm{d}\,\Omega\, \mathrm{e}^{\mathrm{i}\omega\,\mathbf{n}\cdot(\mathbf{x}-\mathbf{x}')}\mathrm{e}^{-\,\mathrm{i}\omega\tau}. \tag{6.1.9}$$

From Eq. (6.12), with $(t - t') = \tau$,

$$\dot{\mathbf{R}}(t)\cdot\dot{\mathbf{R}}(t') = \mathrm{v}^2(\sin\omega_{\mathrm{o}}t\,\sin\omega_{\mathrm{o}}t' + \cos\omega_{\mathrm{o}}t\,\cos\omega_{\mathrm{o}}t) = \mathrm{v}^2\cos\omega_{\mathrm{o}}\tau, \tag{6.1.10}$$

and using, in the process, that $1 - \cos z = 2\sin^2(z/2)$,

$$|\mathbf{R}(t') - \mathbf{R}(t)|^2 = \mathbf{R}^2(t') + \mathbf{R}^2(t) - 2\mathbf{R}(t')\cdot\mathbf{R}(t) = 2\,R^2\big(1 - \cos\omega_{\mathrm{o}}\tau\big), \tag{6.1.11}$$

$$\Rightarrow\;\; |\mathbf{R}(t') - \mathbf{R}(t)| = 2\,R\sin\Big(\frac{\omega_{\mathrm{o}}\tau}{2}\Big). \tag{6.1.12}$$

Also from Eq. (6.1.1),

$$J^{\mu}(x)J_{\mu}(x') = \mathrm{e}^2\left(\dot{\mathbf{R}}(t')\cdot\dot{\mathbf{R}}(t) - 1\right)\delta^{(3)}\big(\mathbf{x} - \mathbf{R}(t)\big)\,\delta^{(3)}\big(\mathbf{x}' - \mathbf{R}(t')\big). \tag{6.1.13}$$

Accordingly, upon integration over $\mathbf{x}$, $\mathbf{x}'$, as well as over the angular integration and using, in the process Eqs. (6.1.10), (6.1.12), the following expression emerges for the power per unit energy about the energy $\omega$,

$$\mathrm{P}(\omega) = \frac{\mathrm{e}^2}{4\pi}\frac{\omega\,\omega_{\mathrm{o}}}{2\pi\,\mathrm{v}}\int_{-\infty}^{\infty}\mathrm{d}\tau\,\big(\mathrm{v}^2\cos\omega_{\mathrm{o}}\tau - 1\big)\sin\left(\frac{2\omega\mathrm{v}}{\omega_{\mathrm{o}}}\sin\Big(\frac{\omega_{\mathrm{o}}\tau}{2}\Big)\right)\frac{\mathrm{e}^{-\mathrm{i}\omega\tau}}{\sin\big(\frac{\omega_{\mathrm{o}}\tau}{2}\big)}, \tag{6.1.14}$$

using the notation $\mathrm{P}(\omega, T) = \mathrm{P}(\omega)$, and, as given in Eq. (6.13), we have used the fact that $R = \mathrm{v}/\omega_{\mathrm{o}}$.

The power $\mathrm{P}(\omega)$ will be expressed in terms of harmonics, involving Bessel functions in Sect. 6.5, which is useful for various considerations of synchrotron radiation.

## 6.2 Explicit Expression for the Average Number of Photon Emission

From Eq. (4.1.36), or directly from Eq. (6.1.14), we may immediately write down the number of photons $N(\omega)$ of energy $\omega$ emitted *per revolution* by removing the overall factor $\omega$ in it, corresponding to the energy of a photon with $\hbar = 1$, *and* multiplying it by $2\pi/\omega_{\mathrm{o}}$. This gives,

$$N(\omega) = \frac{\alpha}{\mathrm{v}} \int_{-\infty}^{\infty} \mathrm{d}\tau \left(\mathrm{v}^2 \cos \omega_o \tau - 1\right) \sin\left(\frac{2\omega \mathrm{v}}{\omega_o} \sin\left(\frac{\omega_o \tau}{2}\right)\right) \frac{\mathrm{e}^{-\mathrm{i}\omega\tau}}{\sin\left(\frac{\omega_o \tau}{2}\right)}, \tag{6.2.1}$$

where

$$\alpha = \frac{\mathrm{e}^2}{4\pi}, \tag{6.2.2}$$

denotes the *fine-structure* constant.

To obtain the average number of photons $< N >$ emitted by the electron per revolution of any energy, we have to integrate $\mathrm{N}(\omega)$ over $\omega$ from 0 to $\infty$. Accordingly, using the convenient variables,

$$z = \frac{\omega}{\omega_o}, \qquad x = \omega_o \tau, \tag{6.2.3}$$

we obtain

$$< N >= \alpha \int_{-\infty}^{\infty} \mathrm{d}x \, \frac{(\mathrm{v}^2 \cos x - 1)}{\mathrm{v} \sin\left(\frac{x}{2}\right)} \int_0^{\infty} \mathrm{d}z \, \sin\left(2\, z \mathrm{v} \sin\left(\frac{x}{2}\right)\right) \mathrm{e}^{-\mathrm{i}z\,x}, \tag{6.2.4}$$

up to satisfying the boundary condition that $< N >$ vanishes for $\mathrm{v} = 0$. If we formally set $\mathrm{v} = 0$, the whole integral, multiplying $\alpha$, becomes

$$-\,2 \int_{-\infty}^{\infty} \mathrm{d}x \int_0^{\infty} \mathrm{d}z \, z \mathrm{e}^{-\mathrm{i}zx}.$$

If we integrate over $x$ first, we get $-4\pi \int_0^{\infty} \mathrm{d}z \, z \, \delta(z)$ which gives zero, as it should for $\mathrm{v} = 0$. On the other hand, one may differentiate $< N >$ with respect to v and then integrate it again, leading one to consider $< N > - < N >|_{\mathrm{v}=0}$, rigorously and unambiguously satisfying the necessary boundary condition of a vanishing $< N >$ for $\mathrm{v} = 0$ as a boundary value problem.

We readily integrate $< N >$ over the $z$ variable first, involving the frequency, then impose this necessary boundary condition to be satisfied.

We note that the imaginary part $-\sin(z\,x)$, of $\exp[-\mathrm{i}z\,x]$, will generate part of the integrand which is an odd function in the variable $x$ and will thus not contribute. This will also ensure the reality of $< N >$. Accordingly, the latter takes the form

$$< N >= \alpha \int_{-\infty}^{\infty} \mathrm{d}x \, \frac{(\mathrm{v}^2 \cos x - 1)}{\mathrm{v} \sin\left(\frac{x}{2}\right)} \int_0^{\infty} \mathrm{d}z \, \sin\left(2\, z \mathrm{v} \sin\left(\frac{x}{2}\right)\right) \cos\left(z\,x\right). \tag{6.2.5}$$

To carry out the $z-$ integral, we proceed as follows, we first write

$$\sin(az)\cos(zx) = \frac{1}{2}\big[\sin(a-x)z + \sin(a+x)z\big], \quad a = 2\mathrm{v}\sin\Big(\frac{x}{2}\Big), \tag{6.2.6}$$

$$= \frac{1}{4\mathrm{i}}\big[\mathrm{e}^{\mathrm{i}z(a-x)} - \mathrm{e}^{-\mathrm{i}z(a-x)} + \mathrm{e}^{\mathrm{i}z(a+x)} - \mathrm{e}^{-\mathrm{i}z(a+x)}\big], \tag{6.2.7}$$

using the notation $a = 2\mathrm{v}\sin(x/2)$. We use the integrals

$$\int_0^\infty \mathrm{d}z\, \mathrm{e}^{\pm\mathrm{i}zx} = \frac{\pm\mathrm{i}}{x \pm \mathrm{i}\epsilon}, \qquad \epsilon \to +0. \tag{6.2.8}$$

These give for the $z-$ integral

$$\int_0^\infty \mathrm{d}z\, \sin\Big(2z\mathrm{v}\sin\Big(\frac{x}{2}\Big)\Big)\cos(zx)$$
$$= \frac{1}{4\mathrm{i}}\int_0^\infty \mathrm{d}z\,\big[\mathrm{e}^{\mathrm{i}z(a-x)} - \mathrm{e}^{-\mathrm{i}z(a-x)} + \mathrm{e}^{\mathrm{i}z(a+x)} - \mathrm{e}^{-\mathrm{i}z(a+x)}\big]. \tag{6.2.9}$$

$$= \frac{1}{4}\Big[\frac{1}{(a-x)+\mathrm{i}\epsilon} + \frac{1}{(a-x)-\mathrm{i}\epsilon} + \frac{1}{(a+x)+\mathrm{i}\epsilon} + \frac{1}{(a+x)-\mathrm{i}\epsilon}\Big]. \tag{6.2.10}$$

We note from Eq. (3.6.16) that no delta singularity arises by considering the imaginary part of the above expression. That is, we have to consider its real part, and we may set $\epsilon = 0$, referred to as taking the principal part of the integral in question, subject to the proper boundary condition that $< N >$ vanishes for $\mathrm{v} = 0$. We may combine the denominators, with the first pair and the second pair separately, to obtain for the latter, and for $\epsilon \to 0$,

$$\frac{1}{2}\Big[\frac{(a-x)}{(a-x)^2+\epsilon^2} + \frac{(a+x)}{(a+x)^2+\epsilon^2}\Big] \to \frac{1}{2}\Big[\frac{1}{(a-x)} + \frac{1}{(a+x)}\Big]$$
$$\to \frac{a}{a^2-x^2} = 2\frac{\mathrm{v}\sin(x/2)}{4\mathrm{v}^2\sin^2(x/2)-x^2}. \tag{6.2.11}$$

Accordingly, from Eq. (6.2.5), we obtain for the properly normalized expression for $< N >$, the explicit expression

$$< N > = 2\alpha\int_{-\infty}^\infty \mathrm{d}x\,\frac{(\mathrm{v}^2\cos x - 1)}{4\mathrm{v}^2\sin^2\big(\frac{x}{2}\big) - x^2} - 2\alpha\int_{-\infty}^\infty \mathrm{d}x\,\frac{(\mathrm{v}^2\cos x - 1)}{4\mathrm{v}^2\sin^2\big(\frac{x}{2}\big) - x^2}\bigg|_{\mathrm{v}=0} \tag{6.2.12}$$

$$= 2\alpha\int_{-\infty}^\infty \mathrm{d}x\,\Big[\frac{(\mathrm{v}^2\cos x - 1)}{4\mathrm{v}^2\sin^2\big(\frac{x}{2}\big) - x^2} - \frac{1}{x^2}\Big], \tag{6.2.13}$$

and there is no question of the existence of the integral.

Upon changing the variable of integration $x \to 2x$, we obtain the simple expression[9]

$$< N >= 2\alpha\,\mathrm{v}^2 \int_0^\infty \frac{\mathrm{d}x}{x^2}\left[\frac{\left(\frac{\sin x}{x}\right)^2 - \cos(2\,x)}{\left(1 - \mathrm{v}^2\left(\frac{\sin x}{x}\right)^2\right)}\right], \qquad (6.2.14)$$

for the average number of photons emitted per revolution in synchrotron radiation. There is no question of satisfying the appropriate boundary condition and of the simultaneous existence of the integral for all $0 \leq \mathrm{v} < 1$. In the above equation, we have used the symmetry in $x$ to integrate from 0 to $\infty$ and multiply by a factor of 2.

In the next section, we carry out a systematic rigorous analysis of the average number of photons emitted for a high energetic electron, $1 - \mathrm{v}^2 \to 0$.

## 6.3 Systematic High Energy Analysis of the Average Number of Photon Emission

To investigate the behavior of the average number of photons $< N >$ emitted per revolution for a high energy electron corresponding to $1 - \mathrm{v}^2 \to 0$, we may rewrite it in the following convenient explicit form:

$$< N >= \alpha \int_0^\infty \mathrm{d}x\,\big[I_1(v,x) + I_2(v,x) + I_3(v,x) + \epsilon(\mathrm{v},x)\big], \qquad (6.3.1)$$

$$\frac{2\mathrm{v}^2}{x^2}\left[\frac{\left(\frac{\sin x}{x}\right)^2 - \cos(2\,x)}{\left(1 - \mathrm{v}^2\left(\frac{\sin x}{x}\right)^2\right)}\right] = \sum_{i=1}^{3} I_i(\mathrm{v},x) + \epsilon(\mathrm{v},x), \qquad (6.3.2)$$

where,

$$I_1(\mathrm{v},x) = \frac{10\,\mathrm{v}}{3(1-\mathrm{v}^2) + \mathrm{v}^2x^2}, \quad I_2(x) = \frac{2}{x^2}\frac{6\left(\frac{\sin x}{x}\right)^2 - \cos(2x) - 5}{1 - \left(\frac{\sin x}{x}\right)^2}, \qquad (6.3.3)$$

$$I_3(\mathrm{v},x) = -2(1-\mathrm{v}^2)\left(\frac{9}{30}\frac{\mathrm{v}}{(1-\mathrm{v}^2) + \mathrm{v}^2x^2/3} - \frac{2}{3}\frac{(1-\mathrm{v}^2)\mathrm{v}}{[(1-\mathrm{v}^2) + \mathrm{v}^2x^2/3]^2}\right), \qquad (6.3.4)$$

$$\epsilon(\mathrm{v},x) = -2(1-\mathrm{v}^2)h(\mathrm{v},x), \qquad (6.3.5)$$

[9] Manoukian and Jearnkulprasert (2000) [9].

where $h(\mathrm{v}, x)$ for $\mathrm{v} \to 1$, is explicitly given by

$$h(1,x) = \Big(\frac{1}{x^2}\Big)\bigg(\frac{\big[\big(\frac{\sin x}{x}\big)^2 - \cos(2x)\big]}{\big[1-\big(\frac{\sin x}{x}\big)^2\big]^2} + \frac{8}{5} - \frac{15}{x^2}\bigg). \tag{6.3.6}$$

For $x \to 0$, we may use the expansions

$$\Big(\frac{\sin x}{x}\Big)^2 - \cos(2x) \simeq \frac{5}{3}x^2 - \frac{28}{45}x^4 + \frac{31}{360}x^6 + \dots, \tag{6.3.7}$$

$$\Big[1-\Big(\frac{\sin x}{x}\Big)^2\Big]^2 \simeq \frac{x^4}{9}\Big(1 - \frac{4}{15}x^2 + \frac{31}{900}x^4 + \dots\Big). \tag{6.3.8}$$

These give, for $x \to 0$

$$h(1,x) \to \frac{1}{x^2}\bigg(\Big[\frac{15}{x^2} - \frac{28}{5} + \frac{31}{40}x^2 + \dots\Big] + \frac{8}{5}\Big[1 - \frac{4}{15}x^2 + \frac{31}{900}x^4 + \dots\Big]$$
$$- \frac{15}{x^2}\Big[1 - \frac{4}{15}x^2 + \frac{31}{900}x^4 + \dots\Big]\bigg), \tag{6.3.9}$$

from which

$$h(1,x) \to -\frac{101}{600}, \quad \text{for } x \to 0. \tag{6.3.10}$$

On the other hand, it directly follows from Eq. (6.3.6), that

$$h(1,x) \to O\Big(\frac{1}{x^2}\Big), \quad \text{for } x \to \infty. \tag{6.3.11}$$

Therefore, there is no question of the existence of the integral,

$$(1-\mathrm{v}^2)\int_0^\infty \mathrm{d}x\, h(\mathrm{v},x) = O(1-\mathrm{v}^2), \tag{6.3.12}$$

corresponding to the term in Eq. (6.3.5).

Regarding the integral of $I_1(\mathrm{v}, x)$ with the latter given in Eq. (6.3.3). Make a change of variable $x \to x\sqrt{1-\mathrm{v}^2}/\mathrm{v}$ to obtain

$$\int_0^\infty \mathrm{d}x\, I_1(\mathrm{v},x) = \frac{10}{\sqrt{1-\mathrm{v}^2}}\int_0^\infty \mathrm{d}x\, \frac{1}{(3+x^2)} = \frac{5\pi}{\sqrt{3(1-\mathrm{v}^2)}}. \tag{6.3.13}$$

The integral of $I_2(x)$ is independent of v and is given by

$$\int_0^\infty \mathrm{d}x\, I_2(x) = 2\int_0^\infty \frac{\mathrm{d}x}{x^2}\, \frac{6\Big(\frac{\sin x}{x}\Big)^2 - \cos(2x) - 5}{1 - \Big(\frac{\sin x}{x}\Big)^2} = -9.55797. \tag{6.3.14}$$

For the integral of $I_3(\mathrm{v}, x)$, make a change of variable $x \to x\sqrt{1-\mathrm{v}^2}/\mathrm{v}$, as before, to obtain

$$\int_0^\infty \mathrm{d}x\, I_3(\mathrm{v}, x) = -2\sqrt{1-\mathrm{v}^2}\Big(\frac{9}{10}\int_0^\infty \frac{\mathrm{d}x}{(3+x^2)} - 6\int_0^\infty \frac{\mathrm{d}x}{(3+x^2)^2}\Big) \tag{6.3.15}$$

$$= -2\sqrt{1-\mathrm{v}^2}\Big(\frac{9}{10}.\frac{\pi}{2\sqrt{3}} - 6.\frac{\pi}{12\sqrt{3}}\Big) = \frac{\pi}{10\sqrt{3}}\sqrt{1-\mathrm{v}^2}. \tag{6.3.16}$$

All told, the average number of photons emitted per revolution for high energetic electron, for which $1 - \mathrm{v}^2 \to 0$, is given by.[10]

$$< N >= \frac{\alpha}{\sqrt{1-\mathrm{v}^2}}\Big[\frac{5\pi}{\sqrt{3}} + a_o\sqrt{1-\mathrm{v}^2} + a_1(1-\mathrm{v}^2) + O\big((1-\mathrm{v}^2)^{3/2}\big)\Big], \tag{6.3.17}$$

$$a_o = -9.55797, \quad a_1 = \frac{\pi}{10\sqrt{3}}. \tag{6.3.18}$$

The first term within the square brackets in Eq. (6.3.13) corresponds to the *truly* asymptotic limit quoted and tabulated in the literature.[11] The second term will certainly contribute even for v close to one.

## 6.4 Power of Radiation Per Unit Angular Frequency

We express the power of radiation per unit angular frequency about $\omega$, derived in Eq. (6.1.14) in terms of harmonics. This is expressed in terms of Bessel functions, with some detailed properties given in Sect. 6.7 which is convenient for numerical work in the study of synchrotron radiation and for carrying asymptotic analysis of the power of radiation in question for various energies of the electron.

[10] Manoukian, Jearnkulprasert, and Suebka (2004) [10].

[11] See, e.g., Navas et al. PDG (2024) [11].

The power of radiation per unit angular frequency about $\omega$, is given by

$$\mathrm{P}(\omega) = \frac{\mathrm{e}^2}{4\pi}\frac{\omega\,\omega_0}{2\pi\,\mathrm{v}}\int_{-\infty}^{\infty}\mathrm{d}\tau\,\mathrm{e}^{-\mathrm{i}\omega\tau}\frac{\left(\mathrm{v}^2\cos\omega_0\tau-1\right)}{\sin\left(\frac{\omega_0\tau}{2}\right)}\sin\left(\frac{2\omega\mathrm{v}}{\omega_0}\sin\left(\frac{\omega_0\tau}{2}\right)\right). \tag{6.4.1}$$

Upon defining the variables,

$$\omega_0\tau = x,\quad \frac{\omega}{\omega_0} = z, \tag{6.4.2}$$

we may rewrite Eq. (6.4.1) as

$$\mathrm{P}(\omega) = \frac{\mathrm{e}^2}{4\pi}\frac{\omega}{2\pi}\int_{-\infty}^{\infty}\mathrm{d}x\,\mathrm{e}^{-\mathrm{i}zx}\frac{\left(\mathrm{v}^2\cos x-1\right)}{\mathrm{v}\,\sin\left(\frac{x}{2}\right)}\sin\left(2z\mathrm{v}\,\sin\left(\frac{x}{2}\right)\right). \tag{6.4.3}$$

The $x-$ integrand, multiplying the $\exp[-\mathrm{i}zx]$ factor, is invariant under the transformation, variable $x\to x+2\pi n$, for all $n=0,\pm1,\pm2,\ldots$. This invites us to carry out a Fourier series as follows;

$$\frac{\left(\mathrm{v}^2\cos x-1\right)}{\sin\left(\frac{x}{2}\right)}\sin\left(2z\mathrm{v}\,\sin\left(\frac{x}{2}\right)\right) = \sum_{\ell=-\infty}^{\infty} a_\ell\,\mathrm{e}^{\mathrm{i}\ell x}. \tag{6.4.4}$$

To obtain the value of the Fourier coefficient $a_\ell$, we multiply the above equation by $\exp[-\mathrm{i}\ell' x]$ and integrate over $x$ from $-\pi$ to $\pi$. This gives,

$$\int_{-\pi}^{\pi}\mathrm{d}x\,\mathrm{e}^{-\mathrm{i}\ell' x}\frac{\left(\mathrm{v}^2\cos x-1\right)}{\sin\left(\frac{x}{2}\right)}\sin\left(2z\mathrm{v}\,\sin\left(\frac{x}{2}\right)\right) = \sum_{\ell=-\infty}^{\infty} a_\ell\int_{-\pi}^{\pi}\mathrm{d}x\,\mathrm{e}^{i(\ell-\ell')x}. \tag{6.4.5}$$

Using the fact that $\sin\big(\pm(\ell-\ell')\pi\big)=0$, $\cos\big((\ell-\ell')\pi\big)=\cos\big(-(\ell-\ell')\pi\big)$, give

$$\int_{-\pi}^{\pi}\mathrm{d}x\,\mathrm{e}^{i(\ell-\ell')x} = 2\pi\,\delta_{\ell\ell'}. \tag{6.4.6}$$

From this, we explicitly have

$$a_\ell = \frac{1}{2\pi}\int_{-\pi}^{\pi}\mathrm{d}x\,\mathrm{e}^{-\mathrm{i}\ell x}\frac{\left(\mathrm{v}^2\cos x-1\right)}{\sin\left(\frac{x}{2}\right)}\sin\left(2z\mathrm{v}\,\sin\left(\frac{x}{2}\right)\right). \tag{6.4.7}$$

Accordingly, from Eq. (6.4.4)

$$\frac{(\mathrm{v}^2\cos x-1)}{\sin\left(\frac{x}{2}\right)}\sin\left(2z\mathrm{v}\sin\left(\frac{x}{2}\right)\right)$$
$$=\sum_{\ell=-\infty}^{\infty}\mathrm{e}^{\mathrm{i}x\ell}\frac{1}{2\pi}\int_{-\pi}^{\pi}\mathrm{d}x'\,\mathrm{e}^{-\mathrm{i}\ell x'}\frac{(\mathrm{v}^2\cos x'-1)}{\sin\left(\frac{x'}{2}\right)}\sin\left(2z\mathrm{v}\sin\left(\frac{x'}{2}\right)\right). \tag{6.4.8}$$

Upon substituting the latter in Eq. (6.4.3), integrating over $x$, using, in the process, $z=\omega/\omega_o$ and

$$\frac{1}{2\pi}\int_{-\infty}^{\infty}\mathrm{d}x\,\mathrm{e}^{-\mathrm{i}(z-\ell)x}=\delta(z-\ell)=\omega_o\delta(\omega-\omega_o\ell), \tag{6.4.9}$$

the following expression emerges

$$\mathrm{P}(\omega)=\frac{\mathrm{e}^2}{4\pi}\omega\,\omega_o\sum_{\ell=-\infty}^{\infty}\delta(\omega-\omega_o\ell)$$
$$\times\frac{1}{2\pi}\int_{-\pi}^{\pi}\mathrm{d}x\,\cos(\ell x)\frac{(\mathrm{v}^2\cos x-1)}{\mathrm{v}\sin\left(\frac{x}{2}\right)}\sin\left(2\ell\mathrm{v}\sin\left(\frac{x}{2}\right)\right), \tag{6.4.10}$$

where we have denoted the other variable of integration $x'$ in Eq. (6.4.8) by $x$ for convenience. We have also used the fact that the odd part $-\mathrm{i}\sin(\ell x)$ of $\exp[-\mathrm{i}\ell x]$ does not contribute to the $x$-integral. Clearly the $\ell=0$ term does not contribute either. Finally, invoking the positivity of $\omega$, we have

$$\mathrm{P}(\omega)=\frac{\mathrm{e}^2}{4\pi}\sum_{\ell=1}^{\infty}\delta(\omega-\omega_o\ell)\,\mathrm{P}_\ell, \tag{6.4.11}$$

$$\mathrm{P}_\ell=\omega\omega_o\frac{1}{2\pi}\int_{-\pi}^{\pi}\mathrm{d}x\,\cos(\ell x)\frac{(\mathrm{v}^2\cos x-1)}{\mathrm{v}\sin\left(\frac{x}{2}\right)}\sin\left(2\ell\mathrm{v}\sin\left(\frac{x}{2}\right)\right). \tag{6.4.12}$$

It is more convenient to split the $x-$ integral in $\mathrm{P}_\ell$ into two parts, by using in the process that $\cos x=1-2\sin^2(x/2)$ in $(\mathrm{v}^2\cos x-1)$, as follows:

$$\mathrm{P}_\ell=f_1(\ell,\mathrm{v})+f_2(\ell,\mathrm{v}), \tag{6.4.13}$$

$$f_1(\ell,\mathrm{v})=-(1-\mathrm{v}^2)\,\omega\,\omega_o\frac{1}{2\pi}\int_{-\pi}^{\pi}\mathrm{d}x\,\cos(\ell x)\frac{\sin\left[2\ell\mathrm{v}\sin\left(\frac{x}{2}\right)\right]}{\mathrm{v}\sin\left(\frac{x}{2}\right)}, \tag{6.4.14}$$

$$f_2(\ell,\mathrm{v})=-2\mathrm{v}\omega\,\omega_o\frac{1}{2\pi}\int_{-\pi}^{\pi}\mathrm{d}x\,\cos(\ell x)\sin\left(\frac{x}{2}\right)\sin\left[2\ell\mathrm{v}\sin\left(\frac{x}{2}\right)\right]. \tag{6.4.15}$$

Now we may use the following integral representation of the Bessel function[12] of order $2\ell$, given by[13]

$$J^{\mathrm{B}}_{2\ell}(x) \;=\; \frac{1}{\pi}\int_{-\pi/2}^{\pi/2} \mathrm{d}y\ \cos(2\ell y)\ \cos(x\sin y). \tag{6.4.16}$$

Upon integrating the latter over $x$, from 0 to $2\ell\mathrm{v}$, gives

$$\int_0^{2\ell\mathrm{v}} \mathrm{d}x\ J^{\mathrm{B}}_{2\ell}(x) \;=\; \frac{1}{\pi}\int_{-\pi/2}^{\pi/2} \mathrm{d}y\ \cos(2\ell y)\,\frac{\sin(2\ell\mathrm{v}\sin y)}{\sin y}, \tag{6.4.17}$$

$$\text{and with } y \to y/2 \text{ gives} \;=\; \frac{1}{2\pi}\int_{-\pi}^{\pi} \mathrm{d}y\ \cos(\ell y)\,\frac{\sin\left(2\ell\mathrm{v}\sin\left(\frac{y}{2}\right)\right)}{\sin\left(\frac{y}{2}\right)}. \tag{6.4.18}$$

Similarly differentiating $J^{\mathrm{B}}_{2\ell}(x)$ in Eq. (6.4.16) with respect to $x$, setting $x = 2\ell\mathrm{v}$, and making a simple change of the variable of integration $y$ to $y/2$, lead to

$$J^{\mathrm{B}'}_{2\ell}(2\ell\mathrm{v}) \;=\; -\,\frac{1}{2\pi}\int_{-\pi}^{\pi} \mathrm{d}y\ \cos(\ell y)\ \sin\left(\frac{y}{2}\right)\ \sin\left[2\ell\mathrm{v}\ \sin\left(\frac{y}{2}\right)\right]. \tag{6.4.19}$$

Upon comparison Eq. (6.4.18) with Eqs. (6.4.14), and (6.4.19) with Eq. (6.4.15), the following expression emerges for $\mathrm{P}_\ell$, from the expression given in Eq. (6.4.13),[14]

$$\mathrm{P}(\omega) \;=\; \frac{\mathrm{e}^2}{4\pi}\sum_{\ell=1}^{\infty} \delta(\omega - \omega_{\mathrm{o}}\ell)\,\mathrm{P}_\ell, \tag{6.4.20}$$

$$\mathrm{P}_\ell \;=\; \frac{\omega\,\omega_{\mathrm{o}}}{\mathrm{v}}\left[2\mathrm{v}^2\,J^{\mathrm{B}'}_{2\ell}(2\ell\mathrm{v}) \;-\; (1-\mathrm{v}^2)\int_0^{2\ell\mathrm{v}} \mathrm{d}x\ J^{\mathrm{B}}_{2\ell}(x)\right], \tag{6.4.21}$$

expanded in terms of harmonics.

The interested reader may refer to the references cited below for numerical estimates[15] of the functions appearing in synchrotron studies, as well as various investigations of asymptotic analysis involved in various energy regions,[16] for further studies.

---

[12] In order not to confuse the Bessel function here with current, we have introduced the letter B.

[13] See, e.g., Gradshteyn and Ryzhik (2000) [4], p. 902.

[14] This was derived by Schwinger in (1949, 1946) [17, 18]. See also Schott (1912) [16], Schwinger, Tsai and Erber (1976) [19].

[15] Fouka and Ouichaoui (2013) [3]. See also the related references: Rybicki and Lightman (1979) [14], Pacholczyk (1970) [12].

[16] Fouka and Ouichaoui (2013) [3], Schwinger (1949) [18], and the monumental work of Schwinger et al. (1976) [19]. See also Rybicki and Lightman (1979) [14], Pacholczyk (1970) [12]. The following references : Abramowitz and Stegun (1964) [1], Watson (1966) [22] are also useful in this respect.

In the next section we carry out another expansion of the power of radiation in terms of harmonics involving polarizations of radiation.

## 6.5 Polarization Aspects

The power of radiation per unit frequency, about $\omega$, now per unit solid angle about a unit vector $\mathbf{n}$, expressed in terms of polarization vectors was given Eq. (4.1.38):

$$\begin{aligned} \mathrm{P}_{\mathbf{n}}(\omega, T) &= \frac{\omega^2}{16\pi^3}\int \mathrm{d}^3\mathbf{x} \int \mathrm{d}^3\mathbf{x}' \int_{-\infty}^{\infty} \mathrm{d}\tau \\ &\times \sum_{\lambda=1,2} \big(J^i(x)\,\mathbf{e}^i_\lambda\big)\big(J^j(x')\,\mathbf{e}^j_\lambda\big)\,\mathrm{e}^{\mathrm{i}\,\omega\mathbf{n}\cdot(\mathbf{x}-\mathbf{x}')}\mathrm{e}^{-\,\mathrm{i}\omega\tau}, \end{aligned} \tag{6.5.1}$$

Consider a unit vector $\mathbf{n}$, and polarization vectors defined by

$$\mathbf{n} = (0, \sin\theta, \cos\theta), \quad \mathbf{e}_1 = (1, 0, 0), \quad \mathbf{e}_2 = (0, \cos\theta, -\sin\theta), \tag{6.5.2}$$

$$\mathbf{n}\cdot\mathbf{e}_1 = 0, \quad \mathbf{n}\cdot\mathbf{e}_2 = 0, \quad \mathbf{e}_1\cdot\mathbf{e}_2 = 0, \tag{6.5.3}$$

referring to Fig. 6.1 in the beginning of the chapter. We set $t = T + \tau/2$, $t' = T - \tau/2$, $t - t' = \tau$.

Concerning currents $\mathbf{J}(x)$, $\mathbf{J}(x')$ in Eq.(6.1.1), and the vectors $\mathbf{R}(t)$, $\mathbf{R}(t')$, $\dot{\mathbf{R}}(t)$, $\dot{\mathbf{R}}(t')$, in Eq. (6.12), we then have

$$\mathbf{n}\cdot\Big[\mathbf{R}\Big(T+\frac{\tau}{2}\Big) - \mathbf{R}\Big(T-\frac{\tau}{2}\Big)\Big] = R\sin\theta\Big[\cos\omega_{\mathrm{o}}\Big(T+\frac{\tau}{2}\Big) - \cos\omega_{\mathrm{o}}\Big(T-\frac{\tau}{2}\Big)\Big]. \tag{6.5.4}$$

Moreover,

$$\big(\dot{\mathbf{R}}(t)\cdot\mathbf{e}_1\big)\big(\dot{\mathbf{R}}(t')\cdot\mathbf{e}_1\big) = \mathrm{v}^2\sin(\omega_{\mathrm{o}}t)\sin(\omega_{\mathrm{o}}t') \tag{6.5.5}$$

$$= \frac{\mathrm{v}^2}{2}\big[\cos\omega_{\mathrm{o}}\tau - \cos 2\omega_{\mathrm{o}}T\big]. \tag{6.5.6}$$

and

$$\big(\dot{\mathbf{R}}(t)\cdot\mathbf{e}_2\big)\big(\dot{\mathbf{R}}(t')\cdot\mathbf{e}_2\big) = \mathrm{v}^2\cos^2\theta\,\cos(\omega_{\mathrm{o}}t)\cos(\omega_{\mathrm{o}}t') \tag{6.5.7}$$

$$= \cos^2\theta\,\frac{\mathrm{v}^2}{2}\big[\cos\omega_{\mathrm{o}}\tau + \cos 2\omega_{\mathrm{o}}T\big]. \tag{6.5.8}$$

Accordingly, upon integration over $\mathbf{x}$, $\mathbf{x}'$, using the Dirac delta distributions in the currents in Eq. (6.1.1), with $R = \mathrm{v}/\omega_{\mathrm{o}}$, as given in Eq. (6.13), lead to

$$
\begin{aligned}
\mathrm{P}_{\mathbf{n}}(\omega, T) &= \frac{\omega^2 \mathrm{v}^2 \mathrm{e}^2}{32\pi^3} \int_{-\infty}^{\infty} \mathrm{d}\tau\, \mathrm{e}^{-\mathrm{i}\omega\tau} \\
\times \exp&\Big[\mathrm{i}\,\frac{\mathrm{v}\,\omega}{\omega_o} \sin\theta \cos\Big(\omega_o\Big(T+\frac{\tau}{2}\Big)\Big)\Big] \exp\Big[-\mathrm{i}\,\frac{\mathrm{v}\,\omega}{\omega_o} \sin\theta \cos\Big(\omega_o\Big(T-\frac{\tau}{2}\Big)\Big)\Big] \\
\times &\Big(\big[\cos\omega_o\tau \;-\; \cos 2\omega_o T\big] + \cos^2\theta\,\big[\cos\omega_o\tau \;+\; \cos 2\omega_o T\big]\Big). \qquad (6.5.9)
\end{aligned}
$$

To average the above expression over one period, and eventually carry out the $\tau-$ integral, we use the following summation formula involving Bessel functions:[17]

$$
\mathrm{e}^{\mathrm{i}z\cos x} \;=\; \sum_{\ell=-\infty}^{\infty} (\mathrm{i})^{\ell}\, J_{\ell}^{\mathrm{B}}(z)\, \mathrm{e}^{\mathrm{i}\ell x}. \qquad (6.5.10)
$$

Therefore, it is convenient to denote

$$
\frac{\mathrm{v}\,\omega}{\omega_o} \sin\theta \;\equiv\; z, \qquad (6.5.11)
$$

to write

$$
\exp\Big[\mathrm{i}z\cos\Big(\omega_o\Big(T+\frac{\tau}{2}\Big)\Big)\Big] = \sum_{\ell=-\infty}^{\infty} (\mathrm{i})^{\ell}\, J_{\ell}^{\mathrm{B}}(z) \exp\Big[\mathrm{i}\omega_o\Big(T+\frac{\tau}{2}\Big)\ell\Big]. \qquad (6.5.12)
$$

$$
\exp\Big[-\mathrm{i}z\cos\Big(\omega_o\Big(T-\frac{\tau}{2}\Big)\Big)\Big] = \sum_{\ell'=-\infty}^{\infty} (\mathrm{i})^{\ell'}(-1)^{\ell'}\, J_{\ell'}^{\mathrm{B}}(z) \exp\Big[\mathrm{i}\omega_o\Big(T-\frac{\tau}{2}\Big)\ell'\Big]. \qquad (6.5.13)
$$

where we have used that $J_{\ell}^{\mathrm{B}}(-z) = (-1)^{\ell} J_{\ell}^{\mathrm{B}}(z)$.

We perform the average over one cycle of the integral in Eq. (6.5.9) using, in the process, Eqs. (6.5.12), (6.5.13). This involves the averaging

$$
\begin{aligned}
\frac{\omega_o}{2\pi} &\int_0^{2\pi/\omega_o} \mathrm{d}T\, \mathrm{e}^{\mathrm{i}\omega_o T(\ell+\ell')} \big[\cos\omega_o\tau \;\mp\; \cos 2\omega_o T\big] \\
&= \delta_{\ell',-\ell}\cos\omega_o\tau \;\mp\; \frac{1}{2}\big[\delta_{\ell',-\ell-2} \;+\; \delta_{\ell',-\ell+2}\big], \qquad (6.5.14)
\end{aligned}
$$

where we have used

$$
\cos(2\omega_o T) \;=\; \frac{1}{2}\big[\mathrm{e}^{2\mathrm{i}\omega_o T} \;+\; \mathrm{e}^{-2\mathrm{i}\omega_o T}\big], \qquad (6.5.15)
$$

in the process.

---

[17] See, e.g., Gradshteyn and Ryzhik (2000) [4], p. 923.

Using the fact that $J^{\mathrm{B}}_{\ell}(z) = (-1)^{\ell} J^{\mathrm{B}}_{\ell}(z)$, the $\tau-$ integral of first term in Eq. (6.5.9), corresponding to the polarization $\mathbf{e}_1$, in Eq. (6.5.6) is from Eqs. (6.5.12)–(6.5.14),

$$\int_{-\infty}^{\infty} \mathrm{d}\tau\, \mathrm{e}^{-\mathrm{i}\omega\tau} \sum_{\ell=-\infty}^{\infty} \big[\cos(\omega_{\mathrm{o}}\tau)\, \mathrm{e}^{\mathrm{i}\ell\omega_{\mathrm{o}}\tau}\, (J^{\mathrm{B}}_{\ell}(z))^2 + \\ \mathrm{e}^{\mathrm{i}(\ell+1)\omega_{\mathrm{o}}\tau}\, J^{\mathrm{B}}_{\ell}(z)\, J^{\mathrm{B}}_{\ell+2}(z) \;+\; \mathrm{e}^{\mathrm{i}(\ell-1)\omega_{\mathrm{o}}\tau}\, J^{\mathrm{B}}_{\ell}(z)\, J^{\mathrm{B}}_{\ell-2}(z)\big]. \tag{6.5.16}$$

Upon using the integrals

$$\int_{-\infty}^{\infty} \mathrm{d}\,\tau\, \mathrm{e}^{\mathrm{i}(\omega-\omega_{\mathrm{o}}\ell)\tau} = 2\pi\,\delta(\omega - \omega_{\mathrm{o}}\ell), \tag{6.5.17}$$

$$\cos(\omega_{\mathrm{o}}\tau) = \frac{1}{2}\big[\mathrm{e}^{\mathrm{i}\omega_{\mathrm{o}}\tau} + \mathrm{e}^{\mathrm{i}\omega_{\mathrm{o}}\tau}\big], \tag{6.5.18}$$

the $\tau-$ integral in Eq. (6.5.16) may be readily carried. If we simultaneously change the parameter $\ell$ to $(\ell - 1)$ in the second term in the summand within the square brackets, and $\ell$ to $(\ell + 1)$ in the third term in the summand, the following expression emerges for the power in question, averaged over one cycle, corresponding to the polarization $\mathbf{e}_1$:

$$\mathrm{P_n}(\omega, \mathbf{e}_1) = \sum_{\ell=1}^{\infty} \delta(\omega - \omega_{\mathrm{o}}\ell)\, \mathrm{P}_{\mathbf{e}_1,\ell}, \tag{6.5.19}$$

$$\begin{aligned} \mathrm{P}_{\mathbf{e}_1,\ell} &= \frac{\mathrm{v}^2\omega_{\mathrm{o}}^2\mathrm{e}^2}{32\pi^2}\, \ell^2\Big(\big(J^{\mathrm{B}}_{\ell-1}(z)\big)^2 - 2\, J^{\mathrm{B}}_{\ell+1}(z) J^{\mathrm{B}}_{\ell-1}(z) + \big(J^{\mathrm{B}}_{\ell+1}(z)\big)^2\Big), \\ &= \frac{\mathrm{v}^2\omega_{\mathrm{o}}^2\mathrm{e}^2}{32\pi^2}\, \ell^2\Big(J^{\mathrm{B}}_{\ell+1}(z) - J^{\mathrm{B}}_{\ell-1}(z)\Big)^2, \quad z = \mathrm{v}\ell\sin\theta,\ \ \ell = \frac{\omega}{\omega_{\mathrm{o}}}, \end{aligned} \tag{6.5.20}$$

where we have invoked the positivity of $\omega$.

Finally, using the recurrence relation,[18]

$$J^{\mathrm{B}'}_{\ell}(z) = \frac{1}{2}\big[J^{\mathrm{B}}_{\ell-1}(z) - J^{\mathrm{B}}_{\ell+1}(z)\big], \tag{6.5.21}$$

gives

$$\mathrm{P}_{\mathbf{e}_1,\ell} = \frac{\mathrm{v}^2\omega_{\mathrm{o}}^2}{2\pi}\frac{\mathrm{e}^2}{4\pi}\, \ell^2\big(J^{\mathrm{B}'}_{\ell}(\mathrm{v}\ell\sin\theta)\big)^2. \tag{6.5.22}$$

For the corresponding expression to the polarization $\mathbf{e}_2$, we note that for $\mathbf{e}_1$, we have a minus sign within the first square brackets on the last line of Eq. (6.5.9), and a minus sign within the round brackets on the last line of Eq. (6.5.20). On the other hand

[18] See, e.g., Gradshteyn and Ryzhik (2000) [4], p. 361.

for $\mathbf{e}_2$, we have a + sign within the square brackets on the last line of Eq. (6.5.9) and the expression within the square brackets is simply multiplied by $\cos^2\theta$. Accordingly, we may infer, directly from Eq. (6.5.20), that

$$\mathrm{P_n}(\omega, \mathbf{e}_2) = \sum_{\ell=1}^{\infty} \delta(\omega - \omega_o \ell)\, \mathrm{P}_{\mathbf{e}_2,\ell}, \tag{6.5.23}$$

$$\mathrm{P}_{\mathbf{e}_2,\ell} = \frac{\mathrm{v}^2\omega_o^2 \mathrm{e}^2}{32\pi^2}\, \ell^2 \cos^2\theta \left( J^{\mathrm{B}}_{\ell+1}(z) + J^{\mathrm{B}}_{\ell-1}(z) \right)^2, \quad z = \mathrm{v}\ell \sin\theta. \tag{6.5.24}$$

Using the recurrence relation,[19]

$$J^{\mathrm{B}}_{\ell+1}(z) + J^{\mathrm{B}}_{\ell-1}(z) = \frac{2\ell}{z} J^{\mathrm{B}}_{\ell}(z), \tag{6.5.25}$$

gives

$$\mathrm{P}_{\mathbf{e}_2,\ell} = \frac{\mathrm{v}^2\omega_o^2}{2\pi} \frac{\mathrm{e}^2}{4\pi}\, \ell^2 \cos^2\theta \left( \frac{J^{\mathrm{B}}_{\ell}(\mathrm{v}\ell \sin\theta)}{\mathrm{v}\sin\theta} \right)^2. \tag{6.5.26}$$

In particular, we learn that for emission of the radiation considered along the orbital plane, referring to Fig. 6.1 in the beginning of the chapter, i.e. $\theta = \pi/2$, it is totally polarized along $\mathbf{e}_1$ as $\mathrm{P}_{\mathbf{e}_2,\ell} = 0$ above. Such polarization is referred to as $\sigma$ polarization lying the orbital plane of the electron..

Upon integrating Eqs. (6.5.22) and (6.5.26) corresponding to the contribution of both polarizations, one obtains the total contribution as a sum of harmonics:

$$\int_0^\infty \mathrm{d}\omega\, [P_{\mathbf{n}}(\omega, \mathbf{e}_1) + P_{\mathbf{n}}(\omega, \mathbf{e}_1)]$$
$$= \frac{\mathrm{v}^2\omega_o^2}{2\pi} \frac{\mathrm{e}^2}{4\pi} \sum_{\ell=1}^{\infty} \left[ \ell^2 \big(J^{\mathrm{B}'}_{\ell}(\mathrm{v}\ell \sin\theta)\big)^2 + \ell^2 \cos^2\theta \left( \frac{J^{\mathrm{B}}_{\ell}(\mathrm{v}\ell \sin\theta)}{\mathrm{v}\sin\theta} \right)^2 \right], \tag{6.5.27}$$

where the contribution of each of the two polarizations in the sum appear separately. The integral of the latter over the solid angle $\mathrm{d}\Omega$ is referred to as *Schott's formula*.[20]

In the following section, we carry out the sum over the harmonics in the above equation to study the profile of radiation for an ultra-relativistic electron.

---

[19] See, e.g., Gradshteyn and Ryzhik (2000) [4], p. 361.

[20] Schott (1912) [16].

## 6.6 Summing Over the Harmonics: Profile of Radiation and the Ultra-Relativistic Electron

It is remarkable that after having integrated over $\omega$ obtaining the expression in Eq. (6.5.27), one may explicitly carry out the sum over the harmonics.

To the above end, we use the following summation formulas[21] involving Bessel functions:

$$\sum_{\ell=1}^{\infty} \ell^2 \big(J_\ell^{\mathrm{B}'}(\ell x)\big)^2 = \frac{4+3x^2}{16(1-x^2)^{5/2}}, \tag{6.6.1}$$

$$\sum_{\ell=1}^{\infty} \ell^2 \big(J_\ell^{\mathrm{B}}(\ell x)\big)^2 = \frac{x^2(4+x^2)}{16(1-x^2)^{7/2}}. \tag{6.6.2}$$

Upon setting $x = \mathrm{v}\sin\theta$, as before, the expression in Eq. (6.5.27) for the total radiation emitted in direction of $\mathbf{n}$ simply becomes the sum of the following two terms:

$$\mathrm{P}_{\mathbf{n}}(\mathbf{e}_1) = \frac{\mathrm{v}^2\omega_{\mathrm{o}}^2}{32\pi}\Big(\frac{\mathrm{e}^2}{4\pi}\Big)\frac{4+3\mathrm{v}^2\sin^2\theta}{(1-\mathrm{v}^2\sin^2\theta)^{5/2}}, \tag{6.6.3}$$

$$\mathrm{P}_{\mathbf{n}}(\mathbf{e}_2) = \frac{\mathrm{v}^2\omega_{\mathrm{o}}^2}{32\pi}\Big(\frac{\mathrm{e}^2}{4\pi}\Big)\cos^2\theta\,\frac{4+\mathrm{v}^2\sin^2\theta}{(1-\mathrm{v}^2\sin^2\theta)^{7/2}}, \tag{6.6.4}$$

where $\theta$ is measured relative to the magnetic field $\mathbf{B}$, as shown in Fig. 6.1 in the beginning of the chapter, with $\theta = \pi/2$ corresponding to the orbital plane of the electron.

In order to describe the radiation emitted by an ultra-relativistic electron in more details, we first introduce the following useful integrals:

$$\int \frac{\mathrm{d}z}{(1+z^2)^{3/2}} = \frac{z}{(1+z^2)^{1/2}}. \tag{6.6.5}$$

$$\int \frac{\mathrm{d}z}{(1+z^2)^{5/2}} = \frac{z(2z^2+3)}{3(1+z^2)^{3/2}}. \tag{6.6.6}$$

$$\int \frac{\mathrm{d}z}{(1+z^2)^{7/2}} = \frac{z(8z^4+20z^2+15)}{15(1+z^2)^{5/2}}. \tag{6.6.7}$$

$$\int \mathrm{d}z\left[\frac{1}{(1+z^2)^{5/2}} - \frac{1}{(1+z^2)^{7/2}}\right] = \frac{z^3(2z^2+5)}{15(1+z^2)^{5/2}}. \tag{6.6.8}$$

Upon introducing the convenient variable $z = \mathrm{v}\cos\theta/\sqrt{1-\mathrm{v}^2}$ and referring to Eq. (6.6.3), we may write

[21] See Schott (1912) [16], pp. 121–125.

$$\frac{4+3\mathrm{v}^2\sin^2\theta}{(1-\mathrm{v}^2\sin^2\theta)^{5/2}} = \frac{4+3\mathrm{v}^2-3\mathrm{v}^2\cos^2\theta}{(1-\mathrm{v}^2+\mathrm{v}^2\cos^2\theta)^{5/2}} \tag{6.6.9}$$

$$= \frac{4+3\mathrm{v}^2+3(1-\mathrm{v}^2)-3(\mathrm{v}^2\cos^2\theta+1-\mathrm{v}^2)}{(1-\mathrm{v}^2+\mathrm{v}^2\cos^2\theta)^{5/2}}$$

$$= \frac{7}{(1-\mathrm{v}^2)^{5/2}}\frac{1}{(1+z^2)^{5/2}} - \frac{3}{(1-\mathrm{v}^2)^{3/2}}\frac{1}{(1+z^2)^{3/2}}. \tag{6.6.10}$$

Similarly, referring to Eq. (6.6.4), we have

$$\cos^2\theta\,\frac{4+\mathrm{v}^2\sin^2\theta}{(1-\mathrm{v}^2\sin^2\theta)^{7/2}} = \cos^2\theta\,\frac{4+\mathrm{v}^2+1-\mathrm{v}^2-(\mathrm{v}^2\cos^2\theta+1+\mathrm{v}^2)}{(1-\mathrm{v}^2+\mathrm{v}^2\cos^2\theta)^{7/2}} \tag{6.6.11}$$

$$= \frac{5\cos^2\theta}{(1-\mathrm{v}^2+\mathrm{v}^2\cos^2\theta)^{7/2}} - \frac{\cos^2\theta}{(1-\mathrm{v}^2+\mathrm{v}^2\cos^2\theta)^{5/2}}$$

$$= \frac{5}{\mathrm{v}^2(1-\mathrm{v}^2)^{5/2}}\left[\frac{1}{(1+z^2)^{5/2}} - \frac{1}{(1+z^2)^{7/2}}\right]$$

$$- \frac{1}{\mathrm{v}^2(1-\mathrm{v}^2)^{3/2}}\left[\frac{1}{(1+z^2)^{3/2}} - \frac{1}{(1+z^2)^{5/2}}\right], \tag{6.6.12}$$

with $z = v\cos\theta/\sqrt{1-\mathrm{v}^2}$.

For considering an application to the ultra-relativistic electron, up to a $(2\pi)$ factor, the solid angle measure $\mathrm{d}(\cos\theta)$, may be expresses in terms of the $z$ variable by the following

$$\mathrm{d}(\cos\theta) = \frac{\sqrt{1-\mathrm{v}^2}}{\mathrm{v}}\,\mathrm{d}z, \quad \text{which} \tag{6.6.13}$$

$$\text{for} \quad -\frac{3}{2}\frac{\sqrt{1-\mathrm{v}^2}}{\mathrm{v}} \leq \cos\theta \leq \frac{3}{2}\frac{\sqrt{1-\mathrm{v}^2}}{\mathrm{v}}, \quad \Rightarrow \quad -\frac{3}{2} \leq z \leq \frac{3}{2}, \tag{6.6.14}$$

$$\text{and} \quad \text{for} \quad -1 \leq \cos\theta \leq 1, \quad \Rightarrow \quad -\frac{v}{\sqrt{1-\mathrm{v}^2}} \leq z \leq \frac{v}{\sqrt{1-\mathrm{v}^2}}. \tag{6.6.15}$$

By referring to Eqs. (6.6.10) and (6.6.12), we note that for both regions of integrations just described, for an ultra-relativistic electron $(1-\mathrm{v}^2) \to 0$, we may make the substitutions:

$$\frac{4+3\mathrm{v}^2\sin^2\theta}{(1-\mathrm{v}^2\sin^2\theta)^{5/2}}\,\mathrm{d}\cos\theta \to \frac{7}{(1-\mathrm{v}^2)^2}\frac{1}{\mathrm{v}(1+z^2)^{5/2}}\,\mathrm{d}z, \tag{6.6.16}$$

$$\cos^2\theta\,\frac{4+\mathrm{v}^2\sin^2\theta}{(1-\mathrm{v}^2\sin^2\theta)^{7/2}}\,\mathrm{d}\cos\theta \to \frac{5}{\mathrm{v}^3(1-\mathrm{v}^2)^2}\left[\frac{1}{(1+z^2)^{5/2}} - \frac{1}{(1+z^2)^{7/2}}\right]\mathrm{d}z. \tag{6.6.17}$$

Now we compare the portion of radiation emitted within the region specified by the angular integration in Eq. (6.6.14) to the corresponding one specified in Eq. (6.6.15). To this end, we use the integrals, as directly obtained from Eqs. (6.6.6), (6.6.8),

$$\int_{-\infty}^{\infty} dz \frac{7}{(1+z^2)^{5/2}} = \frac{28}{3}, \quad \int_{-\infty}^{\infty} dz\, 5\left[\frac{1}{(1+z^2)^{5/2}} - \frac{1}{(1+z^2)^{7/2}}\right] = \frac{4}{3}, \tag{6.6.18}$$

$$\int_{-3/2}^{3/2} dz \frac{7}{(1+z^2)^{5/2}} = 8.96, \tag{6.6.19}$$

$$\int_{-3/2}^{3/2} dz\, 5\left[\frac{1}{(1+z^2)^{5/2}} - \frac{1}{(1+z^2)^{7/2}}\right] = 1.13, \tag{6.6.20}$$

and from Eqs. (6.6.16)/(6.6.3), (6.6.17)/(6.6.4), as well as from the following equation,

$$\frac{[8.96 + 1.13]}{[28/3 + 4/3]} \simeq 0.95,$$

we may infer that about 95 % of contribution of radiation lies within an angle $\theta$ satisfying $-(3/2)\sqrt{1-\mathrm{v}^2}/\mathrm{v} \le \cos\theta \le (3/2)\sqrt{1-\mathrm{v}^2}/\mathrm{v}$, for which $\theta \to \pi/2$ for $\mathrm{v} \to 1$, where the angle $\theta = \pi/2$ corresponds to the orbital plane of the ultra-relativistic electron.[22]

By referring to Eqs. (6.6.3)/(6.6.16), (6.6.4)/(6.6.17) and (6.6.18), we note that radiation is mainly polarized along $\mathbf{e}_1$, referred to as $\sigma$ polarization, rather than along $\mathbf{e}_2$, referred to as $\pi$ polarization.

## 6.7 Appendix: Bessel Functions, Integrals and Estimates Useful in Synchrotron Radiation Studies

In this appendix we provide some useful mathematical aspects of Bessel functions which are relevant to the present chapter.

The special function that is most encountered in synchrotron studies is the Bessel functions of the first kind $J_\nu^{\mathrm{B}}(z)$. We will only consider a real variable $z$. For the Bessel function $J_\nu^{\mathrm{B}}(z)$, we will mainly be interested in $\nu = \ell$, denoting integers. An integral representation of $J_\ell^{\mathrm{B}}(z)$ is given by:[23]

---

[22] That radiation is mostly concentrated in the plane of the orbit of the electron, i.e., for $\theta \approx \pi/2$, is confirmed experimentally, see, e.g., Joos (1960) [7]; Tomboulian and Hartman (1956) [21], Ternov (1995) [20].

[23] Gradshteyn and Ryzhik (2000) [4], p. 902.

$$J_\ell^{\mathrm{B}}(z) = \frac{1}{2\pi} \int_{-\pi}^{\pi} \mathrm{d}\theta \; \mathrm{e}^{\mathrm{i}(z \sin\theta - \ell\theta)}, \tag{6.7.1}$$

In particular, an integral representation for $J_{2\ell}^{\mathrm{B}}(z)$, may be spelled out as

$$J_{2\ell}^{\mathrm{B}}(z) = \frac{1}{\pi} \int_{-\pi/2}^{\pi/2} \mathrm{d}\theta \; \cos(2\ell\theta) \, \cos(z \sin\theta). \tag{6.7.2}$$

Some recurrence relations of the Bessel function $J_\nu^{\mathrm{B}}(z)$, in general, are:

$$J_{\nu-1}^{\mathrm{B}}(z) + J_{\nu+1}^{\mathrm{B}}(z) = \frac{2\nu}{z} J_\nu^{\mathrm{B}}(z), \tag{6.7.3}$$

$$J_{\nu-1}^{\mathrm{B}}(z) - J_{\nu+1}^{\mathrm{B}}(z) = 2 J_\nu^{\mathrm{B}'}(z), \tag{6.7.4}$$

$$J_{\nu-1}^{\mathrm{B}}(z) - \frac{\nu}{z} J_\nu^{\mathrm{B}}(z) = J_\nu^{\mathrm{B}'}(z), \tag{6.7.5}$$

$$\frac{\nu}{z} J_\nu^{\mathrm{B}}(z) - J_{\nu+1}^{\mathrm{B}}(z) = J_\nu^{\mathrm{B}'}(z), \tag{6.7.6}$$

and $J_{-\ell}^{\mathrm{B}}(z) = (-1)^\ell J_\ell^{\mathrm{B}}(z)$.

The Bessel function $J_\ell^{\mathrm{B}}(z)$ enjoys some useful summation relations:[24]

$$\sum_{\ell=1}^{\infty} J_\ell^{\mathrm{B}}(\ell z) = \frac{z}{2(1-z)}, \qquad \sum_{\ell=1}^{\infty} J_{2\ell}^{\mathrm{B}}(2\ell z) = \frac{z^2}{2(1-z^2)}, \tag{6.7.7}$$

$$\sum_{\ell=1}^{\infty} \ell \, J_{2\ell}^{\mathrm{B}'}(2\ell z) = \frac{z}{2(1-z^2)^2}, \qquad \sum_{\ell=1}^{\infty} \ell \int_0^{2\ell z} \mathrm{d}\rho \; J_{2\ell}^{\mathrm{B}}(\rho) = \frac{1}{3} \frac{z^3}{(1-z^2)^3}, \tag{6.7.8}$$

as well as

$$\sum_{\ell=1}^{\infty} \frac{\left(J_\ell^{\mathrm{B}}(\ell z)\right)^2}{\ell^2} = \frac{1}{4} z^2, \qquad \sum_{\ell=1}^{\infty} \frac{J_\ell^{\mathrm{B}}(\ell z) \, J_\ell^{\mathrm{B}'}(\ell z)}{\ell} = \frac{1}{4} z, \tag{6.7.9}$$

$$\sum_{\ell=1}^{\infty} \left(J_\ell^{\mathrm{B}}(\ell z)\right)^2 = \frac{1}{2} \left[ \frac{1}{\sqrt{1-z^2}} - 1 \right], \tag{6.7.10}$$

$$\sum_{\ell=1}^{\infty} \ell^2 \left(J_\ell^{\mathrm{B}}(\ell z)\right)^2 = \frac{z^2(4+z^2)}{16(1-z^2)^{7/2}}, \qquad \sum_{\ell=1}^{\infty} \ell^2 \left(J_\ell^{\mathrm{B}'}(\ell z)\right)^2 = \frac{4+3z^2}{16(1-z^2)^{5/2}}. \tag{6.7.11}$$

[24] Gradshteyn and Ryzhik (2000) [4], Schott (1912) [16], pp. 121–125.

The following expansion formula in terms of Bessel functions should be noted

$$e^{iz\cos\theta} = \sum_{\ell=-\infty}^{\infty} (i)^{\ell} J_{\ell}^{B}(z)\, e^{i\ell\theta}, \tag{6.7.12}$$

which facilitates integrating a function, involving the expression on the left-hand side, over the variable $\theta$. Another expansion formula for the exponential in the above equation is given in Eq. (13.6.6). Some additional properties of the Bessel functions are also spelled out in Chap. 12.

## References


1. Abramowitz, M., & Stegun (Eds.) (1964). *Handbook of mathematical functions, National Bureau of Standards*, Washington, D.C.
2. Bordovitsyn. V. A., & Gushchina, V. S. (1999). *Synchrotron radiation and its development.* Singapore: World Scientific.
3. Fouka, M. & S. Ouichaoui, S. (2013). Analytical fits to the synchrotron functions, Research in Astronomy. *Astrophysics, 13 No. 6*, 680–686.
4. Gradshteyn, I. S., & Ryzhik, I. M. (2000). *Tables of integrals, series and products*, 6th Edition. New York: Academic Press.
5. Heaviside, O. (1902). The waste of energy from a moving electron. *Nature, 67*, 6-8.
6. Ivanenko, D., & Pomeranchuk, I. (1944). On the maximal energy attainable in a Betatron. *Physical Review 65*, 343.
7. Joos, P. (1960). Measurenent of the polarization of synchrotron radiation. *Physical Review Letters, 4 (11)*, 558–559.
8. Liénard, A. (1898). Champ électrique et magnétique produit par une charge électrique contentrée en un point et animée d'un mouvement quelconque,*L'Éclairage Électrique, 16*, 5.
9. Manoukian, E. B., & Jearnkulprasert, N. (2000). Explicit Expression for the photon number emission in synchrotron radiation. *Physics Letters A, 268*, 35–36.
10. Manoukian E. B., Jearnkulprasert, N., & Suebha, P. (2004). Photon number emission in synchrotron radiation: Systematics for high-energy particles. *Il Nuovo Cimento, B, 119 N.1*, 9–15.
11. Navas, S., et al. PDG (2024). *Particle Data Group Collaboration. Physical Review D 110*, 030001.
12. Pacholczyk, A. G. (1970). *Radio astrophysics. Nonthermal processes in galactic and extragalactic sources.* San Francisco: Freeman.
13. Pollock, H. C. (1983), The discovery of synchrotron radiation. *American Journal of Physics, 51*, 278–280.
14. Rybicki, G. B., & Lightman, A. P. (1979). *Radiative processes in astrophysics.* New York: Wiley.
15. Schott, G. A. (1907). On the electron theory of matter and on radiation. *Philosophical Magazine, 13*, 189–213.
16. Schott, G. A. (1912). *Electromagnetic radiation and the mechanical reactions arising from it.* Cambridge.
17. Schwinger, J. (1946). Electron radiation in high energy accelerators. *Physical Review, 70*, 798.
18. Schwinger, J. (1949). On the classical radiation of accelerated electrons. *Physical Review, 75*, 1912–1925. [Erratum: *Physical Review, D, 17*, 1696 (1973).]
19. Schwinger, J. ,Tsai, Wu-Yang, & Erber, T. (1976). Classical and quantum theory of synergic synchrotron-Čerenkov radiation. *Annals of Physics (NY), 96*, 303–332.

20. Ternov, I. M. (1995). Synchrotron radiation. *Physics-Uspekhi 38 (4)*. 409–434.
21. Tomboulian, D. H., & Hartman, P. L. (1956). Spectral and angular distribution of ultraviolet radiation from the 300-Mev Cornell Synchrotron. *Physical Review 102*, 1423–1427.
22. Watson. G. N. (1966). *A treatise on the theory of Bessel functions*. Second Edition. Cambridge.
23. Westfold, K. C. (1959). The polarization of synchrotron radiation. *Astrophysics Jornal, 130*, 241–258.

# Chapter 7
# Vacuum-to-Vacuum Transition Amplitude $< 0_+ \mid 0_- >$ in a Medium

*The vacuum-to-vacuum transition amplitude in a medium emerges simply from a scaling procedure by phenomenological parameters*

In a straightforward manner, we recast the vacuum-to-vacuum transition amplitude, already derived in Chap. 3, by taking into account of the phenomenological parameters that may be needed to describe radiation theory in a medium, as introduced in Chap. 5.

Generally, when the wavelength of light in a given medium is too big to unravel its underlying detailed structure then, as a working hypothesis, phenomenological parameters may be introduced to, hopefully, describe the interaction of light with matter, in a satisfactory way and not dwell on a detailed description of such a structure. On the other hand, when one deals with X-rays, as treated in Chap. 15, the small wavelength of such rays may reveal such a structure and the rays are considered to interact directly with the electrons in atoms making up the medium. In such a problem, phenomenological parameters are not needed to do physics, and, regarding this, Feynman considered this improvement in this theoretical description as[1] "Chemistry becoming Physics". On the other hand, when such parameters are needed to develop a theory, Einstein would be expected to say,[2] they must be "freely invented" and inserted.

The general expression of power of radiation is derived followed by an application to radiation emitted by an electron in a medium of unbounded region of space. This chapter follows with numerous such applications in different media, in subsequent chapters.

---

[1] Feynman (1967) [2].

[2] Schwinger (1957) [9].



E. B. Manoukian, *Radiation*,
https://doi.org/10.1007/978-3-032-21506-2_7

## 7.1 Recasting $< 0_+ \mid 0_- >$ in a Medium

The vacuum-to-vacuum transition amplitude was derived in Chap. 3 and, for example, from Chap. 4, Eqs. (4.1.1)–(4.1.4) is given by

$$\langle 0_+ \mid 0_- \rangle = \exp\left[\frac{\mathrm{i}}{2}\int (\mathrm{d}x)(\mathrm{d}x')J^\mu(x)D_+(x-x')J_\mu(x')\right], \tag{7.1.1}$$

$$\partial_\mu J^\mu(x) = 0, \qquad (\mathrm{d}x) = \mathrm{d}x^0\mathrm{d}^3\mathbf{x}, \quad x^0 = t, \tag{7.1.2}$$

$$D_+(x-x') = \int \frac{(\mathrm{d}k)}{(2\pi)^4}\frac{\mathrm{e}^{\mathrm{i}k(x-x')}}{k^2 - \mathrm{i}\epsilon}, \quad \epsilon \to +0, \quad k^2 = \mathbf{k}^2 - k^{0^2}, \tag{7.1.3}$$

$$(\mathrm{d}k) = \mathrm{d}k^0\mathrm{d}^3\mathbf{k}, \quad k(x'-x) = \mathbf{k}\cdot(\mathbf{x}'-\mathbf{x}) - k^0(t'-t), \tag{7.1.4}$$

and from Eqs. (5.2.2), (5.2.3), up to gauge fixing,

$$\big[-\eta^{\mu\nu}\Box + \partial^\mu\partial^\nu\big]A_\nu(x) = J^\mu(x). \tag{7.1.5}$$

$$-\Box D_+(x-x') = \delta^{(4)}(x-x'), \tag{7.1.6}$$

$\Box \equiv \partial^\mu\partial_\mu = \boldsymbol{\nabla}^2 - (\partial/\partial t)^2$.

Consider the following scalings:

$$(x^0,\mathbf{x}) = (\sqrt{\epsilon\mu}\,\underline{x}^0,\underline{\mathbf{x}}), \quad (\mathrm{d}x) = \sqrt{\mu\epsilon}\,(\mathrm{d}\underline{x}), \quad \partial_0 = \frac{1}{\sqrt{\mu\epsilon}}\underline{\partial}_0, \ \partial_i = \underline{\partial}_i,$$

$$A^i(x) = \left(\frac{\mu}{\epsilon}\right)^{1/4}\underline{A}^i(x), \qquad A^0(x) = \frac{1}{\mu^{1/4}\epsilon^{3/4}}\underline{A}^0(x),$$

$$J^0(x) = \left(\frac{\epsilon}{\mu}\right)^{1/4}\underline{J}^0(x), \qquad J^i(x) = \frac{1}{\mu^{3/4}\epsilon^{1/4}}\underline{J}^i(x). \tag{7.1.7}$$

of Maxwell's equation, in rationalized CGS units, in Eqs. (5.4.1)–(5.4.4), but *not* scaling the arguments of the fields themselves That is, in detail they read:

$$\underline{\boldsymbol{\nabla}}\cdot\underline{\mathbf{E}}(x) = \underline{J}^0(x), \tag{7.1.8}$$

$$\underline{\boldsymbol{\nabla}}\times\underline{\mathbf{E}}(x) = -\underline{\partial}_0\underline{\mathbf{B}}(x), \tag{7.1.9}$$

$$\underline{\boldsymbol{\nabla}}\times\underline{\mathbf{B}}(x) = \underline{\partial}_0\underline{\mathbf{E}}(x) + \underline{\mathbf{J}}(x), \tag{7.1.10}$$

$$\underline{\boldsymbol{\nabla}}\cdot\underline{\mathbf{B}}(x) = 0, \tag{7.1.11}$$

where,

$$\underline{\mathbf{E}}^i(x) = \underline{\partial}^0\underline{A}^i(x) - \underline{\partial}^i\underline{A}^0(x), \tag{7.1.12}$$

$$\underline{\mathbf{B}}^i(x) = \epsilon^{ijk}\underline{\partial}^j\underline{A}^k(x), \tag{7.1.13}$$

with the arguments unchanged.

It is easily verified, from the above, equations, that they lead to

$$\nabla \cdot \varepsilon\, \mathbf{E}(x) = J^0(x), \tag{7.1.14}$$

$$\nabla \times \mathbf{E}(x) = -\frac{\partial \mathbf{B}(x)}{\partial t}, \tag{7.1.15}$$

$$\nabla \times \frac{\mathbf{B}(x)}{\mu} = \frac{\partial \varepsilon\, \mathbf{E}(x)}{\partial t} + \mathbf{J}(x), \tag{7.1.16}$$

$$\nabla \cdot \mathbf{B}(x) = 0, \tag{7.1.17}$$

recovering Maxwell's equations in Eqs. (5.3.19)–(5.3.22), which take into account the phenomenological parameters.

The simple rule for incorporating of phenomenological parameters, is that *all* the variables $x$, the partial derivatives $\partial_\mu$, and $J_\nu$, $A^\mu$, are all to be scaled according to the rules in Eq. (7.1.7), *with the main exception that the arguments of* $J^\nu(x)$, and $A_\mu(x)$ *remain unchanged.*

The above means that in a medium, one is to implement the following replacement in order to account for the phenomenological parameters needed to desribe physics in a medium. In reference to Eq. (7.1.1) of $\langle 0_+ \mid 0_- \rangle$, the following replacement arises:

$$(\mathrm{d}x)(\mathrm{d}x')\, J^\mu(x) D_+(x-x') J_\mu(x') \to (\mathrm{d}\underline{x})(\mathrm{d}\underline{x}')\underline{J}^\mu(x) D_+(\underline{x}-\underline{x}')\underline{J}_\mu(x'). \tag{7.1.18}$$

$$= \Big(\frac{\mu}{\varepsilon}\Big)^{1/2} (\mathrm{d}x)(\mathrm{d}x')\Big[\mathbf{J}(x)\cdot\mathbf{J}(x') - \frac{J^0(x)J^0(x')}{\varepsilon\mu}\Big] D_+(\underline{x}-\underline{x}'), \tag{7.1.19}$$

where, from Eqs. (7.1.3) and (7.1.7), $D_+(\underline{x}-\underline{x}')$ is given by

$$D(\underline{x}-\underline{x}') = \int \frac{(dk)}{(2\pi)^4} \frac{\mathrm{e}^{\mathrm{i}\mathbf{k}\cdot(\mathbf{x}-\mathbf{x}')}\, \mathrm{e}^{-[ik^0(x^0-x'^0)/\sqrt{\varepsilon\mu}]}}{k^2 - \mathrm{i}\epsilon}, \quad \epsilon \to +0. \tag{7.1.20}$$

Also note that

$$\underline{\partial}_\mu \underline{J}^\mu(x) = \mu^{3/4}\varepsilon^{1/4}\partial_\mu J^\mu(x) = 0. \tag{7.1.21}$$

Vacuum-to-vacuum transition amplitude $< 0_+ \mid 0_- >$ in a medium then takes the form,

$$< 0_+ \mid 0_- > = \mathrm{e}^{\frac{\mathrm{i}}{2}\mathcal{M}}, \tag{7.1.22}$$

$$\mathcal{M} = \Big(\frac{\mu}{\varepsilon}\Big)^{1/2} \int (\mathrm{d}x) \int (\mathrm{d}x')\Big[\mathbf{J}(x)\cdot\mathbf{J}(x') - \frac{J^0(x)J^0(x')}{\varepsilon\mu}\Big] D_+(\underline{x}-\underline{x}'). \tag{7.1.23}$$

Needless say, the explicit expression of $< 0_+ \mid 0_- >$ involving the phenomenological parameters would lead to us, as we have done in Chap. 5, to a derivation of the Maxwell's equations including the parameters. With $< 0_+ \mid 0_- >$ now available at hand, we do not have to solve Maxwell's equations, and we may apply it directly to consider physical problems of electrodynamics in a medium as done next, and as will be done again in following chapters.

## 7.2 Power of Radiation in a Medium

To derive the expressions of power radiation in a medium, we follow the procedure given in Chap. 4. As in the latter chapter, Eq. (4.1.5), we first take the absolute value square of the vacuum-to-vacuum transition amplitude in Eq. (7.1.22). We use the symmetry over the $x$ and $x'$ variables in the integral in Eq. (7.1.23), as well as make use of Eqs. (4.1.6)–(4.1.9), to infer that the average number $< N >$ emitted by $J^\mu(x)$ is given by

$$< N > = \Big(\frac{\mu}{\varepsilon}\Big)^{1/2} \int (\mathrm{d}x) \int (\mathrm{d}x') \Big[\mathbf{J}(x)\cdot\mathbf{J}(x') \;-\; \frac{J^0(x)J^0(x')}{\varepsilon\mu}\Big]\, Im\big[D_+(\underline{x}-\underline{x}')\big]. \tag{7.2.1}$$

From Eq. (7.1.20), and Eqs. (4.1.6)–(4.1.9), as mentioned above,

$$Im\ D_+(\underline{x}-\underline{x}') \to \int \frac{(\mathrm{d}k)}{(2\pi)^4} \mathrm{e}^{\mathrm{i}\mathbf{k}\cdot(\mathbf{x}-\mathbf{x}')}\, \mathrm{e}^{-[\mathrm{i}k^0(x^0-x'^0)/\sqrt{\varepsilon\mu}]}\, Im\Big(\frac{1}{k^2-\mathrm{i}\epsilon}\Big) \tag{7.2.2}$$

$$= \pi \int \frac{(\mathrm{d}k)}{(2\pi)^4}\, \mathrm{e}^{\mathrm{i}\mathbf{k}\cdot(\mathbf{x}-\mathbf{x}')}\, \mathrm{e}^{-[\mathrm{i}k^0(x^0-x'^0)/\sqrt{\varepsilon\mu}]}\, \delta(k^2), \tag{7.2.3}$$

where

$$\delta(k^2) = \frac{\delta(k^0- \mid \mathbf{k} \mid) + \delta(k^0+ \mid \mathbf{k} \mid)}{2 \mid \mathbf{k} \mid}. \tag{7.2.4}$$

Accordingly,

$$< N > = \pi \Big(\frac{\mu}{\varepsilon}\Big)^{1/2} \int (\mathrm{d}x) \int (\mathrm{d}x') \Big[\mathbf{J}(x)\cdot\mathbf{J}(x') \;-\; \frac{J^0(x)J^0(x')}{\varepsilon\mu}\Big]$$
$$\times \int \frac{(\mathrm{d}k)}{(2\pi)^4}\, \mathrm{e}^{\mathrm{i}\mathbf{k}\cdot(\mathbf{x}-\mathbf{x}')}\, \mathrm{e}^{-[\mathrm{i}k^0(x^0-x'^0)/\sqrt{\varepsilon\mu}]}\, \delta(k^2). \tag{7.2.5}$$

We may proceed by integrating over the variable $\mathbf{k}$ buy using

$$(\mathrm{d}k) \equiv \mathrm{d}k^0\mathrm{d}^3\mathbf{k}, \quad \mathrm{d}^3\mathbf{k} = |\mathbf{k}|^2\mathrm{d}|\mathbf{k}|\,\mathrm{d}\Omega, \quad \mathbf{k} = |\mathbf{k}|\mathbf{n}, \quad \mathbf{k}\cdot\mathbf{n} = |\mathbf{k}|. \tag{7.2.6}$$

to obtain

$$< N > = \pi\left(\frac{\mu}{\varepsilon}\right)^{1/2}\int(\mathrm{d}x)\int(\mathrm{d}x')\Big[\mathbf{J}(x)\cdot\mathbf{J}(x') - \frac{J^0(x)J^0(x')}{\varepsilon\mu}\Big]\int_{-\infty}^{\infty}\frac{\mathrm{d}k^0}{2\pi}$$
$$\times\,\mathrm{e}^{-[ik^0(x^0-x'^0)/\sqrt{\varepsilon\mu}]}\int_0^{\infty}\frac{|\mathbf{k}|^2\mathrm{d}|\mathbf{k}|}{(2\pi)^3}\delta(k^2)\int\mathrm{d}\Omega\,\mathrm{e}^{\mathrm{i}|\mathbf{k}|\mathbf{n}\cdot(\mathbf{x}-\mathbf{x}')}. \tag{7.2.7}$$

We are interested in the average number of photons, denoted by $N(\omega)$, emitted with a given energy $\hbar\omega \equiv \omega$ (with $\hbar$ set equal to one). Moreover, we have learnt in Sect. 5.5, that the speed of light in a medium is changed according to the rule $\mathrm{c} \to \mathrm{c}/\sqrt{\epsilon\mu}$. From energy conservation, and the simple relations

$$\frac{\mathrm{c}}{\sqrt{\epsilon\mu}} = \frac{\omega}{2\pi}\lambda = \frac{\omega}{|\mathbf{k}|}, \quad \Rightarrow \quad |\mathbf{k}| = \frac{\omega}{\mathrm{c}}\sqrt{\epsilon\mu} \;\to\; \omega\sqrt{\epsilon\mu}, \tag{7.2.8}$$

with c set equal to one, in the latter expression, we get a relation of the magnitude of the momentum of a photon in the medium in terms of angular frequency $\omega$.

To above end of obtaining the expression of $N(\omega)$ we insert the identity

$$1 = \int_0^{\infty}\mathrm{d}\omega\,\delta\Big(\omega - \frac{|\mathbf{k}|}{\sqrt{\epsilon\mu}}\Big), \tag{7.2.9}$$

in the overall integrand in Eq. (7.2.7) and use, in the process, that

$$\delta\Big(\omega - \frac{|\mathbf{k}|}{\sqrt{\epsilon\mu}}\Big)\delta(k^2) = \sqrt{\epsilon\mu}\,\delta\big(\sqrt{\epsilon\mu}\omega - |\mathbf{k}|\big)\frac{1}{2|\mathbf{k}|}\big[\delta(k^0 - |\,\mathbf{k}\,|) + \delta(k^0 + |\,\mathbf{k}\,|)\big]$$
$$= \frac{\delta\big(\sqrt{\epsilon\mu}\omega - |\mathbf{k}|\big)}{\omega}\,\frac{\big[\delta(k^0 - \sqrt{\epsilon\mu}\omega) + \delta(k^0 + \sqrt{\epsilon\mu}\omega)\big]}{2}. \tag{7.2.10}$$

making use of Eq. (7.2.4).

For the $|\mathbf{k}|$-integral we have

$$\int_0^{\infty}\frac{|\mathbf{k}|^2\mathrm{d}|\mathbf{k}|}{(2\pi)^3}\,\frac{\delta\big(\sqrt{\epsilon\mu}\omega - |\mathbf{k}|\big)}{\omega}\,\mathrm{e}^{\mathrm{i}|\mathbf{k}|\mathbf{n}\cdot(\mathbf{x}-\mathbf{x}')} = \frac{(\epsilon\mu)\,\omega}{(2\pi)^3}\,\mathrm{e}^{\mathrm{i}\sqrt{\epsilon\mu}\omega\,\mathbf{n}\cdot(\mathbf{x}-\mathbf{x}')}. \tag{7.2.11}$$

On the other hand, the $k^0$-integral gives

$$\int_{-\infty}^{\infty}\frac{\mathrm{d}k^0}{2\pi}\,\frac{\big[\delta(k^0 - \sqrt{\epsilon\mu}\omega) + \delta(k^0 + \sqrt{\epsilon\mu}\omega)\big]}{2}\,\mathrm{e}^{-[\mathrm{i}\,k^0(x^0-x'^0)/\sqrt{\varepsilon\mu}]}$$
$$= \frac{1}{4\pi}\big[\mathrm{e}^{-\,\mathrm{i}\omega(x^0-x'^0)} + \mathrm{e}^{\mathrm{i}\omega(x^0-x'^0)}\big]. \tag{7.2.12}$$

That is

$$< N > = \int_0^\infty \mathrm{d}\omega\, N(\omega), \tag{7.2.13}$$

$$N(\omega) = \frac{\omega}{32\pi^3}\Big(\frac{\mu}{\varepsilon}\Big)^{1/2}(\epsilon\mu)\int(\mathrm{d}x)\int(\mathrm{d}x')\Big[\mathbf{J}(x)\cdot\mathbf{J}(x') - \frac{J^0(x)J^0(x')}{\varepsilon\mu}\Big] \times \int \mathrm{d}\Omega\, \mathrm{e}^{\mathrm{i}\sqrt{\epsilon\mu}\omega\,\mathbf{n}\cdot(\mathbf{x}-\mathbf{x}')}\Big[\mathrm{e}^{-\mathrm{i}\omega(x^0-x'^0)} + \mathrm{e}^{\mathrm{i}\omega(x^0-x'^0)}\Big]. \tag{7.2.14}$$

Note the consistency of the multiplicative factor $(\epsilon\mu)$. If we integrate over $\omega$ from 0 to $\infty$, we have an integral involving $\sqrt{\epsilon\mu}\,\omega\sqrt{\epsilon\mu}\,\mathrm{d}\omega$, which is the same as $|\mathbf{k}|\mathrm{d}|\mathbf{k}|$, as it should be.

We will see below in Eq. (7.2.19), that the $\Omega-$ integral is a function of $|\mathbf{x}-\mathbf{x}'|$, i.e., is even in the interchange $\mathbf{x} \leftrightarrow \mathbf{x}'$. Accordingly, we may interchange

$$\mathrm{e}^{\mathrm{i}\sqrt{\epsilon\mu}\omega\,\mathbf{n}\cdot(\mathbf{x}-\mathbf{x}')}\mathrm{e}^{\mathrm{i}\omega(x^0-x'^0)} \to \mathrm{e}^{\mathrm{i}\sqrt{\epsilon\mu}\omega\,\mathbf{n}\cdot(\mathbf{x}-\mathbf{x}')}\mathrm{e}^{-\mathrm{i}\omega(x^0-x'^0)},$$

in Eq. (7.2.14), and combine it with the first term. This allows us to rewrite Eqs. (7.2.13), (7.2.14) in the more convenient form consisting one term and multiply the resulting overall integral by 2. That is,

$$< N > = \int_0^\infty \mathrm{d}\omega\, N(\omega), \tag{7.2.15}$$

$$N(\omega) = \frac{\omega}{16\pi^3}\Big(\frac{\mu}{\varepsilon}\Big)^{1/2}(\epsilon\mu)\int(\mathrm{d}x)\int(\mathrm{d}x')\Big[\mathbf{J}(x)\cdot\mathbf{J}(x') - \frac{J^0(x)J^0(x')}{\varepsilon\mu}\Big] \times \int \mathrm{d}\Omega\, \mathrm{e}^{\mathrm{i}\sqrt{\epsilon\mu}\omega\,\mathbf{n}\cdot(\mathbf{x}-\mathbf{x}')}\mathrm{e}^{-\mathrm{i}\omega(x^0-x'^0)}. \tag{7.2.16}$$

Since each photon has an energy $\hbar\omega \equiv \omega$, the energy of radiation of (angular) frequency $\omega$ emitted is then given by

$$E(\omega) = \frac{\omega^2}{16\pi^3}\Big(\frac{\mu}{\varepsilon}\Big)^{1/2}(\epsilon\mu)\int(\mathrm{d}x)\int(\mathrm{d}x')\Big[\mathbf{J}(x)\cdot\mathbf{J}(x') - \frac{J^0(x)J^0(x')}{\varepsilon\mu}\Big] \times \int \mathrm{d}\Omega\, \mathrm{e}^{\mathrm{i}\sqrt{\epsilon\mu}\omega\,\mathbf{n}\cdot(\mathbf{x}-\mathbf{x}')}\mathrm{e}^{-\mathrm{i}\omega(x^0-x'^0)}. \tag{7.2.17}$$

We note, in passing, that the integration over the solid angle may be explicitly carried out by using

$$\mathrm{d}\Omega = 2\pi d(\cos\theta), \quad -1 \leq \cos\theta \leq 1, \quad \mathbf{n}\cdot(\mathbf{x}-\mathbf{x}') = |\mathbf{x}-\mathbf{x}'|\cos\theta, \tag{7.2.18}$$

$$\Rightarrow \int \mathrm{d}\Omega\, \mathrm{e}^{\mathrm{i}\sqrt{\epsilon\mu}\omega\,\mathbf{n}\cdot(\mathbf{x}-\mathbf{x}')} = \frac{-2\pi\,\mathrm{i}}{\omega\sqrt{\epsilon\mu}|\mathbf{x}-\mathbf{x}'|}\Big[\mathrm{e}^{\mathrm{i}\sqrt{\epsilon\mu}\omega|\mathbf{x}-\mathbf{x}'|} - \mathrm{e}^{-\mathrm{i}\sqrt{\epsilon\mu}\omega|\mathbf{x}-\mathbf{x}'|}\Big]. \tag{7.2.19}$$

Using the conservation law[3] $\partial_\mu J^\mu = 0$, allows us to use the expression

$$\Big[\mathbf{J}(x)\cdot\mathbf{J}(x') \;-\; \frac{J^0(x)J^0(x')}{\varepsilon\mu}\Big] = \Big[\delta_{ij} \;-\; \frac{\partial_i\,\partial_j{}'}{\varepsilon\mu\,\partial_0\,\partial_0^{'}}\Big]J^i(x)J^j(x'). \qquad (7.2.20)$$

and integrate Eq. (7.2.17) by parts over $x, x'$, to replace the second term involving derivatives within the square brackets above simply by $\mathbf{n}_i\mathbf{n}_j$, and obtain

$$\begin{aligned} E(\omega) \;&=\; \frac{\mu\sqrt{\epsilon\mu}}{16\pi^3}\,\omega^2\int(\mathrm{d}x)\int(\mathrm{d}x')J^i(x)J^j(x') \\ &\times\int\mathrm{d}\Omega\,\big[\delta_{ij}-\mathbf{n}_i\mathbf{n}_j\big]\mathrm{e}^{\mathrm{i}\sqrt{\epsilon\mu}\omega\,\mathbf{n}\cdot(\mathbf{x}-\mathbf{x}')}\mathrm{e}^{-\,\mathrm{i}\omega(t-t')}, \end{aligned} \qquad (7.2.21)$$

obtaining an expression involving the 3-vector components of the current.

We may introduce variables of integration $T$, $\tau$ as follows

$$t \;=\; T + \frac{\tau}{2}, \quad t' \;=\; T \;-\; \frac{\tau}{2} \quad t-t' = \tau, \qquad (7.2.22)$$

to obtain the power of radiation per unit frequency about $\omega$

$$\begin{aligned} \mathrm{P}(\omega,T) \;&=\; \frac{\omega^2}{16\pi^3}\mu\sqrt{\varepsilon\mu}\int\mathrm{d}^3\mathbf{x}\int\mathrm{d}^3\mathbf{x}'\int_{-\infty}^{\infty}\mathrm{d}\tau\; J^i(x)J^j(x') \\ &\times\int\mathrm{d}\Omega\,\big[\delta_{ij}-\mathbf{n}_i\mathbf{n}_j\big]\mathrm{e}^{\mathrm{i}\sqrt{\epsilon\mu}\omega\,\mathbf{n}\cdot(\mathbf{x}-\mathbf{x}')}\mathrm{e}^{-\,\mathrm{i}\omega\tau}. \end{aligned} \qquad (7.2.23)$$

Needless to say, we may equivalently rewrite the above expression in an equivalent form,

$$\begin{aligned} \mathrm{P}(\omega,T) \;&=\; \frac{\omega^2}{16\pi^3}\mu\sqrt{\varepsilon\mu}\int\mathrm{d}^3\mathbf{x}\int\mathrm{d}^3\mathbf{x}'\int_{-\infty}^{\infty}\mathrm{d}\tau\,\Big[\mathbf{J}(x)\cdot\mathbf{J}(x') \;-\; \frac{J^0(x)J^0(x')}{\varepsilon\mu}\Big] \\ &\times\int\mathrm{d}\Omega\,\mathrm{e}^{\mathrm{i}\sqrt{\epsilon\mu}\omega\,\mathbf{n}\cdot(\mathbf{x}-\mathbf{x}')}\mathrm{e}^{-\,\mathrm{i}\omega\tau}. \end{aligned} \qquad (7.2.24)$$

Equation (7.2.23), for example, also gives the power of radiation per unit frequency per solid unit angle with radiation emitted along a unit vector $\mathbf{n}$:

$$\begin{aligned} \mathrm{P}_{\mathbf{n}}(\omega,T) \;&=\; \frac{\omega^2}{16\pi^3}\mu\sqrt{\varepsilon\mu}\int\mathrm{d}^3\mathbf{x}\int\mathrm{d}^3\mathbf{x}'\int_{-\infty}^{\infty}\mathrm{d}\tau\; J^i(x)J^j(x') \\ &\times\,\big[\delta_{ij}-\mathbf{n}_i\mathbf{n}_j\big]\mathrm{e}^{\mathrm{i}\sqrt{\epsilon\mu}\omega\,\mathbf{n}\cdot(\mathbf{x}-\mathbf{x}')}\mathrm{e}^{-\,\mathrm{i}\omega\tau}. \end{aligned} \qquad (7.2.25)$$

We note from Eq. (7.2.21), that $E(\omega)$ in invariant under the simultaneous interchange $(x, x', \omega) \to (x', x, -\,\omega)$. Accordingly, to obtain the total energy of radiation

[3] See also Eq. (7.1.21).

$E$, we may instead of integrating $E(\omega)$ in Eq. (7.2.21) from 0 to $\infty$, we may instead integrate over $\omega$ from $-\infty$ to $\infty$ and divide the resulting integral by 2, to obtain

$$E = \frac{\mu\sqrt{\epsilon\mu}}{32\pi^3}\int_{-\infty}^{\infty}\mathrm{d}\omega\,\omega^2\int(\mathrm{d}x)\int(\mathrm{d}x')J^i(x)J^j(x') \times\int\mathrm{d}\Omega\,\big[\delta_{ij}-\mathbf{n}_i\mathbf{n}_j\big]\mathrm{e}^{\mathrm{i}\sqrt{\epsilon\mu}\omega\,\mathbf{n}\cdot(\mathbf{x}-\mathbf{x}')}\mathrm{e}^{-\,\mathrm{i}\omega(t-t')}. \tag{7.2.26}$$

We may introduce the following Fourier transforms:

$$J^{i*}(\omega,\mathbf{x}) = \int_{-\infty}^{\infty}\mathrm{d}t\,\mathrm{e}^{-\mathrm{i}\omega t}\,J^i(t,\mathbf{x}), \tag{7.2.27}$$

$$J^{j}(\omega,\mathbf{x}') = \int_{-\infty}^{\infty}\mathrm{d}t'\,\mathrm{e}^{\mathrm{i}\omega t'}\,J^j(t',\mathbf{x}'), \tag{7.2.28}$$

as well as the Dirac distribution

$$\delta\big(\omega-\omega'\big) = \frac{1}{2\pi}\int_{-\infty}^{\infty}\mathrm{d}T\,\mathrm{e}^{\mathrm{i}\,(\omega-\omega')T}, \tag{7.2.29}$$

to obtain for the energy of radiation emitted per unit time, i.e., the power of radiation

$$\mathrm{P}(T) = \frac{\mu\sqrt{\epsilon\mu}}{64\pi^4}\int\mathrm{d}\Omega\,\Big|\int\mathrm{d}^3\mathbf{x}\,\mathrm{e}^{-\mathrm{i}\sqrt{\varepsilon\mu}\,\mathbf{n}\cdot\mathbf{x}}\int_{-\infty}^{\infty}\mathrm{d}\omega\,\omega\,\mathrm{e}^{-\mathrm{i}\omega T}\,\hat{\mathbf{J}}(\omega,\mathbf{x})\Big|^2, \tag{7.2.30}$$

where

$$\hat{J}^j(\omega,\mathbf{x}) = J^i(\omega,\mathbf{x})\big[\delta^{ij}-\mathbf{n}^i\mathbf{n}^j\big],\ \text{and we have used that} \tag{7.2.31}$$

$$\big[\delta^{ik}-\mathbf{n}^i\mathbf{n}^k\big]\,\big[\delta^{kj}-\mathbf{n}^k\mathbf{n}^j\big] = \big[\delta^{ij}-\mathbf{n}^i\mathbf{n}^j\big]. \tag{7.2.32}$$

Finally for the power of radiation emitted within a solid angle in the direction of a unit vector $\mathbf{n}$, the following expression emerges

$$\mathrm{P}_{\mathbf{n}}(T) = \frac{\mu\sqrt{\epsilon\mu}}{64\pi^4}\Big|\int\mathrm{d}^3\mathbf{x}\,\mathrm{e}^{-\mathrm{i}\sqrt{\varepsilon\mu}\,\mathbf{n}\cdot\mathbf{x}}\int_{-\infty}^{\infty}\mathrm{d}\omega\,\omega\,\mathrm{e}^{-\mathrm{i}\omega T}\,\hat{\mathbf{J}}(\omega,\mathbf{x})\Big|^2. \tag{7.2.33}$$

## 7.3 Electron in a Medium in an Unbounded Region: The Classic Čerenkov Radiation

We consider an electron moving with a certain velocity, say, v in the $z \equiv x_3$—direction of a given coordinate system in a homogeneous and isotropic medium of permeability $\mu$ and permittivity $\varepsilon$. In order to work out the power of radiation emitted by the electron of given energy $\omega$, in certain direction, specified by a unit vector $\mathbf{n}$, we introduce the four-current components:

$$J^{\mu}(x) = (J^0, 0, 0, J_3(x)), \tag{7.3.1}$$

$$J_3(x) = \mathrm{e}\mathrm{v}\delta(x_3 - vt)\delta(x_2)\delta(x_1), \qquad J^0(x) = \mathrm{e}\delta(x_3 - vt)\delta(x_2)\delta(x_1), \tag{7.3.2}$$

$$t = T + \frac{\tau}{2}, \quad t' = T - \frac{\tau}{2}, \quad t - t' = \tau, \tag{7.3.3}$$

$$\mathbf{n} = \big(\sin\theta\cos\phi, \sin\theta\sin\phi, \cos\theta), \qquad \mathbf{n}\cdot(\mathbf{x} - \mathbf{x}'\big) = \cos\theta\, \mathrm{v}(t - t'). \tag{7.3.4}$$

Accordingly, from Eq. (7.2.24), for example, and, in the process, by carrying out the relevant integrals involving the Dirac delta distributions in Eq. (7.3.2), we obtain

$$\begin{aligned} \mathrm{P}(\omega, T) = {} & \frac{\omega^2}{16\pi^3}\mu\sqrt{\varepsilon\mu}\int \mathrm{d}^3\mathbf{x}\int \mathrm{d}^3\mathbf{x}'\int_{-\infty}^{\infty}\mathrm{d}\tau\left[\mathbf{J}(x)\cdot\mathbf{J}(x') - \frac{J^0(x)J^0(x')}{\varepsilon\mu}\right] \\ & \times \int \mathrm{d}\Omega\, \mathrm{e}^{\mathrm{i}\sqrt{\epsilon\mu}\omega\,\mathbf{n}\cdot(\mathbf{x}-\mathbf{x}')}\mathrm{e}^{-\,\mathrm{i}\omega\tau}. \end{aligned} \tag{7.3.5}$$

$$\mathrm{P}(\omega, T) = \frac{\mathrm{e}^2\omega^2}{16\pi^3}\mu\sqrt{\varepsilon\mu}\Big(\mathrm{v}^2 - \frac{1}{\epsilon\mu}\Big)(2\pi)\int_{-1}^{1}\mathrm{d}(\cos\theta)\int_{-\infty}^{\infty}\mathrm{d}\tau\, \mathrm{e}^{\mathrm{i}\left(\sqrt{\epsilon\mu}\mathrm{v}\cos\theta - 1\right)\omega\tau} \tag{7.3.6}$$

$$= \frac{\mathrm{e}^2\mathrm{v}^2\omega}{8\pi^2}\mu\sqrt{\varepsilon\mu}\Big(1 - \frac{1}{\mathrm{v}^2\epsilon\mu}\Big)\int_{-1}^{1}(2\pi)\,\delta\big(\sqrt{\epsilon\mu}\mathrm{v}\cos\theta - 1\big)\,\mathrm{d}(\cos\theta). \tag{7.3.7}$$

Or using the fact that $|\cos\theta| \leq 1$,

$$\mathrm{P}(\omega, T) = \mathrm{P}(\omega) = \begin{cases} 0, & \text{for } \mathrm{v} < \frac{1}{\sqrt{\varepsilon\mu}} \\ \frac{\mathrm{e}^2}{4\pi}\mu\mathrm{v}\omega\Big(1 - \frac{1}{\mathrm{v}^2\epsilon\mu}\Big), & \text{for } \mathrm{v} > \frac{1}{\sqrt{\varepsilon\mu}} \end{cases}. \tag{7.3.8}$$

That is radiation may be emitted by the electron only if its speed is greater than that of the speed of light *in* the medium. This is the classical Čerenkov radiation formula. It is important to realize that the *boundary condition* that $\mathrm{P}(\omega)$ vanishes for $\mathrm{v} = 0$, is as expected. Due to the fact $\mathrm{P}(\omega)$ is linear in $\omega$, one cannot integrate over

$\omega$, to arbitrary large values, to obtain the total power. A quantum treatment of the problem, however, produces naturally a "cut-off" for large $\omega$,[4] expressed in terms of the speed of the electron itself given by

$$\mathrm{v}\sqrt{\varepsilon\mu} > \Big(1 + (\varepsilon\mu - 1)\frac{\omega}{2E}\Big), \tag{7.3.9}$$

which replaces condition $\mathrm{v}\sqrt{\varepsilon\mu} > 1$ in case 2 in Eq. (7.3.8), where $E$ is the energy of the electron.

It is interesting that Apollo Astronauts during Apollo Missions have reported of "seeing" flashes of light even with their eyes closed. An explanation of this was generally attributed to high energy cosmic charged particles, encountered freely in outer space, that would pass one's eyelids causing Čerenkov radiation within one's eye itself.[5]

## 7.4 Power of Radiation for Complex Harmonic Fields

We reconsider the expression of the total energy radiated by a given current distribution $J^\mu(x)$, given in Eq. (7.2.26)

$$\begin{aligned} E &= \frac{\mu\sqrt{\epsilon\mu}}{32\pi^3}\int_{-\infty}^{\infty}\mathrm{d}\omega'\,\omega'^2\int(\mathrm{d}x)\int(\mathrm{d}x')J^i(x)J^j(x') \\ &\times\int\mathrm{d}\Omega'\left[\delta_{ij} - \mathbf{n}_i\mathbf{n}_j\right]\mathrm{e}^{\mathrm{i}\sqrt{\epsilon\mu}\omega'\,\mathbf{n}\cdot(\mathbf{x}-\mathbf{x}')}\mathrm{e}^{-\mathrm{i}\omega'(t-t')}. \end{aligned} \tag{7.4.1}$$

We consider harmonic fields by defining the time dependence of a current as follows

$$J^i(x) = J^i(\omega,\mathbf{x})\,\mathrm{e}^{-\mathrm{i}\omega t}. \tag{7.4.2}$$

Due to the reality condition of the currents, we set a current $J^i(x)$ as the real part of the right-hand side of the above expression

$$\mathcal{Re}\left(\mathrm{e}^{-\mathrm{i}\omega t}J^i(\omega,\mathbf{x})\right) = \frac{1}{2}\Big(\mathrm{e}^{-\mathrm{i}\omega t}J^i(\omega,\mathbf{x}) + \mathrm{e}^{\mathrm{i}\omega t}J^{*i}(\omega,\mathbf{x})\Big), \tag{7.4.3}$$

where * denotes complex conjugation.

Upon introducing the variables $\tau$, $T$ through the equations

$$t = T + \frac{\tau}{2}, \quad t' = T - \frac{\tau}{2} \quad t - t' = \tau, \tag{7.4.4}$$

[4] See, e.g., Sect. 2 of Manoukian and Charuchittapan (2000) [3].

[5] See, e.g., Fazio et al. (1970) [1]; Pinsky et al. (1974) [8]; McNulty et al. (1976) [7].

the product of the currents in Eq. (7.4.1), becomes

$$\begin{aligned}
&\mathcal{Re}\left(\mathrm{e}^{-\mathrm{i}\omega t} J^i(\omega,\mathbf{x})\right)\mathcal{Re}\left(\mathrm{e}^{-\mathrm{i}\omega t'} J^j(\omega,\mathbf{x}')\right) = \\
&= \frac{1}{4}\Big(J^i(\omega,\mathbf{x})J^j(\omega,\mathbf{x}')\mathrm{e}^{-2\mathrm{i}\omega T} + J^{*i}(\omega,\mathbf{x})J^{*j}(\omega,\mathbf{x}')\mathrm{e}^{+2\mathrm{i}\omega T} \\
&+ J^i(\omega,\mathbf{x})J^{*j}(\omega,\mathbf{x}')\mathrm{e}^{-\mathrm{i}\omega\tau} + J^{*i}(\omega,\mathbf{x})J^j(\omega,\mathbf{x}')\mathrm{e}^{+\mathrm{i}\omega\tau}\Big). \qquad (7.4.5)
\end{aligned}$$

We perform an average over $T$ over one cycle

$$\frac{\omega}{2\pi}\int_0^{2\pi/\omega} \mathrm{d}T\, \mathrm{e}^{\mp 2\omega \mathrm{i}T} = 0, \qquad (7.4.6)$$

from which only the last two terms in Eq. (7.4.5) survive, leading to a time-averaged power of radiation per unit cycle given by

$$\begin{aligned}
\mathrm{P}(\omega) &= \frac{\mu\sqrt{\epsilon\mu}}{128\pi^3}\int_{-\infty}^{\infty}\mathrm{d}\omega'\,\omega'^2\int \mathrm{d}^3\mathbf{x}\int \mathrm{d}^3\mathbf{x}'\int_{-\infty}^{\infty}\mathrm{d}\tau\left(\mathrm{e}^{-\mathrm{i}\omega\tau}J^i(\omega,\mathbf{x})J^{*j}(\omega,\mathbf{x}')\right. \\
&\left.+\ \mathrm{e}^{\mathrm{i}\omega\tau}J^{*i}(\omega,\mathbf{x})J^j(\omega,\mathbf{x}')\right)\int \mathrm{d}\Omega\,[\delta^{ij}-\mathbf{n}^i\mathbf{n}^j]\mathrm{e}^{\mathrm{i}\sqrt{\epsilon\mu}\omega'\,\mathbf{n}\cdot(\mathbf{x}-\mathbf{x}')}\mathrm{e}^{-\mathrm{i}\omega'\tau}. \qquad (7.4.7)
\end{aligned}$$

Due to symmetry of $[\delta^{ij}-\mathbf{n}^i\mathbf{n}^j]$ in the interchange of $i$ and $j$, in the above equation, we may interchange $i$ and $j$ in the product of the two currents in the second term within the round brackets, and simultaneously interchange $\mathbf{x}$ and $\mathbf{x}'$ in the corresponding term in the integral, as follows

$$J^{*i}(\omega,\mathbf{x})J^j(\omega,\mathbf{x}')\mathrm{e}^{\mathrm{i}\sqrt{\epsilon\mu}\omega'\,\mathbf{n}\cdot(\mathbf{x}-\mathbf{x}')} \rightarrow J^i(\omega,\mathbf{x})J^{*j}(\omega,\mathbf{x}')\mathrm{e}^{-\mathrm{i}\sqrt{\epsilon\mu}\omega'\,\mathbf{n}\cdot(\mathbf{x}-\mathbf{x}')}. \qquad (7.4.8)$$

This allows us to combine the products of the two currents within the round brackets in Eq. (7.4.7) to rewrite

$$\begin{aligned}
\mathrm{P}(\omega) &= \frac{\mu\sqrt{\epsilon\mu}}{128\pi^3}\int \mathrm{d}^3\mathbf{x}\int \mathrm{d}^3\mathbf{x}'J^i(\omega,\mathbf{x})J^{*j}(\omega,\mathbf{x}')\int \mathrm{d}\Omega\,[\delta^{ij}-\mathbf{n}^i\mathbf{n}^j]\times \\
&\int_{-\infty}^{\infty}\mathrm{d}\omega'\omega'^2\int_{-\infty}^{\infty}\mathrm{d}\tau\Big[\mathrm{e}^{-\mathrm{i}(\omega+\omega')\tau}\mathrm{e}^{\mathrm{i}\sqrt{\epsilon\mu}\omega'\,\mathbf{n}\cdot(\mathbf{x}-\mathbf{x}')}+\mathrm{e}^{\mathrm{i}(\omega-\omega')\tau}\mathrm{e}^{-\mathrm{i}\sqrt{\epsilon\mu}\omega'\,\mathbf{n}\cdot(\mathbf{x}-\mathbf{x}')}\Big]. \qquad (7.4.9)
\end{aligned}$$

We may readily integrate over $\tau$, recalling the definition of the Dirac Delta distribution, and then over $\omega'$, to obtain

$$\begin{aligned}
\mathrm{P}(\omega) &= \frac{\omega^2\mu\sqrt{\epsilon\mu}}{32\pi^2}\int \mathrm{d}^3\mathbf{x}\int \mathrm{d}^3\mathbf{x}'J^i(\omega,\mathbf{x})J^{*j}(\omega,\mathbf{x}') \\
&\times\int \mathrm{d}\Omega\,[\delta^{ij}-\mathbf{n}^i\mathbf{n}^j]\mathrm{e}^{-\mathrm{i}\sqrt{\epsilon\mu}\omega\,\mathbf{n}\cdot(\mathbf{x}-\mathbf{x}')}. \qquad (7.4.10)
\end{aligned}$$

We may also use the fact that

$$\big[\delta^{ij} - \mathbf{n}^i\mathbf{n}^j\big] = \big[\delta^{ik} - \mathbf{n}^i\mathbf{n}^k\big]\big[\delta^{kj} - \mathbf{n}^k\mathbf{n}^j\big], \tag{7.4.11}$$

and the vector identity, for a unit vector $\mathbf{n}$

$$[J^k - n^k\mathbf{n}\cdot\mathbf{J}] = -[\mathbf{n}\times(\mathbf{n}\times\mathbf{J})]^k, \tag{7.4.12}$$

in components form, to write,

$$\begin{aligned} J^i\big[\delta^{ij} - \mathbf{n}^i\mathbf{n}^j\big]J^{*j} &= \big[\delta^{ik} - \mathbf{n}^i\mathbf{n}^k\big]J^i\,\big[\delta^{kj} - \mathbf{n}^k\mathbf{n}^j\big]J^{*j}, && (7.4.13)\\ &= \big[\mathbf{n}\times(\mathbf{n}\times\mathbf{J})\cdot\big[\mathbf{n}\times(\mathbf{n}\times\mathbf{J}^*)\big]. && (7.4.14)\end{aligned}$$

According the time averaged power of radiation per unit (angular) frequency, unit angular solid angle is, from Eqs. (7.4.10), (7.4.14), given by

$$\frac{\mathrm{d}\,\mathrm{P}(\omega)}{\mathrm{d}\Omega} = \frac{\omega^2\mu\sqrt{\epsilon\mu}}{32\pi^2}\Big|\int \mathrm{d}^3\mathbf{x}\,[\mathbf{n}\times(\mathbf{n}\times\mathbf{J}(\omega,\mathbf{x})]\mathrm{e}^{-\mathrm{i}\omega\sqrt{\epsilon\mu\omega}\,\mathbf{n}\cdot\mathbf{x}})\Big|^2. \tag{7.4.15}$$

## 7.5 The Role of the Electric Field in the Expression of Power of Radiation

In the temporal gage, $\mathbf{E} = \partial^0\mathbf{A}$, we may infer from Eq. (5.4.50) that

$$\Big[\big(-\nabla^2 + \varepsilon\mu\,\partial_0^2\big)\delta^{ij} + \partial^i\partial^j\Big]A^j(x) = \mu\,J^i(x). \tag{7.5.1}$$

Upon multiplying the latter by $\partial_i$, and using $\partial^i\partial^i \equiv \nabla^2$, we obtain

$$\varepsilon\mu\,\partial_0^2\partial_j A^j(x) = \mu\,\partial_i J^i(x), \;\Rightarrow\; \partial_j A^j(x) = \frac{\partial_i}{\varepsilon\partial_0^2}J^i(x), \tag{7.5.2}$$

which, from Eq. (7.5.1), allows us to spell out the field equation for the electric field components $E^i = \partial^0 A^i$ as

$$\big(-\nabla^2 + \varepsilon\mu\,\partial_0^2\big)E^i(x) = \mu\Big[\delta^{ij} - \frac{\partial^i\partial^j}{\varepsilon\mu\,\partial_0^2}\Big]\partial^0 J^j(x). \tag{7.5.3}$$

For harmonic fields with a time dependence factor $\exp[-\mathrm{i}\omega t]$, we have

$$\big(-\nabla^2 - \varepsilon\mu\,\omega^2\big)E^i(\omega,\mathbf{x}) = \mathrm{i}\mu\omega\Big[\delta^{ij} + \frac{\partial^i\partial^j}{\varepsilon\mu\,\omega^2}\Big]J^j(\omega,\mathbf{x}). \tag{7.5.4}$$

This may be conveniently re-written as

$$\big(-\nabla^2 - \varepsilon\mu\,\omega^2\big)E^i(\omega,\mathbf{x}) = \mathrm{i}\,\mu\,\omega\Big[\delta^{ij} + \frac{\partial^i\partial^j}{\varepsilon\mu\,\omega^2}\Big]\int \mathrm{d}^3\mathbf{x}'\,\delta^{(3)}(\mathbf{x}-\mathbf{x}')\,J^j(\omega,\mathbf{x}') \tag{7.5.5}$$

We will show below that

$$(\nabla^2 + k^2)\Big[\frac{1}{4\pi}\frac{\mathrm{e}^{\mathrm{i}\,k|\mathbf{x}-\mathbf{x}'|}}{|\mathbf{x}-\mathbf{x}'|}\Big] = -\,\delta^{(3)}(\mathbf{x}-\mathbf{x}'), \quad \text{for e.g., } k = \omega\sqrt{\varepsilon\mu}. \tag{7.5.6}$$

From this, we may, from Eq. (7.5.5), infer that

$$E^i(\omega,\mathbf{x}) = \frac{\mathrm{i}\,\mu\,\omega}{4\pi}\Big[\delta^{ij} + \frac{\partial^i\partial^j}{\varepsilon\mu\,\omega^2}\Big]\int \mathrm{d}^3\mathbf{x}'\frac{\mathrm{e}^{\mathrm{i}\sqrt{\varepsilon\mu}\omega|\mathbf{x}-\mathbf{x}'|}}{|\mathbf{x}-\mathbf{x}'|}\,J^j(\omega,\mathbf{x}'). \tag{7.5.7}$$

We are interested in the leading behavior of the electric field for $|\mathbf{x}| \to \infty$. To this end,

$$|\mathbf{x}-\mathbf{x}'| = \big(|\mathbf{x}|^2 - 2\,\mathbf{x}\cdot\mathbf{x}' + |\mathbf{x}'|^2\big)^{1/2} \to |\mathbf{x}| - \frac{\mathbf{x}}{|\mathbf{x}|}\cdot\mathbf{x}', \tag{7.5.8}$$

$$\frac{\mathrm{e}^{\mathrm{i}\sqrt{\varepsilon\mu}\omega|\mathbf{x}-\mathbf{x}'|}}{|\mathbf{x}-\mathbf{x}'|} \to \frac{\mathrm{e}^{\mathrm{i}\sqrt{\varepsilon\mu}\omega|\mathbf{x}|}}{|\mathbf{x}|}\,\mathrm{e}^{-\mathrm{i}\sqrt{\varepsilon\mu}\omega\mathbf{n}\cdot\mathbf{x}'}, \quad \mathbf{n} \equiv \frac{\mathbf{x}}{|\mathbf{x}|}, \tag{7.5.9}$$

as well

$$\Big[\delta^{ij} + \frac{\partial^i\partial^j}{\varepsilon\mu\,\omega^2}\Big]\frac{\mathrm{e}^{\mathrm{i}\sqrt{\varepsilon\mu}\omega|\mathbf{x}-\mathbf{x}'|}}{|\mathbf{x}-\mathbf{x}'|} \to [\delta^{ij} - \mathbf{n}^i\mathbf{n}^j]\frac{\mathrm{e}^{\mathrm{i}\sqrt{\varepsilon\mu}\omega|\mathbf{x}|}}{|\mathbf{x}|}\mathrm{e}^{-\mathrm{i}\sqrt{\varepsilon\mu}\omega\mathbf{n}\cdot\mathbf{x}'}. \tag{7.5.10}$$

Hence from Eq. (7.5.7), for $|\mathbf{x}| \to \infty$,

$$E^i(\omega,\mathbf{x}) \to \frac{\mathrm{i}\,\mu\,\omega}{4\pi}\frac{\mathrm{e}^{\mathrm{i}\sqrt{\varepsilon\mu}\omega|\mathbf{x}|}}{|\mathbf{x}|}\int \mathrm{d}^3\mathbf{x}'\,[\delta^{ij} - \mathbf{n}^i\mathbf{n}^j]\,\mathrm{e}^{-\mathrm{i}\sqrt{\varepsilon\mu}\omega\mathbf{n}\cdot\mathbf{x}'}\,J^j(\omega,\mathbf{x}'). \tag{7.5.11}$$

Upon comparing this expression with Eqs. (7.4.12)–(7.4.15), we learn that

$$\frac{1}{2}\sqrt{\frac{\varepsilon}{\mu}}\lim_{|\mathbf{x}|\to\infty}\big[|\mathbf{x}|^2\,\mathbf{E}^*(\omega,\mathbf{x})\cdot\mathbf{E}(\omega,\mathbf{x})\big] = \frac{\mathrm{d}\,\mathrm{P}(\omega)}{\mathrm{d}\Omega}. \tag{7.5.12}$$

Also from the expression of the average power of radiation per unit frequency $\mathrm{P}(\omega)$ in Eqs. (7.4.10), and (7.4.11), and the expression of $|\mathbf{x}|\mathbf{E}(\omega,\mathbf{x})$ as obtained from Eq. (7.5.11), we have

$$\mathrm{P}(\omega) = \frac{1}{2}\sqrt{\frac{\varepsilon}{\mu}}\lim_{|\mathbf{x}|\to\infty}\Big[\,|\mathbf{x}|^2\!\int \mathrm{d}\Omega\,|\mathbf{E}(\omega,\mathbf{x})|^2\Big]. \tag{7.5.13}$$

The proof of Eq. (7.5.6) follows. To establish this equation, consider the following relevant steps:

$$\partial^i \frac{\mathrm{e}^{\mathrm{i}\,k|\mathbf{r}-\mathbf{r}'|}}{|\mathbf{r}-\mathbf{r}'|} = \Big[\mathrm{i}k\frac{\partial^i|\mathbf{r}-\mathbf{r}'|}{|\mathbf{r}-\mathbf{r}'|} + \partial^i\frac{1}{|\mathbf{r}-\mathbf{r}'|}\Big]\mathrm{e}^{\mathrm{i}\,k|\mathbf{r}-\mathbf{r}'|}, \tag{7.5.14}$$

$$\begin{aligned}\partial^j\partial^i \frac{\mathrm{e}^{\mathrm{i}\,k|\mathbf{r}-\mathbf{r}'|}}{|\mathbf{r}-\mathbf{r}'|} =& \Big[\mathrm{i}k\frac{\partial^j\partial^i|\mathbf{r}-\mathbf{r}'|}{|\mathbf{r}-\mathbf{r}'|} + \mathrm{i}k(\,\partial^i|\mathbf{r}-\mathbf{r}'|\,)\partial^j\frac{1}{|\mathbf{r}-\mathbf{r}'|} + \partial^j\partial^i\frac{1}{|\mathbf{r}-\mathbf{r}'|}\Big]\mathrm{e}^{\mathrm{i}\,k|\mathbf{r}-\mathbf{r}'|}\\ &+ \Big[\mathrm{i}k\frac{\partial^i|\mathbf{r}-\mathbf{r}'|}{|\mathbf{r}-\mathbf{r}'|} + \partial^i\frac{1}{|\mathbf{r}-\mathbf{r}'|}\Big]\big(\mathrm{i}\,k\,\partial^j|\mathbf{r}-\mathbf{r}'|\,\big)\mathrm{e}^{\mathrm{i}\,k|\mathbf{r}-\mathbf{r}'|}.\end{aligned} \tag{7.5.15}$$

In particular, by setting $j = i$, and summing over $i$, lead to

$$\begin{aligned}\nabla^2\frac{\mathrm{e}^{\mathrm{i}\,k|\mathbf{r}-\mathbf{r}'|}}{|\mathbf{r}-\mathbf{r}'|} =& \Big[\mathrm{i}k\frac{\nabla^2|\mathbf{r}-\mathbf{r}'|}{|\mathbf{r}-\mathbf{r}'|} + 2\mathrm{i}k(\,\partial^i|\mathbf{r}-\mathbf{r}'|\,)\partial^i\frac{1}{|\mathbf{r}-\mathbf{r}'|}\\ &- k^2\frac{\partial^i|\mathbf{r}-\mathbf{r}'|\,)\partial^i|\mathbf{r}-\mathbf{r}'|\,)}{|\mathbf{r}-\mathbf{r}'|} + \nabla^2\frac{1}{|\mathbf{r}-\mathbf{r}'|}\Big]\mathrm{e}^{\mathrm{i}\,k|\mathbf{r}-\mathbf{r}'|}.\end{aligned} \tag{7.5.16}$$

Using the equalities

$$|\mathbf{r}-\mathbf{r}'| = \big(\mathbf{r}^2 - 2\mathbf{r}\cdot\mathbf{r}' + \mathbf{r}'^2\big)^{1/2}, \tag{7.5.17}$$

$$\partial^i|\mathbf{r}-\mathbf{r}'| = \frac{\mathbf{r}^i - \mathbf{r}'^i}{|\mathbf{r}-\mathbf{r}'|}, \qquad \nabla^2|\mathbf{r}-\mathbf{r}'| = \frac{2}{|\mathbf{r}-\mathbf{r}'|}, \tag{7.5.18}$$

$$\partial^i\frac{1}{|\mathbf{r}-\mathbf{r}'|} = -\frac{\mathbf{r}^i - \mathbf{r}'^i}{|\mathbf{r}-\mathbf{r}'|^3}, \qquad \nabla^2\frac{1}{|\mathbf{r}-\mathbf{r}'|} = -4\pi\,\delta^{(3)}(\mathbf{r}-\mathbf{r}'), \tag{7.5.19}$$

where the last relation was proved in Sect. 3.6, below Eq. (3.6.46), lead to

$$\begin{aligned}\nabla^2\frac{\mathrm{e}^{\mathrm{i}\,k|\mathbf{r}-\mathbf{r}'|}}{|\mathbf{r}-\mathbf{r}'|} &= \Big[-\frac{k^2}{|\mathbf{r}-\mathbf{r}'|} - 4\pi\,\delta^{(3)}(\mathbf{r}-\mathbf{r}')\Big]\mathrm{e}^{\mathrm{i}\,k|\mathbf{r}-\mathbf{r}'|}\\ &= -\frac{k^2}{|\mathbf{r}-\mathbf{r}'|}\mathrm{e}^{\mathrm{i}\,k|\mathbf{r}-\mathbf{r}'|} - 4\pi\,\delta^{(3)}(\mathbf{r}-\mathbf{r}')\mathrm{e}^{\mathrm{i}\,k|\mathbf{r}-\mathbf{r}'|}\\ &= -k^2\frac{\mathrm{e}^{\mathrm{i}\,k|\mathbf{r}-\mathbf{r}'|}}{|\mathbf{r}-\mathbf{r}'|} - 4\pi\,\delta^{(3)}(\mathbf{r}-\mathbf{r}').\end{aligned} \tag{7.5.20}$$

which establishes the result in Eq. (7.5.6), by finally dividing by $4\pi$.

Some references to this chapter are given below.[6]

[6] See, e.g., Manoukian (1993), (2015), (2016) [4–6]. See also the pioneering monumental work of Schwinger et al. (1976) [10].

# References

1. Fazio, G. G., et al (1970). Generation of Čerenkov light flashes by cosmic radiation within the eyes of the Apollo Astronauts. *Nature, 228*, 260-264.
2. Feynman, R. (1967). *The character of physical law*. Cambridge, MA: MIT Press.
3. Manoukian, E. B., & Charuchittapan D. (2000). Quantum electrodynamics of Čerenkov radiation at finite temperature. *International Journal of Theoretical Physics, 19*, 2197-2205.
4. Manoukian E. B. (1993). Discrete Čerenkov power spectrum. *Physical Review B, 48*, 543-547.
5. Manoukian E. B. (2015). Vacuum-to-vacuum transition amplitude in a medium. *Radiation Physics and Chemistry, 112*, 104-107.
6. Manoukian, E. B. (2016). *Quantum field theory I: Foundations and abelian and non-abelian gauge theories*. Switzerland: Springer.
7. McNulty, P. J., et al. (1976). Role of Čerenkov radiation in the eye-flashes observed by Apollo Astronauts. *Life Science and Space Research, 14*, 205-217.
8. Pinsky, L. S., et al. (1974). Light flashes observed by Astronauts on Apollo 11 through Apollo 17. *Science,183*, 957-959.
9. Schwinger, J. (1957). A theory of the fundamental interactions. *Annals of Physics (NY), 2, 407*, 407-434.
10. Schwinger, J. ,Tsai, Wu-Yang, & Erber, T. (1976). Classical and quantum theory of synergic synchrotron-Čerenkov radiation. *Annals of Physics (NY), 96*, 303-332.

# Chapter 8
# Propagator Theory and Scattering Processes

*Follow a photon from the time it is emitted to the time it is detected in a time evolutionary process*

The propagator theory approach derived and exploited in electrodynamics as applied to various physical situations, provides a physically appealing approach to scattering theory of electromagnetic phenomena, in a time evolutionary process in spacetime, in complex situations in the presence of obstacles, and emphasizes the dynamical aspects of the formalism.

It generalizes the expression of the propagator $D_+^{ij}(x - x')$ in Eq. (3.3.14) set up in an infinite region, as described in Sect. 3.3 of Chap. 3, by simply satisfying the appropriate boundary conditions, as one crosses, in general, an intervening surface between two regions. The advantage of working with a propagator is that one avoids dealing with plane waves. It generalizes the well publicized treatment by Feynman,[1] of adding-up time-dependent arrows, to a rigorous computational formalism described in the manner carried out experimentally. The propagator describes the time development of a process from the time a photon is emitted to the time it is detected. This present method of describing *scattering* processes, in classical electrodynamics, was given in the reference cited below.[2]

As seen in Sect. 3.4 of Chap. 3, the fate of photons, as they meet an obstacle in the present, general cases, may be elegantly described by considering the combinations $J_2^i(x)D_+^{ij}(x, x')J_1^j(x)$ of, causally arranged, emission and detection current distributions $J_1^i(x)$, $J_2^j(x)$, set up appropriately in spacetime as encountered in experimental situations.

We first recast the content of Sects. 3.3, 3.4 of Chap. 3, in a form convenient to describe the propagation of a photon in an unbounded region in free space, which will provide a clear-cut way, in subsequent sections, for generalising the corresponding

---

[1] Feynman (1985) [1], pp. 17–35, 61–72.

[2] Manoukian (1992) [4].



E. B. Manoukian, *Radiation*,
https://doi.org/10.1007/978-3-032-21506-2_8

propagator $D_+^{ij}(x, x')$ to cases when more than one region is present in the physical situation being considered, and actually describe a scattering process, as well as to extract information on polarization aspects of the problem. We describe, in particular, the propagator approach to the classical problem of reflection off a perfect conductor. This will follow by applications to other several physical situations, far more complex, in forthcoming chapters.

## 8.1 Setting-Up the Scattering Problem in Propagator Theory

To see how propagator theory describes the scattering problem, we re-iterate part of the development in Sects. 3.3, 3.4 of Chap. 3 to this end.

The vacuum-to-vacuum transition amplitude, taking into account of current conservation, was given in Eqs. (3.3.13)–(3.3.15), which in detail reads,

$$< 0_+ \mid 0_- > = \exp\Big[\frac{\mathrm{i}}{2}\int(\mathrm{d}x)\int(\mathrm{d}x' J^i(x) D_+^{ij}(x-x') J^j(x')\Big], \tag{8.1.1}$$

$$D_+^{ij}(x-x') = \int \frac{(\mathrm{d}k)}{(2\pi)^4}\Big[\delta^{ij} - \frac{k^i k^j}{k^{02}}\Big]\frac{\mathrm{e}^{\mathrm{i}k(x-x')}}{k^2 - \mathrm{i}\,\epsilon}, \qquad \epsilon \to +0, \tag{8.1.2}$$

$$(\mathrm{d}k) \equiv \mathrm{d}k^0 \mathrm{d}^3\mathbf{k},\;\; k^2 = \mathbf{k}^2 - k^{02},\;\; k(x-x') = \mathbf{k}\cdot(\mathbf{x}-\mathbf{x}') - k^0(x^0 - k'^0), \tag{8.1.3}$$

$$\big[(-\,\partial^2 + \partial^{02})\delta^{i\ell} + \partial^i\partial^\ell\big]D_+^{\ell j}(x-x') = \delta^{ij}\,\delta^{(4)}(x-x'). \tag{8.1.4}$$

Note that the structure $[\delta^{ij} - (k^i k^j/k^{02})]$ arises in the propagator as a consequence of current conservation.

Upon writing the current as

$$J^\mu(x) = J_2^\mu(x) + J_1^\mu(x), \tag{8.1.5}$$

where $J_1^\mu(x)$, $J_2^\mu(x)$ are causally arranged such that the current $J_2^\mu(x)$ is switched on after $J_1^\mu(x)$ is switched off, i.e., removed, allows us to write $< 0_+ \mid 0_- >$ explicitly as:

$$< 0_+ \mid 0_- >^J = < 0_+ \mid 0_- >^{J_2}\; T\; < 0_+ \mid 0_- >^{J_1}, \tag{8.1.6}$$

$$T = \exp\Big[\int \frac{\mathrm{d}^3\mathbf{k}}{(2\pi)^3 2|\mathbf{k}|}\,(\mathrm{i})J_2^{*i}(k)\big[\delta^{ij} - \frac{\mathbf{k}^i}{|\mathbf{k}|}\frac{\mathbf{k}^j}{|\mathbf{k}|}\big](\mathrm{i})J_1^j(k)\Big], \tag{8.1.7}$$

where we have used the expression for $D_+^{ij}(x-x')$, for $t > t'$, as obtained from Eq. (3.2.9), and note the (i) factors multiplying the currents. Using the completeness relation

$$\left[\delta^{ij} - \frac{\mathbf{k}^i}{|\mathbf{k}|}\frac{\mathbf{k}^j}{|\mathbf{k}|}\right] = \sum_{\lambda=1,2} \mathbf{e}^i_\lambda \mathbf{e}^j_\lambda, \quad i = 1, 2, 3, \tag{8.1.8}$$

as given in Eq. (3.4.4), the process that a photon of momentum/polarization ($\mathbf{k}'$, $\lambda'$) was emitted by $J_1$, and detected by $J_2$ with momentum/polarization ($\mathbf{k}$, $\lambda$) is given by

$$< 0_+ \mid \mathbf{k}, \lambda >< \mathbf{k}, \lambda \mid \mathbf{k}', \lambda' >< \mathbf{k}', \lambda' \mid 0_- >, \tag{8.1.9}$$

where $< \mathbf{k}, \lambda \mid \mathbf{k}', \lambda' >$ denotes the amplitude that a photon of momentum/polarization ($\mathbf{k}'$, $\lambda'$) ends up with momentum/polarization ($\mathbf{k}$, $\lambda$),

$$< \mathbf{k}, \lambda \mid \mathbf{k}', \lambda' > = \sqrt{\frac{\mathrm{d}^3\mathbf{k}}{(2\pi)^3 2|\mathbf{k}|}}\,\delta^{(3)}(\mathbf{k} - \mathbf{k}')\delta_{\lambda\lambda'}\sqrt{\frac{\mathrm{d}^3\mathbf{k}'}{(2\pi)^3 2|\mathbf{k}'|}}, \tag{8.1.10}$$

$$< 0_+ \mid \mathbf{k}, \lambda >^{J_2} = < 0_+ \mid 0_- >^{J_2} \sqrt{\frac{\mathrm{d}^3\mathbf{k}}{(2\pi)^3 2|\mathbf{k}|}}\ (\mathrm{i})\, \mathbf{J}_2^* \cdot \mathbf{e}_\lambda, \tag{8.1.11}$$

$$< \mathbf{k}', \lambda' \mid 0_- >^{J_1} = \mathbf{e}_{\lambda'} \cdot (\mathrm{i})\, \mathbf{J}_1 \sqrt{\frac{\mathrm{d}^3\mathbf{k}'}{(2\pi)^3 2|\mathbf{k}'|}} < 0_+ \mid 0_- >^{J_1}. \tag{8.1.12}$$

as described in Sect. 3.4 in Chap. 3.

Now we have to see what happens to the propagator $D_+^{ij}(x - x')$ when we introduce a perfectly reflecting surface in the problem, with proper boundary conditions, and eventually extract the amplitude $< \mathbf{k}, \lambda \mid \mathbf{k}', \lambda' >$ of interest.

## 8.2 Propagator and the Classical Reflection Problem

Let the $x^1 - x^2$ plane denote a reflecting surface. The physical process occurs in the region $x^3 \equiv z \geq 0$ in free space with $\epsilon$, $\mu = 1$. A vector $\mathbf{V}$ parallel to, or lying in, the reflecting surface will be denoted by $\mathbf{v}_\parallel$. Here the process to be described is limited to half-space.

The propagator, in general, satisfies the equation

$$\left(-\Box\delta^{ij} + \partial^i\partial^j\right)D_+^{jk}(x, x') = \delta^{ik}\delta^4(x, x'), \qquad \Box = \boldsymbol{\partial}^2 - \partial^{02}, \tag{8.2.1}$$

where due to the presence of the reflecting surface, the variable $z$ is limited only to $z \geq 0$, i.e., the Dirac delta distribution is, with $z \geq 0$, $z' \geq 0$, to be written as

$$\delta^4(x, x') = \delta^{(2)}\big(\mathbf{x}_\parallel - \mathbf{x}'_\parallel\big)\,\delta\big(z, z'\big)\,\delta\big(x^0 - x'^0\big). \tag{8.2.2}$$

with $\delta(z, z')$ as given in Eqs. (3.6.40), (3.6.41) in Chap. 3, and will be spelled out below.

In particular note that

$$D_+^{3k}(x, x') = D_+^{k3}(x', x). \tag{8.2.3}$$

We spell out Eq. (8.2.1) in terms of various components. With $a, b, c$ taking the values 1, 2, we have

$$-\square D_+^{ab} + \partial^a \partial^c D_+^{cb} + \partial^a \partial^3 D_+^{3b} = \delta^{ab}\, \delta^4(x, x'), \tag{8.2.4}$$

$$-\square D_+^{3a} + \partial^3 \partial^c D_+^{ca} + \partial^3 \partial^3 D_+^{3a} = 0, \tag{8.2.5}$$

$$-\square D_+^{a3} + \partial^a \partial^c D_+^{c3} + \partial^a \partial^3 D_+^{33} = 0, \tag{8.2.6}$$

$$-\square D_+^{33} + \partial^3 \partial^c D_+^{c3} + \partial^3 \partial^3 D_+^{33} = \delta^4(x, x'). \tag{8.2.7}$$

In the temporal gauge $A^0 = 0$, specified in (5.1.15), the field equations are

$$-\square A^i(x) + \partial^i \partial^j A^j(x) = J^i(x). \tag{8.2.8}$$

as given in Eq. (5.1.26). In terms of the propagator $D^{jk}(x, x')$, we may write

$$A^j(x) = \int (\mathrm{d}x') D_+^{jk}(x, x') J^k(x'), \tag{8.2.9}$$

which easily follows by applying the operator $(-\square\delta^{ij} + \partial^i \partial^j)$ to it thus leading to Eq. (8.2.8), by using, in the process, the property of the $\delta^{(4)}(x, x')$ distribution.

In the temporal gauge, Eqs. (5.3.4), (5.3.5), in particular, for $a = 1, 2$ give

$$\mathbf{E}^a = \partial^0 A^a, \quad \mathbf{B}^3 = \partial^1 A^2 - \partial^2 A^1. \tag{8.2.10}$$

We use boundary of an uncharged perfect conductor which from Eqs. (5.4.32), (5.4.34) imply that $\mathbf{E}^a|_{z=+0} = 0$, $\mathbf{B}^3|_{z=+0} = 0$, respectively, where the unit vector $\mathbf{n}$ there is perpendicular to the surface. This will be the case if

$$\partial^0 D_+^{aj}(x, x')\Big|_{z=+0} = 0, \quad a = 1, 2, \ j = 1, 2, 3 \tag{8.2.11}$$

$$\big[\partial^1 D_+^{2j}(x, x') - \partial^2 D^{1j}(x, x')\big]\Big|_{z=+0} = 0, \quad j = 1, 2, 3. \tag{8.2.12}$$

Since the partial derivatives applied to $D^{aj}(x, x')$, in the above two equations, do not involve derives with respect to $z$, we just have the boundary condition that

$$D_+^{aj}(x, x')\Big|_{z=+0} = 0, \quad a = 1, 2, \ j = 1, 2, 3. \tag{8.2.13}$$

In Chap. 3, Eq. (3.6.41), $\delta(z, z')$ for $z, z' \geq 0$, which vanishes for $z = 0$ is given by

$$\delta(z, z') = \frac{1}{\pi}\int_{-\infty}^{\infty} \mathrm{d}q \, \sin qz \sin qz'. \tag{8.2.14}$$

In particular, from Eqs. (8.2.4), (8.2.5), we may then infer that for consistency with the boundary condition in Eq. (8.2.13),

$$D_+^{ab}(x, x') = \int \frac{\mathrm{d}^2\mathbf{K}}{(2\pi)^2}\int_{-\infty}^{\infty}\frac{\mathrm{d}k^0}{2\pi}\int_{-\infty}^{\infty}\frac{\mathrm{d}q}{\pi}\Big(\delta^{ab} - \frac{K^a K^b}{k^{02}}\Big)\frac{\sin(qz)\sin(qz')}{k^2 - \mathrm{i}\epsilon} \times \mathrm{e}^{\mathrm{i}\mathbf{K}\cdot(\mathbf{x}_\parallel - \mathbf{x}'_\parallel)}\,\mathrm{e}^{-\mathrm{i}k^0(x^0 - x'^0)}, \tag{8.2.15}$$

$$D_+^{3b}(x, x') = \int \frac{\mathrm{d}^2\mathbf{K}}{(2\pi)^2}\int_{-\infty}^{\infty}\frac{\mathrm{d}k^0}{2\pi}\int_{-\infty}^{\infty}\frac{\mathrm{d}q}{\pi}\Big(\frac{\mathrm{i}q K^b}{k^{02}}\Big)\frac{\cos(qz)\sin(qz')}{k^2 - \mathrm{i}\epsilon} \times \mathrm{e}^{\mathrm{i}\mathbf{K}\cdot(\mathbf{x}_\parallel - \mathbf{x}'_\parallel)}\,\mathrm{e}^{-\mathrm{i}k^0(x^0 - x'^0)}, \tag{8.2.16}$$

where

$$k = (k^0, \mathbf{K}, q), \quad x = (x^0, \mathbf{x}_\parallel, z), \tag{8.2.17}$$

and $\mathbf{K}$ is the component of $\mathbf{k}$ parallel to the reflecting surface.

On the other hand, from Eqs. (8.2.3) and (8.2.16), we also obtain

$$D_+^{b3}(x, x') = \int \frac{\mathrm{d}^2\mathbf{K}}{(2\pi)^2}\int_{-\infty}^{\infty}\frac{\mathrm{d}k^0}{2\pi}\int_{-\infty}^{\infty}\frac{\mathrm{d}q}{\pi}\Big(\frac{\mathrm{i}K^b q}{k^{02}}\Big)\frac{\sin(qz)\cos(qz')}{k^2 - \mathrm{i}\epsilon} \times \mathrm{e}^{\mathrm{i}\mathbf{k}\cdot(\mathbf{x}'_\parallel - \mathbf{x}_\parallel)}\,\mathrm{e}^{-\mathrm{i}k^0(x'^0 - x^0)}. \tag{8.2.18}$$

Upon making a change of integration variables $\mathbf{K} \to -\mathbf{K}$, $k^0 \to -k^0$, we get

$$D_+^{b3}(x, x') = \int \frac{\mathrm{d}^2\mathbf{K}}{(2\pi)^2}\int_{-\infty}^{\infty}\frac{\mathrm{d}k^0}{2\pi}\int_{-\infty}^{\infty}\frac{\mathrm{d}q}{\pi}\Big(\frac{-\mathrm{i}K^b q}{k^{02}}\Big)\frac{\sin(qz)\cos(qz')}{k^2 - \mathrm{i}\epsilon} \times \mathrm{e}^{\mathrm{i}\mathbf{K}\cdot(\mathbf{x}_\parallel - \mathbf{x}'_\parallel)}\,\mathrm{e}^{-\mathrm{i}k^0(x^0 - x'^0)}. \tag{8.2.19}$$

Using the fact that $E^3$ does not vanish for $z = 0$, i.e. $D^{3j}$ does not vanish for $z = 0$, we may use a Fourier-cosine transform for $\delta(z, z')$ in half space, as given in Chap. 3, Eq. (3.6.40),

$$\delta(z, z') = \frac{1}{\pi}\int_{-\infty}^{\infty} \mathrm{d}q \cos(qz)\, \cos(qz'), \tag{8.2.20}$$

which, from Eq. (8.2.7), as well as Eq. (8.2.19) for $D^{b3}$, lead to the explicit expression for $D^{33}$:

$$D_+^{33}(x,x') = \int \frac{\mathrm{d}^2\mathbf{K}}{(2\pi)^2} \int_{-\infty}^{\infty} \frac{\mathrm{d}k^0}{2\pi} \int_{-\infty}^{\infty} \frac{\mathrm{d}q}{\pi} \Big(1 - \frac{q^2}{k^{02}}\Big) \frac{\cos(qz)\cos(qz')}{k^2 - \mathrm{i}\,\epsilon} \times \mathrm{e}^{\mathrm{i}\mathbf{K}\cdot(\mathbf{x}_\parallel - \mathbf{x}'_\parallel)}\,\mathrm{e}^{-\mathrm{i}k^0(x^0 - x'^0)}. \tag{8.2.21}$$

From Eqs. (8.2.15)/(8.2.16), and (8.2.19)/(8.2.21), we have determined all the components of $D_+^{ij}(x,x')$ for $i, j = 1, 2, 3$.

As in Eq. (8.1.5), we consider two, causally arranged, current sources $J_1^\mu(x)$, $J_2^\mu(x)$ such that the current $J_2^\mu(x)$, as a detector, is switched on after $J_1^\mu(x)$, as an emitter, is switched off, and, in turn, determine the basic structure

$$\mathrm{i}\int(\mathrm{d}x)\int(\mathrm{d}x')\,J_2^i(x)D_+^{ij}(x,x')J_1^j(x') \equiv T. \tag{8.2.22}$$

In order to compute the structure of interest in the above equation, we first evaluate the $k^0-$ integral in $D_+^{ij}(x,x')$. This is easily done from Sect. 3.3, Eqs. (3.3.1), (3.3.6) for $x^0 > x'^0$, due to the denominator $(k^2 - \mathrm{i}\epsilon)$, where $k^2 = [(\mathbf{K}^2 + q^2) - k^{02}]$, giving

$$\int_{-\infty}^{\infty} \frac{\mathrm{d}k^0}{2\pi} \frac{\mathrm{e}^{-\mathrm{i}k^0(x^0 - x'^0)}}{k^2 - \mathrm{i}\,\epsilon} = \frac{\mathrm{i}}{2|\mathbf{k}|}\,\mathrm{e}^{-\mathrm{i}|\mathbf{k}|(x^0 - x'^0)}, \qquad |\mathbf{k}| \equiv \sqrt{\mathbf{K}^2 + q^2}. \tag{8.2.23}$$

where note that $(k^0 - |\mathbf{k}|)(k^0 + |\mathbf{k}|) = -k^2$.

We also use,

$$\sin(qz)\,\sin(qz') = -\frac{1}{4}\big[\mathrm{e}^{\mathrm{i}q(z+z')} + \mathrm{e}^{-\mathrm{i}q(z+z')} - \mathrm{e}^{\mathrm{i}q(z-z')} - \mathrm{e}^{-\mathrm{i}q(z-z')}\big], \tag{8.2.24}$$

$$\sin(qz)\,\cos(qz') = -\frac{\mathrm{i}}{4}\big[\mathrm{e}^{\mathrm{i}q(z+z')} - \mathrm{e}^{-\mathrm{i}q(z+z')} + \mathrm{e}^{\mathrm{i}q(z-z')} - \mathrm{e}^{-\mathrm{i}q(z-z')}\big], \tag{8.2.25}$$

$$\cos(qz)\,\cos(qz') = +\frac{1}{4}\big[\mathrm{e}^{\mathrm{i}q(z+z')} + \mathrm{e}^{-\mathrm{i}q(z+z')} + \mathrm{e}^{\mathrm{i}q(z-z')} + \mathrm{e}^{-\mathrm{i}q(z-z')}\big], \tag{8.2.26}$$

$$\cos(qz)\,\sin(qz') = -\frac{\mathrm{i}}{4}\big[\mathrm{e}^{\mathrm{i}q(z+z')} - \mathrm{e}^{-\mathrm{i}q(z+z')} - \mathrm{e}^{\mathrm{i}q(z-z')} + \mathrm{e}^{-\mathrm{i}q(z-z')}\big], \tag{8.2.27}$$

and the Fourier transforms

$$J^i(k^0, \mathbf{K}, \pm q) = \int(\mathrm{d}x)\,J^i(x)\,\mathrm{e}^{-\mathrm{i}\mathbf{K}\cdot\mathbf{x}_\parallel}\,\mathrm{e}^{\mp\mathrm{i}qz}\,\mathrm{e}^{\mathrm{i}\sqrt{\mathbf{K}^2+q^2}x^0}. \tag{8.2.28}$$

All told, the structure of the scattering process is, from Eq. (8.2.22), extracted from the following expression

$$\big[(\mathrm{i})J_2^{i*}(k)A^{ij}(\mathrm{i})J_1^i(k) + (\mathrm{i})J_2^{i*}(k)B^{ij}(\mathrm{i})J_1^j(k')\big], \tag{8.2.29}$$

where

$$\mathbf{k} = (\mathbf{K}, q) \quad \mathbf{k}' = (\mathbf{K}, -q), \quad k^0 = |\mathbf{k}| = |\mathbf{k}'|. \tag{8.2.30}$$

$$A^{ij} = \left(\delta^{ij} - \frac{k^i k^j}{\mathbf{k}^2}\right), \qquad B^{ij} = -\delta^{ij} + \frac{k^i k'^j}{\mathbf{k}^2} + 2\,\delta^{i3}\delta^{j3}, \tag{8.2.31}$$

and the vector $\mathbf{K}$ is parallel to the reflecting surface. Here the contribution to $A^{ij}$ comes from the $\exp[\pm \mathrm{i}(q(z - z')]$ terms in Eqs. (8.2.24)–(8.2.27), while the contribution to $B^{ij}$ comes from those terms with $\exp[\pm \mathrm{i}(q(z + z')]$, with the latter leading to a scattering with a change of sign in the third component of the incident momentum. It is interesting to note, here, how a solution arises where the third component of a photon, on it's way to be detected has an opposite sign to a photon which was initially emitted. This was, of course, expected and gives the elementary law of reflection.

Note the transversality properties:

$$k^i\,A^{ij} = 0, \quad A^{ij}\,k^j = 0, \qquad k^i\,B^{ij} = 0, \quad B^{ij}k'^j = 0. \tag{8.2.32}$$

Given any two sets of polarization vectors $\mathbf{e}_\lambda$, $\mathbf{e}'_\lambda$ satisfying

$$\mathbf{e}_\lambda \cdot \mathbf{e}_{\lambda'} = \delta_{\lambda\lambda'}, \quad \mathbf{k} \cdot \mathbf{e}_\lambda = 0, \quad \lambda, \lambda' = 1, 2, \tag{8.2.33}$$

$$\delta^{ij} = \frac{k^i\,k^j}{|\mathbf{k}|^2} + \sum_{\lambda=1,2} \mathrm{e}^i_\lambda\,\mathrm{e}^j_\lambda, \tag{8.2.34}$$

and

$$\mathbf{e}'_\lambda \cdot \mathbf{e}'_{\lambda'} = \delta_{\lambda\lambda'}, \quad \mathbf{k}' \cdot \mathbf{e}'_\lambda = 0, \quad \lambda, \lambda' = 1, 2, \tag{8.2.35}$$

$$\delta^{ij} = \frac{k'^i\,k'^j}{|\mathbf{k}|^2} + \sum_{\lambda=1,2} \mathrm{e}'^i_\lambda\,\mathrm{e}'^j_\lambda, \tag{8.2.36}$$

where $\mathbf{k}, \mathbf{k}'$ are defined in Eq. (8.2.30), and referring to Eq. (8.2.31), concerning $A^{ij}$, we have

$$A^{ij} = \sum_{\lambda=1,2} \mathrm{e}^i_\lambda\,\mathrm{e}^j_\lambda, \tag{8.2.37}$$

which corresponds to a non-scattering part, i.e., for a photon which simply goes from the emitter to the detector not approaching the reflecting surface.

The scattering part, is readily obtained from the expression of $B^{ij}$ in Eq. (8.2.31) by using, in the process, the explicit expression of $B^{ij}$ and the completeness relations in Eqs. (8.2.34), (8.2.36),

$$\begin{aligned} B^{ij} &= \delta^{i\ell}\, B^{\ell m}\, \delta^{mj} \\ &= \Big(\frac{k^i k^\ell}{|\mathbf{k}|^2} + \mathrm{e}^i_\lambda \mathrm{e}^\ell_\lambda\Big)\Big( - \delta^{\ell m} + \frac{k^\ell k'^m}{|\mathbf{k}|^2} + 2\delta^{\ell 3}\delta^{m3}\Big)\Big(\frac{k'^m k'^j}{|\mathbf{k}|^2} + \mathrm{e}'^m_{\lambda'} \mathrm{e}'^j_{\lambda'}\Big), \end{aligned} \tag{8.2.38}$$

where summations, separately, over $\lambda$, $\lambda'$ are understood. Upon using the transversality conditions in Eq. (8.2.32), now, on $B^{\ell m}$, the following remarkably simple expression emerges

$$B^{ij} = \mathrm{e}^i_\lambda \Big( - \mathbf{e}_\lambda \cdot \mathbf{e}'_{\lambda'} + 2\, \mathrm{e}^3_\lambda \mathrm{e}'^3_{\lambda'} \Big)\, \mathrm{e}'^j_{\lambda'}. \tag{8.2.39}$$

The importance of this equation in classical electrodynamics cannot be over emphasized.

Accordingly, from Eq. (8.2.29), the reflection process, originating from the emission to detection, for two sets of polarization states described in a unified manner is given by

$$\big[\, (\mathrm{i}) J_2^{i*}(k)\, \mathrm{e}^i_\lambda \,\big]\Big( - \mathbf{e}_\lambda \cdot \mathbf{e}'_{\lambda'} + 2\, \mathrm{e}^3_\lambda \mathrm{e}'^3_{\lambda'} \Big)\big[\, \mathrm{e}'^j_{\lambda'}\, (\mathrm{i}) J_1^j(k') \,\big], \tag{8.2.40}$$

with the amplitude of scattering by reflection given by

$$- \mathbf{e}_\lambda \cdot \mathbf{e}'_{\lambda'} + 2\, \mathrm{e}^3_\lambda \mathrm{e}'^3_{\lambda'}. \tag{8.2.41}$$

We note that given a polarization vector, say, $\mathbf{e}$, then a second polarization vector, say, $\hat{\mathbf{e}}$ must be perpendicular to $\mathbf{e}$ as well as to the unit vector, say, $\hat{\mathbf{n}}$, along the momentum of propagation, with $\mathbf{e} \times \hat{\mathbf{e}} = \hat{\mathbf{n}}$.

*Given* directions of propagations, one can readily introduce associated polarization vectors satisfying a set of properties as given below. To this end, we define unit vectors $\mathbf{n}' = \mathbf{k}'/|\mathbf{k}|$, $\mathbf{n} = \mathbf{k}/|\mathbf{k}|$, and polarization vectors given by

$$\mathbf{e}'_1 = \frac{(\mathbf{n} \times \mathbf{n}') \times \mathbf{n}'}{|(\mathbf{n} \times \mathbf{n}') \times \mathbf{n}'|} = \frac{\mathbf{n}'(\mathbf{n} \cdot \mathbf{n}') - \mathbf{n}}{|\mathbf{n} \times \mathbf{n}'|}, \qquad \mathbf{e}'_2 = \frac{\mathbf{n} \times \mathbf{n}'}{|\mathbf{n} \times \mathbf{n}'|}, \tag{8.2.42}$$

$$\mathbf{e}_1 = \frac{(\mathbf{n} \times \mathbf{n}') \times \mathbf{n}}{|(\mathbf{n} \times \mathbf{n}') \times \mathbf{n}|} = \frac{\mathbf{n}' - \mathbf{n}(\mathbf{n} \cdot \mathbf{n}')}{|\mathbf{n} \times \mathbf{n}'|}, \qquad \mathbf{e}_2 = \frac{\mathbf{n} \times \mathbf{n}'}{|\mathbf{n} \times \mathbf{n}'|}. \tag{8.2.43}$$

They satisfy the following properties:

$$\mathbf{e}'_1 \times \mathbf{e}'_2 = \mathbf{n}', \qquad \mathbf{e}'_{\lambda'} \cdot \mathbf{e}'_\lambda = \delta_{\lambda'\lambda}, \qquad \mathbf{n}' \cdot \mathbf{e}'_\lambda = 0, \tag{8.2.44}$$

$$\mathbf{e}_1 \times \mathbf{e}_2 = \mathbf{n}, \qquad \mathbf{e}_{\lambda'} \cdot \mathbf{e}_\lambda = \delta_{\lambda'\lambda}, \qquad \mathbf{n} \cdot \mathbf{e}_\lambda = 0. \tag{8.2.45}$$

The following inherited aspects of these polarization should be observed,

$$\mathrm{e}'^a_1 = - \mathrm{e}^a_1, \ \text{for } a = 1, 2, \quad \mathrm{e}'^3_1 = \mathrm{e}^3_1, \quad \mathbf{e}'_2 = \mathbf{e}_2, \quad \mathrm{e}^3_2 = 0 = \mathrm{e}'^3_2, \tag{8.2.46}$$

$$\mathrm{e}^3_\lambda \mathrm{e}'^3_{\lambda'} = 0, \ \textit{and}\ \mathbf{e}_\lambda \cdot \mathbf{e}'_{\lambda'} = 0, \ \textit{for}\ \lambda \neq \lambda'. \tag{8.2.47}$$

Accordingly, the scattering amplitude in Eq. (8.2.41) may be written as

$$-\mathbf{e}_{\lambda}\cdot\mathbf{e}'_{\lambda'}+2\,\mathrm{e}^3_{\lambda}\,\mathrm{e}'^3_{\lambda'}=0\quad\text{for}\ \lambda\neq\lambda',\quad\text{from Eq. (8.2.47).}\tag{8.2.48}$$

On the other hand for $\lambda=\lambda'$, with *no* sum over $\lambda$,

$$-\mathbf{e}_{\lambda}\cdot\mathbf{e}'_{\lambda}+2\,\mathrm{e}^3_{\lambda}\,\mathrm{e}'^3_{\lambda}=\mathbf{e}_{\lambda}\cdot\mathbf{e}_{\lambda}=1,\tag{8.2.49}$$

for a *fixed* $\lambda$, using Eqs. (8.2.46).

That is, the following scattering processes occur:

$$\big(\mathbf{K},-q,\mathbf{e}'_1\big)\ \rightarrow\ \big(\mathbf{K},q,\mathbf{e}_1\big),\tag{8.2.50}$$

$$\big(\mathbf{K},-q,\mathbf{e}'_2\big)\ \rightarrow\ \big(\mathbf{K},q,\mathbf{e}_2\big).\tag{8.2.51}$$

These processes are depicted in Fig. 8.1 working in the incident plane.

We may equally define the following pairs of polarization states:

$$\overline{\mathbf{e}}_1=\frac{\mathbf{n}'\times(\mathbf{n}\times\mathbf{n}')}{|\mathbf{n}'\times(\mathbf{n}'\times\mathbf{n})|}=\frac{\mathbf{n}-\mathbf{n}'(\mathbf{n}\cdot\mathbf{n}')}{|\mathbf{n}'\times\mathbf{n}|},\qquad \overline{\mathbf{e}}_2=\frac{\mathbf{n}'\times\mathbf{n}}{|\mathbf{n}'\times\mathbf{n}|},\tag{8.2.52}$$

$$\underline{\mathbf{e}}_1=\frac{\mathbf{n}\times(\mathbf{n}'\times\mathbf{n})}{|\mathbf{n}\times(\mathbf{n}'\times\mathbf{n}|}=\frac{\mathbf{n}'-\mathbf{n}(\mathbf{n}\cdot\mathbf{n}')}{|\mathbf{n}'\times\mathbf{n}|},\qquad \underline{\mathbf{e}}_2=\frac{\mathbf{n}\times\mathbf{n}'}{|\mathbf{n}\times\mathbf{n}'|}.\tag{8.2.53}$$

They satisfy the following properties:

$$\overline{\mathbf{e}}_1\times\overline{\mathbf{e}}_2=\mathbf{n}'\equiv\frac{\mathbf{k}'}{|\mathbf{k}'|},\qquad \overline{\mathbf{e}}_{\lambda}\cdot\overline{\mathbf{e}}_{\lambda'}=\delta_{\lambda\lambda'},\qquad \mathbf{n}'\cdot\overline{\mathbf{e}}_{\lambda}=0,\tag{8.2.54}$$

$$\underline{\mathbf{e}}_1\times\underline{\mathbf{e}}_2=\mathbf{n}\equiv\frac{\mathbf{k}}{|\mathbf{k}|},\qquad \underline{\mathbf{e}}_{\lambda}\cdot\underline{\mathbf{e}}_{\lambda'}=\delta_{\lambda\lambda'},\qquad \mathbf{n}\cdot\underline{\mathbf{e}}_{\lambda}=0,\tag{8.2.55}$$

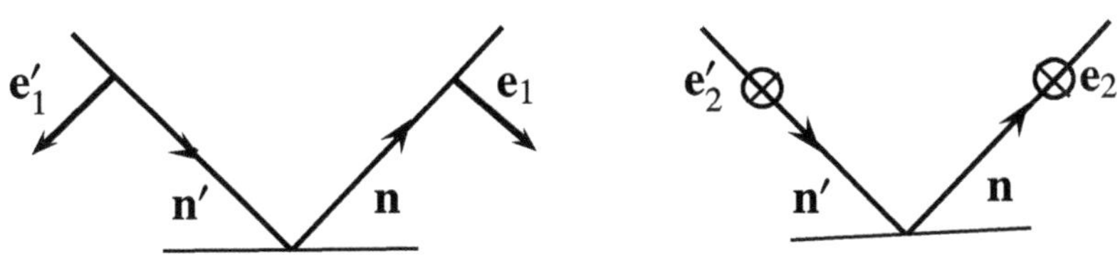


**Fig. 8.1** From left to right, depicting the reflection processes $\big(\mathbf{K},-q,\mathbf{e}'_1\big)\rightarrow\big(\mathbf{K},q,\mathbf{e}_1\big)$, with $\mathbf{e}'_1$ and $\mathbf{e}_1$ in the plane of incidence, and $\big(\mathbf{K},-q,\mathbf{e}'_2\big)\rightarrow\big(\mathbf{K},q,\mathbf{e}_2\big)$, with $\mathbf{e}'_2$ and $\mathbf{e}_2$ normal to it, respectively. The crosses indicate vectors that point into the page. Remember that if light of a given momentum, say, $\mathbf{p}$, is polarized along a direction specified by a polarization vector, say, $\mathbf{e}$, then the direction of the magnetic field $\mathbf{B}$ is given by the relation $\mathbf{e}\times\mathbf{B}\propto\mathbf{p}$

as well as

$$\overline{\mathrm{e}}^{3}_{\lambda}\,\underline{\mathrm{e}}^{3}_{\lambda'}, \quad \textit{and} \quad \overline{\mathbf{e}}_{\lambda}\cdot\underline{\mathbf{e}}_{\lambda'} = 0, \quad \textit{for } \lambda \neq \lambda', \tag{8.2.56}$$

$$\overline{\mathrm{e}}^{a}_{\lambda} = \underline{\mathrm{e}}^{a}_{\lambda}, \quad \text{for } a = 1, 2, \qquad \overline{\mathbf{e}}_{2} = -\,\underline{\mathbf{e}}_{2}, \quad \overline{\mathrm{e}}^{3}_{1} = -\,\underline{\mathrm{e}}^{3}_{1}, \quad \overline{\mathrm{e}}^{3}_{2} = 0 = \underline{\mathrm{e}}^{3}_{2}. \tag{8.2.57}$$

In this case Eq. (8.2.40) is replaced by

$$\big[\,(\mathrm{i})\,J_2^{i\,*}(k)\,\underline{\mathrm{e}}^{i}_{\lambda}\,\big]\Big(-\,\underline{\mathbf{e}}_{\lambda}\cdot\overline{\mathbf{e}}_{\lambda'} + 2\,\underline{\mathrm{e}}^{3}_{\lambda}\overline{\mathrm{e}}^{3}_{\lambda'}\Big)\big[\,\overline{\mathrm{e}}^{j}_{\lambda'}\,(\mathrm{i})\,J_1^{j}(k')\,\big]. \tag{8.2.58}$$

with the amplitude for scattering by reflection given by

$$-\,\underline{\mathbf{e}}_{\lambda}\cdot\overline{\mathbf{e}}_{\lambda'} + 2\,\underline{\mathrm{e}}^{3}_{\lambda}\overline{\mathrm{e}}^{3}_{\lambda'}. \tag{8.2.59}$$

From Eq. (8.2.56),

$$-\,\underline{\mathbf{e}}_{\lambda}\cdot\overline{\mathbf{e}}_{\lambda'} + 2\,\underline{\mathrm{e}}^{3}_{\lambda}\overline{\mathrm{e}}^{3}_{\lambda'} = 0, \quad \text{for } \lambda \neq \lambda', \tag{8.2.60}$$

as before.

On the other hand for $\lambda = \lambda'$, with *no* sum over $\lambda$,

$$\begin{aligned} -\,\underline{\mathbf{e}}_{\lambda}\cdot\overline{\mathbf{e}}_{\lambda} + 2\,\underline{\mathrm{e}}^{3}_{\lambda}\,\overline{\mathrm{e}}^{3}_{\lambda} &= \underline{\mathrm{e}}^{3}_{\lambda}\,\overline{\mathrm{e}}^{3}_{\lambda} - \sum_{a=1,2}\underline{\mathrm{e}}^{a}_{\lambda}\,\overline{\mathrm{e}}^{a}_{\lambda} \\ &= -\,\underline{\mathrm{e}}^{3}_{\lambda}\,\underline{\mathrm{e}}^{3}_{\lambda} - \sum_{a=1,2}\underline{\mathrm{e}}^{a}_{\lambda}\,\underline{\mathrm{e}}^{a}_{\lambda} = -\,\underline{\mathbf{e}}_{\lambda}\cdot\underline{\mathbf{e}}_{\lambda} = -\,1, \end{aligned} \tag{8.2.61}$$

for a *fixed* $\lambda$, using Eq. (8.2.57).

That is, the following scattering processes occur:

$$\big(\mathbf{K}, -q, \overline{\mathbf{e}}_1\big) \rightarrow \big(\mathbf{K}, q, \underline{\mathbf{e}}_1\big), \tag{8.2.62}$$

$$\big(\mathbf{K}, -q, \overline{\mathbf{e}}_2\big) \rightarrow \big(\mathbf{K}, q, \underline{\mathbf{e}}_2\big). \tag{8.2.63}$$

These processes are depicted in Fig. 8.2 working in the incident plane.

By a time evolutionary process, and with no recourse to plane waves, the propagator formalism shows, as we integrate over the energy variable for given fixed momenta, that the photon retains its energy $|\mathbf{k}'| = |\mathbf{k}|$, with processes occurring as described above. We learn, that with the currents $J_1$, $J_2$, causally arranged, respectively, as cites of emissions and detections mimicking the experimental situation, the scattering process may be formulated in terms of momenta and polarizations, and no plane waves, with the propagator involving a summation over phase factors in Eq. (8.2.23), as we integrate over the energy, from the time of emission by $J_1$ to the time of detection by $J_2$.

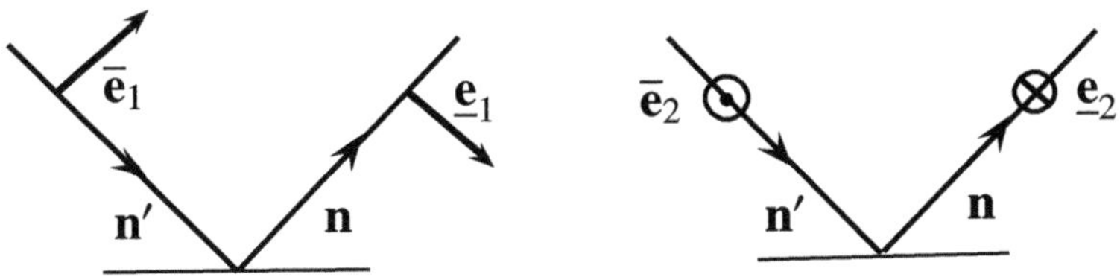


**Fig. 8.2** From left to right, depicting the reflection processes $(\mathbf{K}, -q, \bar{\mathbf{e}}_1) \to (\mathbf{K}, q, \underline{\mathbf{e}}_1)$, with $\bar{\mathbf{e}}_1$ and $\underline{\mathbf{e}}_1$ in the plane of incidence, and $(\mathbf{K}, -q, \bar{\mathbf{e}}_2) \to (\mathbf{K}, q, \underline{\mathbf{e}}_2)$, with $\bar{\mathbf{e}}_2$ and $\underline{\mathbf{e}}_2$ normal to it, respectively. The cross indicates a vector that points into the page, while a dot of a vector that comes out of it. Needless to day, the same situation arises if the directions of *all* the polarization vectors are simultaneously reversed in the figure. Remember that if light of a given momentum, say, $\mathbf{p}$, is polarized along a direction specified by a polarization vector, say, $\mathbf{e}$, then the direction of the magnetic field $\mathbf{B}$ is given by the relation $\mathbf{e} \times \mathbf{B} \propto \mathbf{p}$

## 8.3 Polarization of Light by a Polarizer

We set up the polarizer to constitute the $x^2 - x^3$ plane, such that it polarizes light along the $z = x^3-$ axis, that is, it rejects its $x^2$ component at $x^1 = 0$. Working in the temporal gauge, and using the notation $x = (t, x^1, x^2, z)$, $\mathbf{x}_{\|} = (x^2, x^3)$, the appropriate boundary condition to be applied is then:

$$\partial^0 D_+^{2j}(x, x')\Big|_{x^1=+0} = 0, \quad \Rightarrow \quad D_+^{2j}(x, x')\Big|_{x^1=+0} = 0, \tag{8.3.1}$$

where this boundary condition simply means that the polarization along the $x^2-$ axis has been rejected by the polarizer. This is supplemented by a continuity condition

$$D_+^{aj}(x, x')\Big|_{x^1=-0} = D_+^{aj}(x, x')\Big|_{x^1=+0}, \quad \text{for} \quad a = 1, 3. \tag{8.3.2}$$

This defines our polarizer.

According to Eqs. (5.1.29)/(5.1.34), in Chap. 5, $D_+^{ij}(x, x')$ satisfies the following partial differential equations:

$$\big[(-\boldsymbol{\partial}^2 + \partial^{02})\delta^{i\ell} + \partial^i\partial^\ell\big]D_+^{\ell j}(x, x') = \delta^{ij}\,\delta^{(4)}(x, x'), \tag{8.3.3}$$

$$\big[(-\boldsymbol{\partial}'^2 + \partial'^{02})\delta^{\ell i} + \partial'^\ell\partial'^i\big]D_+^{j\ell}(x, x') = \delta^{ij}\,\delta^{(4)}(x, x'), \tag{8.3.4}$$

where $\delta^{(4)}(x, x') = \delta(x^0 - x'^0)\delta(x^1, x'^1)\delta^{(2)}(\mathbf{x}_{\|} - \mathbf{x}'_{\|})$, and translational invariance is broken along, and only along, the $x^1-$ axis (see Fig. 8.3), where $\mathbf{x}_{\|}$ is parallel to the surface of the polarizer.

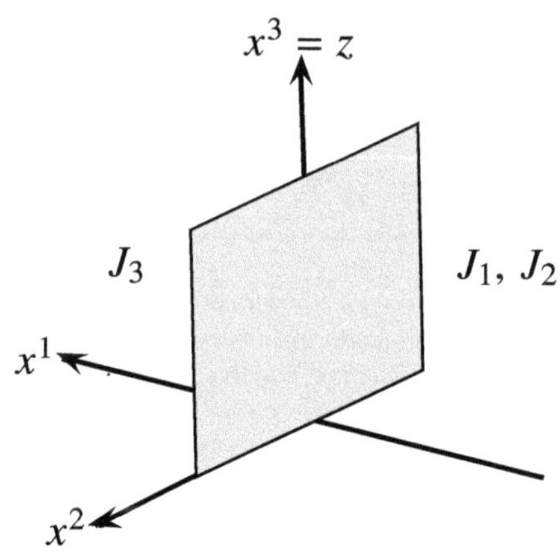


**Fig. 8.3** A Polarizer that polarizes light in the $z$ direction, i.e., it kills the component along the $x^2$– axis. $J_1$ is identified as an emitter of a photon, while $J_2$, $J_3$ as detectors of a reflected one and a transmitted one, respectively. $J_2$, $J_3$ are switched on after $J_1$ is switched off, i.e., removed

In the experimental situation, three things arise. A photon is first emitted which, in turn, may be reflected, i.e., wouldn't go through the polarizer, or it may go through it polarized. That is, we need three current sources: one emitter and two detectors,

$$J(x) = J_1(x) + J_2(x) + J_3(x), \tag{8.3.5}$$

causally arranged, with $J_1$, identified as an emitter, while $J_2$, $J_3$ as detectors. In the regions of 3-space in which $J_3$, $J_2$ reside, $x^1 > 0$, $x^1 < 0$ respectively. That is, $J_2$ is identified as a detector of a reflected photon, while $J_3$ as a detector of a transmitted one. $J_2$, $J_3$ are switched on after $J_1$ is switched off, i.e., removed.

We have, in turn, to consider the propagator $D_+^{ij}(x, x')$ as it emerges from region $x^1 < 0$ into the region $x^1 > 0$:

**The $x^1 < 0$ Region**:

The particular solution of Eq. (8.3.3) is the one you are familiar with, by now, for $t > t'$. Here it is denoted by $D_0^{ij}(x, x')$, and is conveniently written in the form

$$D_0^{ij}(x, x') = \mathrm{i}\int \frac{\mathrm{d}^2\mathbf{K}}{(2\pi)^2}\int \frac{\mathrm{d}q'}{2\pi}\, \frac{\mathrm{e}^{\mathrm{i}\mathbf{K}\cdot(\mathbf{x}_\parallel - \mathbf{x}'_\parallel)}\mathrm{e}^{\mathrm{i}q'(x^1 - x'^1)}}{2\,|\mathbf{k}'|}\mathrm{e}^{-\mathrm{i}|\mathbf{k}'|(t-t')}\Big(\delta^{ij} - \frac{k'^i k'^j}{|\mathbf{k}'|^2}\Big), \tag{8.3.6}$$

$$\text{where}\quad \mathbf{k}' = (q', \mathbf{K}), \quad |\mathbf{k}'| = \sqrt{q'^2 + \mathbf{K}^2}. \tag{8.3.7}$$

where $\mathbf{x}_\parallel$, $\mathbf{x}'_\parallel$ are parallel to the polarizer. This is the *particular* solution. It doesn't include the reflection part. The latter may be obtained from the *homogeneous* solution, here denoted by $\underline{D}^{ij}(x, x')$, which may be written as:

$$\underline{D}^{ij}(x,x') = \mathrm{i}\int \frac{\mathrm{d}^2\mathbf{K}}{(2\pi)^2}\int\frac{\mathrm{d}q\mathrm{d}q'}{2\pi}\int\frac{\mathrm{d}k^0}{2\pi}\,\mathrm{e}^{\mathrm{i}\mathbf{K}\cdot(\mathbf{x}_\parallel-\mathbf{x}'_\parallel)}\mathrm{e}^{\mathrm{i}(qx^1-q'x'^1)} \times \mathrm{e}^{-\mathrm{i}k^0(t-t')}\underline{G}^{ij}(q,q',\mathbf{K},k^0), \tag{8.3.8}$$

with translational invariance broken only along the $x^1-$ axis.

$\underline{D}^{ij}(x,x')$ must satisfy the homogeneous equations

$$\big[(-\boldsymbol{\partial}^2+\partial^{02})\delta^{i\ell}+\partial^i\partial^\ell\big]\underline{D}^{\ell j}(x,x') = 0. \tag{8.3.9}$$

$$\big[(-\boldsymbol{\partial}'^2+\partial'^{02})\delta^{\ell i}+\partial'^\ell\partial'^i\big]\underline{D}^{j\ell}(x,x') = 0. \tag{8.3.10}$$

With $\mathbf{k}=(q,\mathbf{K}),\mathbf{k}'=(q',\mathbf{K})$, Eq. (8.3.9) gives

$$(|\mathbf{k}|^2-k^{02})\underline{G}^{ij}-k^ik^\ell\underline{G}^{\ell j} = 0 \tag{8.3.11}$$

which upon multiplying by $k^i$, and summing over $i$, give, for $k^0\neq 0$,

$$k^{02}k^i\underline{G}^{\ell j} = 0, \;\Rightarrow\; k^i\underline{G}^{ij} = 0, \;\text{ and }\; (|\mathbf{k}|^2-k^{02})\underline{G}^{ij} = 0, \tag{8.3.12}$$

also note that $\underline{G}^{ij}k'^j=0$ as well, The last relation in above equation, for $k^0>0$, gives

$$(|\mathbf{k}|-k^0)\underline{G}^{ij} = 0, \;\Rightarrow\; \underline{G}^{ij} = \delta(k^0-|\mathbf{k}|)G^{ij}, \tag{8.3.13}$$

using the last relation of a property of the delta function in Eq. (3.6.11), Sect. 3.6, Chap. 3, where $G^{ij}$ is to be determined.

On the other hand, upon substituting Eq. (8.3.8) in Eq. (8.3.10) gives

$$(|\mathbf{k}'|^2-k^{02})\underline{G}^{ij} = 0 \;\Rightarrow\; (q'^2-q^2)\underline{G}^{ij} = 0. \tag{8.3.14}$$

using, in the process, the fact that $|\mathbf{k}|^2-k^{02}=0$, from Eq. (8.3.13). This provides the solution $q=\pm q'$. But if we take $q'>0$ in this region, $q$ must be taken to be negative to describe reflection. Accordingly, from Eq. (8.3.13) we may write

$$\underline{G}^{ij}(q,q',\mathbf{K},k^0) = \frac{\pi}{|\mathbf{k}|}\delta(q+q')\delta(k^0-|\mathbf{k}|)G^{ij}_<(q',\mathbf{K}), \tag{8.3.15}$$

where $G^{ij}_<(q',\mathbf{K})$ is to be determined. Note that with $\mathbf{k}'=(q',\mathbf{K}),\mathbf{k}=(q,\mathbf{K})$, that $|\mathbf{k}'|=|\mathbf{k}|$.

Therefore, the solution in this region is given by

$$D_<^{ij}(x,x') = D_0^{ij} + \mathrm{i}\int \frac{\mathrm{d}^2\mathbf{K}}{(2\pi)^2}\int_{-\infty}^{\infty} \frac{\mathrm{d}q\,\mathrm{d}q'}{2\pi}\delta(q+q')\,\frac{\mathrm{e}^{\mathrm{i}\mathbf{K}\cdot(\mathbf{x}_\parallel-\mathbf{x}'_\parallel)}\,\mathrm{e}^{-\mathrm{i}|\mathbf{k}|(t-t')}}{2\,|\mathbf{k}|}\times$$
$$\times\ \mathrm{e}^{\mathrm{i}\,(qx^1-q'x'^1)}G_<^{ij}(q',\mathbf{K}), \tag{8.3.16}$$

with $k^iG_<^{ij}(q',\mathbf{K})=0,\quad k'^jG_<^{ij}(q',\mathbf{K})=0,\quad \mathbf{k}=(-q',\mathbf{K}),\quad \mathbf{k}'=(q',\mathbf{K}).$ (8.3.17)

Through the chosen boundary conditions, the analysis leads to a reflection of a photon which was not transmitted through the polarizer.

**The $x^1 > 0$ Region:**

The solution is immediately obtained to have the general structure:

$$D_>^{ij}(x,x') = \mathrm{i}\int \frac{\mathrm{d}^2\mathbf{K}}{(2\pi)^2}\int_{-\infty}^{\infty} \frac{\mathrm{d}q''}{2\pi}\,\frac{\mathrm{e}^{\mathrm{i}\mathbf{K}\cdot(\mathbf{x}_\parallel-\mathbf{x}'_\parallel)}\mathrm{e}^{\mathrm{i}q''(x^1-x'^1)}}{2\,|\mathbf{k}|''}\mathrm{e}^{-\mathrm{i}|\mathbf{k}''|(t-t')}G_>^{ij}(q'',\mathbf{K}), \tag{8.3.18}$$

$$\text{and}\quad \partial^i D_>^{ij}(x,x') = 0,\quad \partial'^j D_>^{ij}(x,x') = 0.\quad t>t'. \tag{8.3.19}$$

We have to satisfy the boundary conditions. To do this, we now consider, the physical situation involving three basic vectors: $\mathbf{k}'=(\mathbf{K},q')$, $\mathbf{k}=(\mathbf{K},-q')$, $\mathbf{k}''=(\mathbf{K},q')$, corresponding, respectively, to incidence, reflection, and transmission. That is, in turn, we consider $q'>0$, implying, from Eq. (8.3.15), $q<0$. Therefore, the three basic vectors, in question, to consider are:

$$\mathbf{k}' = (|q|,\mathbf{K}),\quad \mathbf{k} = (-|q|,\mathbf{K}),\quad \mathbf{k}'' = (|q|,\mathbf{K}),\quad |\mathbf{k}'| = |\mathbf{k}| = |\mathbf{k}''|. \tag{8.3.20}$$

The boundary condition in Eq. (8.3.1), and the continuity condition in (8.3.2) may be now combined, from the integrands of the integrals in Eqs. (8.3.8), (8.3.16), (8.3.18), at $x^1=0$, to read

$$\left(\delta^{ij} - \frac{k'^ik'^j}{|\mathbf{k}'|^2}\right) + G_<^{ij}(|q|,\mathbf{K}) = G_>^{ij}(|q|,\mathbf{K}), \tag{8.3.21}$$

and

$$G_>^{2j} = 0, \tag{8.3.22}$$

In addition to these relations, $G^{ij}_{<}$, and $G^{ij}_{>}$, from Eqs. (8.3.17), (8.3.19), also satisfy the following equations

$$k^i G^{ij}_{<} = 0, \quad G^{ij}_{<} k'^j = 0, \qquad k''^i G^{ij}_{>} = 0, \quad G^{ij}_{>} k'^j = 0, \tag{8.3.23}$$

where from Eq. (8.3.20), $\mathbf{k} = (-|q|, \mathbf{K})$, $\mathbf{k}'' = (|q|, \mathbf{K}) = \mathbf{k}'$.

From Eqs. (8.3.21), (8.3.22)

$$G^{2j}_{<} = -\Big(\delta^{2j} - \frac{K^2 k'^j}{|\mathbf{k}|^2}\Big). \tag{8.3.24}$$

While the transversality conditions in Eq. (8.3.17), with $q < 0, q' = |q|$, give, respectively, at the boundary

$$-|q| G^{1j}_{<} + K^2 G^{2j}_{<} + K^3 G^{3j}_{<} = 0, \quad |q| G^{1j}_{<} + K^2 G^{2j}_{<} + K^3 G^{3j}_{<} = 0, \tag{8.3.25}$$

$$\Rightarrow \quad G^{1j}_{<} = 0, \tag{8.3.26}$$

and from Eq. (8.3.21), with $i = 1$, we obtain

$$G^{1j}_{>} = \Big(\delta^{1j} - \frac{|q| k'^j}{|\mathbf{k}|^2}\Big). \tag{8.3.27}$$

On the other hand, the transversality condition in Eq. (8.3.19), using, in the process, that $G^{2j}_{>} = 0$, gives

$$|q|\, G^{1j}_{>} + K^3 G^{3j}_{>} = 0, \quad \Rightarrow \qquad G^{3j}_{>} = -\frac{|q|}{K^3}\Big(\delta^{1j} - \frac{|q| k'^j}{|\mathbf{k}|^2}\Big). \tag{8.3.28}$$

Finally, from Eq. (8.3.21) for $i = 3$, and Eq. (8.3.28), we have

$$G^{3j}_{<} = -\frac{|q|}{K^3}\Big(\delta^{1j} - \frac{|q| k'^j}{|\mathbf{k}|^2}\Big) - \Big(\delta^{3j} - \frac{K^3 k'^j}{|\mathbf{k}|^2}\Big). \tag{8.3.29}$$

Therefore, all the components for the reflection and transmission processes are, respectively, given by

$$G^{1j}_{<} = 0, \; G^{2j}_{<} = -\Big(\delta^{2j} - \frac{K^2 k'^j}{|\mathbf{k}|^2}\Big), \; G^{3j}_{<} = -\frac{|q|}{K^3}\Big(\delta^{1j} - \frac{|q| k'^j}{|\mathbf{k}|^2}\Big) - \Big(\delta^{3j} - \frac{K^3 k'^j}{|\mathbf{k}|^2}\Big), \tag{8.3.30}$$

$$G^{1j}_{>} = \Big(\delta^{1j} - \frac{|q| k'^j}{|\mathbf{k}|^2}\Big), \qquad G^{2j}_{>} = 0, \quad G^{3j}_{>} = -\frac{|q|}{K^3}\Big(\delta^{1j} - \frac{|q| k'^j}{|\mathbf{k}|^2}\Big). \tag{8.3.31}$$

To describe the scattering processes we consider an incident and reflected photon with momenta $\mathbf{k}' = (|q|, 0, K^3)$, and $\mathbf{k} = (-|q|, 0, K^3)$, respectively. Note that

$|\mathbf{k}'| = |\mathbf{k}|$. Given the incidence vector $\mathbf{k}'$, we may introduce two polarization vectors $\mathbf{e}'_1$, $\mathbf{e}'_2$, and unit vectors $\mathbf{n}'$, $\mathbf{n}$ satisfying the basic conditions of polarization vectors,

$$\mathbf{n}' = \frac{\mathbf{k}'}{|\mathbf{k}|}, \ \mathbf{n} = \frac{\mathbf{k}}{|\mathbf{k}|}, \qquad \mathbf{k}' \cdot \mathbf{e}'_1 = 0, \ \mathbf{k}' \cdot \mathbf{e}'_2 = 0, \ \mathbf{e}'_1 \cdot \mathbf{e}'_2 = 0, \ \mathbf{e}'_1 \times \mathbf{e}'_2 = \mathbf{n}', \tag{8.3.32}$$

given by

$$\mathbf{e}'_1 = \frac{\mathbf{n}' \times \mathbf{n}}{|\mathbf{n}' \times \mathbf{n}|} = (0, -1, 0), \quad \mathbf{e}'_2 = \frac{\mathbf{n}' \times (\mathbf{n}' \times \mathbf{n})}{|\mathbf{n}' \times (\mathbf{n}' \times \mathbf{n})|} = (K^3, 0, -|q|)\frac{1}{|\mathbf{k}|}, \tag{8.3.33}$$

$$\mathbf{e}'_1 \times \mathbf{e}'_2 = (|q|, 0, K^3)\frac{1}{|\mathbf{k}|} \equiv \mathbf{n}', \tag{8.3.34}$$

and by now the familiar completeness relation,

$$\delta^{ij} - \frac{k'^i k'^j}{|\mathbf{k}'|^2} = \mathrm{e}'^i_\lambda \mathrm{e}'^j_\lambda, \tag{8.3.35}$$

with a summation over $\lambda = 1, 2$, understood.

Similarly for the reflection vector $\mathbf{k} = (-|q|, 0, K^3)$, we have polarization vectors

$$\mathbf{e}_1 = \frac{\mathbf{n} \times \mathbf{n}'}{|\mathbf{n}' \times \mathbf{n}'|} = (0, 1, 0), \quad \mathbf{e}_2 = \frac{\mathbf{n} \times (\mathbf{n} \times \mathbf{n}')}{|\mathbf{n} \times (\mathbf{n} \times \mathbf{n}')|} = -(K^3, 0, |q|)\frac{1}{|\mathbf{k}|}, \tag{8.3.36}$$

$$\mathbf{e}_1 \times \mathbf{e}_2 = (|q|, 0, K^3)\frac{1}{|\mathbf{k}|} \equiv \mathbf{n}, \tag{8.3.37}$$

with corresponding completeness relations as in Eq. (8.3.35).

Finally, for the vector $\mathbf{k}'' = (|q|, 0, K^3)$, one may simply introduce polarization vectors as in Eq. (8.3.33),

$$\mathbf{e}''_1 = \mathbf{e}'_1, \qquad \mathbf{e}''_2 = \mathbf{e}'_2. \tag{8.3.38}$$

with corresponding completeness relations as in Eq. (8.3.35).

The challenge, now, is to see how the emerging photon is polarized along the $z-$ axis.

To the above end, we note From Eq. (8.3.31) that

$$G^{ij}_> = \delta^{i1} G^{1j}_> + \delta^{i3} G^{3j}_> \tag{8.3.39}$$

$$= \Big(\delta^{i1} - \frac{|q|}{K^3}\delta^{i3}\Big)\Big(\delta^{1j} - \frac{|q| k'^j}{|\mathbf{k}|^2}\Big). \tag{8.3.40}$$

The transversality conditions are readily verified

$$k''^i\, G^{ij}_> = 0, \quad G^{ij}_>\, k'^j = 0, \quad \mathbf{k}'' = (|q|, \mathbf{K}) = \mathbf{k}'. \tag{8.3.41}$$

The completeness relations, as in Eq. (8.3.35), should be noted

$$\delta^{ij} = \frac{k''^i k''^j}{|\mathbf{k}''|^2} + \mathrm{e}''^i_{\lambda}\mathrm{e}''^j_{\lambda}, \quad \delta^{ij} = \frac{k^i k^j}{|\mathbf{k}|^2} + \mathrm{e}^i_{\lambda}\mathrm{e}^j_{\lambda}, \tag{8.3.42}$$

where summations over $\lambda = 1, 2$ are understood.

The transmission amplitude is obtained explicitly from Eq. (8.3.5), and, e.g., from Eq.(8.2.29), with $J_1$ and $J_3$ denoting emission and detection sources, as discussed above, to be given by

$$\mathrm{i}\, J^i_3(|q|, \mathbf{K})\, G^{ij}_>\, \mathrm{i}\, J^j_1(|q|, \mathbf{K}) \equiv \mathrm{i}\, J^i_3(|q|, \mathbf{K})\big[\delta^{i\ell}\big]\, G^{\ell m}_>\big[\delta^{mj}\big]\, \mathrm{i}\, J^j_1(|q|, \mathbf{K}) \tag{8.3.43}$$

$$= \big(\mathrm{i}\, J^i_3(|q|, \mathbf{K})\, \mathrm{e}''^i_{\alpha}\big)\Big[\Big(\mathrm{e}''^1_{\alpha} - \frac{|q|}{K^3}\mathrm{e}''^3_{\alpha}\Big)\, \mathrm{e}'^1_{\lambda}\Big]\, \big(\mathrm{e}'^j_{\lambda}\, \mathrm{i}\, J^j_1(|q|\mathbf{K})\big), \tag{8.3.44}$$

where in the second line, we have used the completeness relations in Eq. (8.3.42), the transversality conditions in Eq.(8.3.41) to eliminate the $k''^i k''^\ell$, $k'^m k'^j$ terms, and used the explicit expression of $G^{ij}_>$ in Eq. (8.3.40).

From this, we may extract the amplitude of transmission of a photon of momentum/polarization $(|q|, \mathbf{K}, \mathbf{e}'_{\lambda})$ to momentum/polarization $(|q|, \mathbf{K}, \mathbf{e}_{\alpha})$ to be given by:

$$\big(|q|, \mathbf{K};\, \mathbf{e}''_{\alpha}\big|\, |q|, \mathbf{K};\ \mathbf{e}'_{\lambda}\big) = \Big(\mathrm{e}''^1_{\alpha} - \frac{|q|}{K^3}\mathrm{e}''^3_{\alpha}\Big)\, \mathrm{e}'^1_{\lambda}. \tag{8.3.45}$$

Given the polarizations in Eq. (8.3.33), associated with an incident photon, we are interested in determining the amplitude that it is transmitted through the polarizer with specific polarization states using Eq. (8.3.45).

In reference to Eq. (8.3.45), we have, from Eqs. (8.3.33), (8.3.38), that

$$\mathrm{e}'^1_{\lambda} = \begin{cases} 0, & \text{if } \lambda = 1 \\ K^3/|\mathbf{k}|, & \text{if } \lambda = 2 \end{cases}, \tag{8.3.46}$$

$$\Big(\mathrm{e}''^1_{\alpha} - \frac{q}{K^3}\mathrm{e}''^3_{\alpha}\Big) = \begin{cases} 0, & \text{if } \alpha = 1 \\ |\mathbf{k}|/K^3, & \text{if } \alpha = 2 \end{cases}, \tag{8.3.47}$$

and hence,

$$\big(|q|, \mathbf{K};\, \mathbf{e}''_{\alpha}\big|\, |q|, \mathbf{K};\, \mathbf{e}'_{\lambda}\big) = \begin{cases} 1, & \text{if } \lambda = 2,\ \alpha = 2 \\ 0, & \text{if } \lambda = 1,\ \alpha = 1, \text{ or } 2 \end{cases}. \tag{8.3.48}$$

That is, a photon with polarization $\mathbf{e}_1 = (0, -1, 0)$, having a component along the $x^2 = y-$ axis, cannot get through the polarizer, as expected, since such a polarizer is to eliminate this component. Where does it go ?. It was rejected by the polarizer. It must have been reflected. We should explicitly verify this, in turn. Only the photon with the polarization $\mathbf{e}_2$ emerges from the polarizer with 100 % probability. That is, the photon had no option where to go, it was transmitted.

For the reflection process, we may, in general write,

$$G^{ij}_{<} = \delta^{i\ell} G^{\ell m}_{<} \delta^{mj} = \Big(\frac{k^i k^\ell}{|\mathbf{k}|^2} + \mathrm{e}^i_\alpha\, \mathrm{e}^\ell_\alpha\Big) G^{\ell m}_{<} \Big(\frac{k'^m k'^j}{|\mathbf{k}|^2} + \mathrm{e}'^m_\lambda\, \mathrm{e}'^j_\lambda\Big) \tag{8.3.49}$$

$$= \mathrm{e}^i_\alpha \Big[\mathrm{e}^\ell_\alpha\, G^{\ell m}_{<}\, \mathrm{e}'^m_\lambda\Big] \mathrm{e}'^j_\lambda, \tag{8.3.50}$$

where we have, in the process, used Eq. (8.3.17) to set $k^\ell G^{\ell m}_{<} = 0,\ k'^m G^{\ell m}_{<} = 0$.

Accordingly, the reflection process is described by

$$\Big(\mathrm{i}\, J^{i*}(-|q|, \mathbf{K}) \mathrm{e}^i_\alpha\Big) \Big[\mathrm{e}^\ell_\alpha\, G^{\ell m}_{<}\, \mathrm{e}'^m_\lambda\Big] \Big(\mathrm{e}'^j_\lambda\, \mathrm{i}\, J^j(|q|, \mathbf{K})\Big), \tag{8.3.51}$$

where the amplitude of reflection is given by

$$\big( - |q|, \mathbf{K}; \mathbf{e}_\alpha \big| |q|, \mathbf{K}; \mathbf{e}'_\lambda \big) = \mathrm{e}^\ell_\alpha\, G^{\ell m}_{<}\, \mathrm{e}'^m_\lambda. \tag{8.3.52}$$

To evaluate this expression, we note from Eq. (8.3.30) that

$$G^{\ell m}_{<} = \delta^{\ell 2} G^{2m}_{<} + \delta^{\ell 3} G^{3m}_{<} \tag{8.3.53}$$

$$= -\delta^{\ell 2}\Big(\delta^{2m} - \frac{K^2 k'^m}{|\mathbf{k}|^2}\Big) + \delta^{\ell 3}\Big[\Big(-\frac{|q|}{K^3}\Big)\Big(\delta^{1m} - \frac{|q| k'^m}{|\mathbf{k}|^2}\Big) - \Big(\delta^{3m} - \frac{K^3 k'^m}{|\mathbf{k}|^2}\Big)\Big]. \tag{8.3.54}$$

Moreover $k'^m \mathrm{e}'^m_\lambda = 0$. Therefore, the amplitude for the reflection simply becomes

$$\begin{aligned} \mathrm{e}^\ell_\alpha\, G^{\ell m}_{<}\, \mathrm{e}'^m_\lambda &= -\,\mathrm{e}^2_\alpha\, \mathrm{e}'^2_\lambda + \mathrm{e}^3_\alpha\, \mathrm{e}'^1_\lambda \Big(-\frac{|q|}{K^3}\Big) - \mathrm{e}^3_{\lambda'}\, \mathrm{e}^3_\lambda, \\ &= -\,\mathrm{e}^2_\alpha\, \mathrm{e}'^2_\lambda - \mathrm{e}^3_\alpha\Big(\Big(\frac{|q|}{K^3}\Big)\mathrm{e}'^1_\lambda + \mathrm{e}'^3_\lambda\Big) = -\,\mathrm{e}^2_\alpha\, \mathrm{e}'^2_\lambda, \end{aligned} \tag{8.3.55}$$

where the very last equality follows by considering the first and third components of each of the polarization vectors $\mathbf{e}'_1$ and $\mathbf{e}'_2$ in Eq. (8.3.33).

From Eqs. (8.3.33), (8.3.36), we have

$$\mathrm{e}_{\lambda}^{\prime 2} = \begin{cases} -1, & \text{if } \lambda = 1 \\ 0, & \text{if } \lambda = 2 \end{cases}, \tag{8.3.56}$$

$$\mathrm{e}_{\alpha}^{2} = \begin{cases} 1, & \text{if } \alpha = 1 \\ 0, & \text{if } \alpha = 2 \end{cases}, \tag{8.3.57}$$

That is, from Eqs. (8.3.55)–(8.3.57), the amplitude of reflection is given by

$$\big( -|q|, \mathbf{K}; , \mathbf{e}_{\alpha} \big| \, |q|, \mathbf{K}; \mathbf{e}'_{\lambda} \big) = \begin{cases} 1, & \text{if } \lambda = 1, \alpha = 1 \\ 0, & \text{if } \lambda = 2, \alpha = 1 \text{ or } 2 \end{cases}. \tag{8.3.58}$$

Equation (8.3.58) tells us that we have hove recovered the photon with polarization $\mathbf{e}_1 = (0, 1, 0)$, which was rejected by the polarizer, it has been reflected with 100 % probability.

What about an unpolarized photon. Here, we have to average the probabilities over the initial polarization, not having information of the state of the photon, giving probabilities of 1/2 transmission and 1/2 of reflection.

Let us now treat a more general case in which a photon has an option where to go and we cannot tell how a photon makes up it mind. Consider an incident photon of momentum $\mathbf{k}' = (|q|, 0, K^3)$, impinging on our polarizer, theoretically described above, but now in either of the following polarization states:

$$\mathbf{e}'_1 = (0, -1, 0) \cos\theta + (K^3, 0, -|q|) \frac{\sin\theta}{|\mathbf{k}|}, \tag{8.3.59}$$

$$\mathbf{e}'_2 = (0, 1, 0) \, \sin\theta + (K^3, 0, -|q|) \frac{\cos\theta}{|\mathbf{k}|}, \tag{8.3.60}$$

$$\mathbf{e}'_1 \cdot \mathbf{e}'_2 = 0, \;\; \mathbf{e}'_1 \times \mathbf{e}'_2 = (|q|, 0, K^3) \frac{1}{|\mathbf{k}|}, \;\; \mathbf{k}' \cdot \mathbf{e}'_1, = 0, \;\; \mathbf{k}' \cdot \mathbf{e}'_2, = 0, \tag{8.3.61}$$

where $\theta$ is arbitrary. And, inquire, about the fate of such a photon with transmission or reflection polarization possibilities given, respectively, by

$$\mathbf{e}''_1 = (0, -1, 0), \qquad \mathbf{e}''_2 = (K^3, 0, -|q|) \frac{1}{|\mathbf{k}|}, \tag{8.3.62}$$

$$\mathbf{e}_1 = (0, 1, 0), \qquad \mathbf{e}_2 = -(K^3, 0, |q|) \frac{1}{|\mathbf{k}|}, \tag{8.3.63}$$

as before.

Hence from Eqs. (8.3.45), (8.3.55), the amplitudes of transmissions and reflections read:

$$\big(|q|,\mathbf{K};\mathbf{e}''_{\alpha}\mid |q|,\mathbf{K};\mathbf{e}'_{\lambda}\big) = \Big(\mathrm{e}''^{1}_{\alpha} - \frac{|q|}{K^3}\mathrm{e}''^{3}_{\alpha}\Big)\,\mathrm{e}'^{1}_{\lambda}, \tag{8.3.64}$$

$$(-|q|,\mathbf{K};,\mathbf{e}_{\alpha}\mid |q|,\mathbf{K};\mathbf{e}'_{\lambda}) = -\,\mathrm{e}^{2}_{\alpha}\,\mathrm{e}'^{2}_{\lambda}. \tag{8.3.65}$$

with

$$\mathrm{e}'^{1}_{\lambda} = \begin{cases} K^3\sin\theta/|\mathbf{k}|, & \text{if}\,\lambda = 1\\ K^3\cos\theta/|\mathbf{k}|, & \text{if}\,\lambda = 2\end{cases}, \quad \Big(\mathrm{e}''^{1}_{\alpha} - \frac{q}{K^3}\mathrm{e}''^{3}_{\alpha}\Big) = \begin{cases} 0, & \text{if}\,\alpha = 1\\ |\mathbf{k}|/K^3, & \text{if}\,\alpha = 2\end{cases}, \tag{8.3.66}$$

and

$$\mathrm{e}'^{2}_{\lambda} = \begin{cases} -\cos\theta, & \text{if }\lambda = 1\\ \sin\theta, & \text{if }\lambda = 2\end{cases}, \quad \mathrm{e}^{2}_{\alpha} = \begin{cases} 1, & \text{if }\alpha = 1\\ 0, & \text{if }\alpha = 2\end{cases}, \tag{8.3.67}$$

Accordingly, from Eqs. (8.3.64)–(8.3.67), the amplitudes of transmissions and reflections, for an incident photon with polarizations $\mathbf{e}'_{\lambda}$ are given by

$$\mathbf{e}'_1 \to \begin{cases} \text{transmission amplitude} & \sin\theta, \text{ and polarization } \mathbf{e}''_2\\ \text{reflection amplitude} & \cos\theta, \text{ and polarization } \mathbf{e}_1\end{cases}, \tag{8.3.68}$$

and

$$\mathbf{e}'_2 \to \begin{cases} \text{transmission amplitude} & \cos\theta, \text{ and polarization } \mathbf{e}''_2\\ \text{reflection amplitude} & \sin\theta, \text{ and polarization } \mathbf{e}_1\end{cases}, \tag{8.3.69}$$

That is, a photon with polarization $\mathbf{e}'_1$ gets reflected with a probability of $\cos^2\theta$ and gets transmitted (polarized) with probability $\sin^2\theta$, while a photon with polarization $\mathbf{e}'_2$ gets reflected with a probability of $\sin^2\theta$ and gets transmitted (polarized) with probability $\cos^2\theta$, and probability is conserved for each photon with the given polarization. Most importantly the transmitted photon is polarized with polarization vector $\mathbf{e}''_2$, given in Eq.(8.3.62), involving no component along the $x^2 = y-$ axis. The polarizer has passed the test of a polarizer with flying colors. The answer to the question as how a photon makes up its mind is unknown. The probability laws of $\cos^2\theta/\sin^2\theta$, encountered here, are referred to as the *Malus law*.

It is interesting that the exploration of Feynman[3] of this mystery of the strange behavior of light and his statement as how does a photon decide where to go, has

[3] Feynman (1985) [1].

even inspired a playwright named Peter Parnell to generate a play at Lincoln Center Theater, NY in 2001-2002 called "QED".[4]

For an unpolarized photon we have to average the probabilities corresponding to the incident photon possible polarization states, for each case, giving $(\cos^2\theta + \sin^2\theta)/2$ leading to a probability of 1/2 of transmission and 1/2 of reflection, with transmitted always having no $x^2 = y$ component. The mystery of how a photon makes up its mind still remains.

This chapter is based on the references cited in the footnote below.[5]

## References

1. Feynman, R. P. (1985). *QED: The strange theory of light and matter*. Princeton, New Jersey: Princeton University Press.
2. Manoukian, E. B. (1987a). Reflection-off a reflecting surface in quantum mechanics. Where do the reflections actuall occur ?. *Il Nuovo Cimento B, 99 (2)*, 133–143.
3. Manoukian, E. B. (1987b). Reflection-off a reflecting surface in quantum mechanics. Where do the reflections actuall occur ? II. The law of reflection. *Il Nuovo Cimento B, 100 (2)*, 185–193.
4. Manoukian, E. B. (1992). Quantum field theory of reflection of light. *Hadronic Journal,20*, 253–265.
5. Manoukian, E. B. (1996). Quantum field theory of polarization of light by a polarizer. *International Journal of Theoretical Physics,35 (8)*, 1735–1744.
6. Manoukian, E. B. (2006). *Quantum theory: A Wide Spectrum*. Dordrecht: Springer.
7. Parnell, P. (2001). QED. *Play at Lincoln Center Theater*, October 21, 2001-June 10, 2002.

[4] Parnell (2001) [7].

[5] Manoukian (1992, 1996) [4, 5]. See also Manoukian (2006), Sect. 1.6 [6], as well as Manoukian (1987a,1987b) [2, 3] for further gain and profit.

# Chapter 9
# Propagators: Scattering in Different Media

*Follow a photon when it encounters a different medium during its entire history from its emission to its detection*

The present chapter addresses the classic problem of refraction and reflection off dielectric media as well as off magnetic ones of permeability different from one. The propagators are explicitly solved in a time-dependent evolutionary process as physically imposed by the causal arrangements of an emitter and a detector, avoiding plane waves, in the process. This provides a rigorous treatment, via time-dependent propagators, of adding up amplitudes with time dependent phases, labeled by the initial time a photon is emitted and the final time when a photon is detected. It is a reminder of Feynman's[1] fascinating approach of adding up amplitudes with time-dependent phase factors during the whole process of scattering, which is what a propagator $D_+(t\mathbf{x}, t'\mathbf{x}')$ does from the initial time $t'$ of emission to the final time $t$ of detection. Via the propagator, detailed information is also extracted on polarization aspects of the scattering process. This chapter is based on references to the author cited below.[2] This present method of describing *scattering* processes, via the propagator, in classical electrodynamics, was given in the reference cited below.[3]

---

[1] Feynman (1985) [6].

[2] Manoukian (1996, 1997) [4, 5].

[3] Manoukian (1992) [3]. See also Manoukian (1987a, b) [1, 2] for further gain and profit.



E. B. Manoukian, *Radiation*,
https://doi.org/10.1007/978-3-032-21506-2_9

## 9.1 Refraction and Reflection for a Dielectric Medium in Propagator Theory

The process of scattering is described by introducing three current sources in two semi-infinite regions: One emitter $J_1$ in the free-space region, and two detectors $J_2$, $J_3$, respectively, with one in the free-space region and one in the medium of dielectric constant $\varepsilon$:

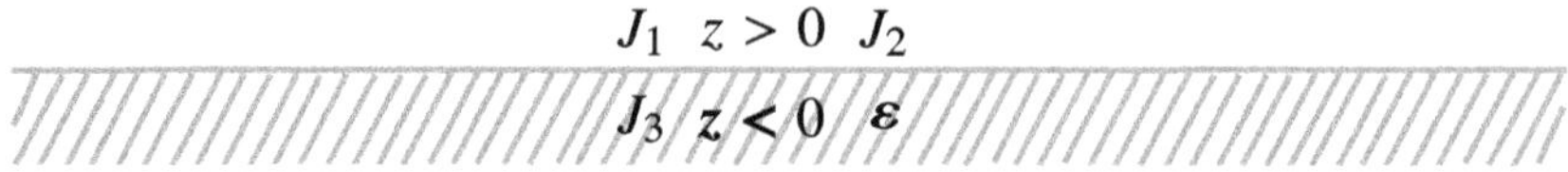


$$J(x) = J_1(x) + J_2(x) + J_3(x), \tag{9.1.1}$$

where $J_2(x)$, $J_3(x)$ are switched on after $J_1(x)$ is switched off.

We may conveniently define:

$$\varepsilon(z) = \begin{cases} 1, & z > 0 \\ \varepsilon, & z < 0 \end{cases}. \tag{9.1.2}$$

In reference to the propagator $D_+(x, x')$, as the emitter lies in the $z > 0$ region, we always have $z' > 0$, and $t > t'$. Here, e.g., $x' = (t', \mathbf{x}'_{\|}, z')$, and $\mathbf{x}'_{\|}$ denotes the component of $\mathbf{x}'$ parallel to intervening surface.

We first write a general expression for the propagator $D_+(x, x')$ in the region $z > 0$, i.e., before it crosses the intervening surface, and then write a general expression for it after it crosses the intervening surface, i.e., for $0 > z$. We than apply the appropriate boundary conditions at $z = 0$, by comparing the two expressions just mentioned, respectively, at $z = +0$ and $z = -0$, to describe the theory of reflection off, and transmission into a dielectric medium.

**The Reflection Region**: $z > 0$, $z' > 0$

Here $\varepsilon(z) = 1$. According, to Eqs. (5.1.29), (5.1.34), the propagator, in question, here denoted by $D_+^{>ij}$ satisfies, in convenient notation, the equations:

$$\big[(-\boldsymbol{\partial}^2 + \partial^{02})\delta^{i\ell} + \partial^i\partial^\ell\big]D_+^{>\ell j}(x, x') = \delta^{ij}\,\delta^{(4)}(x - x'), \tag{9.1.3}$$

$$\big[(-\boldsymbol{\partial}'^2 + \partial'^{02})\delta^{\ell i} + \partial'^\ell\partial'^i\big]D^{>j\ell}(x, x') = \delta^{ij}\,\delta^{(4)}(x - x'), \tag{9.1.4}$$

with $z > 0$, $z' > 0$.

A particular solution, e.g., of Eq. (9.1.3) for $t > t'$ is the one you are familiar with by now, given by

$$D_0^{ij}(x,x') = \mathrm{i}\int \frac{\mathrm{d}^2\mathbf{K}}{(2\pi)^2}\int \frac{\mathrm{d}q}{2\pi}\, \frac{\mathrm{e}^{\mathbf{K}\cdot(\mathbf{x}_\parallel-\mathbf{x}'_\parallel)}\mathrm{e}^{iq(z-z')}}{|\mathbf{k}|}\,\mathrm{e}^{-\mathrm{i}|\mathbf{k}|(t-t')}\Big(\delta^{ij}-\frac{k^ik^j}{|\mathbf{k}|^2}\Big), \tag{9.1.5}$$

$$\text{where}\quad \mathbf{k} = (\mathbf{K},q),\quad |\mathbf{k}| = \sqrt{\mathbf{K}^2+q^2}, \tag{9.1.6}$$

where $\mathbf{x}_\parallel$, $\mathbf{x}'_\parallel$ denote vectors parallel to the surface of the dielectric medium. This provides a solution which doesn't encounter the reflecting surface and is thus not very interesting.

The one describing a reflection is provided by the *homogeneous* solution of the equations, corresponding to Eqs. (9.1.3), (9.1.4),

$$\big[(-\boldsymbol{\partial}^2+\partial^{02})\delta^{i\ell}+\partial^i\partial^\ell\big]H^{>\ell j}(x,x') = 0, \tag{9.1.7}$$

$$\big[(-\boldsymbol{\partial}'^2+\partial'^{02})\delta^{\ell i}+\partial'^\ell\partial'^i\big]H^{>j\ell}(x,x') = 0. \tag{9.1.8}$$

The corresponding solution may be written as

$$H^{>ij}(x,x') = \mathrm{i}\int \frac{\mathrm{d}^2\mathbf{K}}{(2\pi)^2}\int\frac{\mathrm{d}q\,\mathrm{d}q'}{2\pi}\int\frac{\mathrm{d}k^0}{2\pi}\,\mathrm{e}^{\mathbf{K}\cdot(\mathbf{x}_\parallel-\mathbf{x}'_\parallel)}\mathrm{e}^{iqz}\mathrm{e}^{-iq'z'} \\ \times\ \mathrm{e}^{-ik^0(t-t')}\,H^{ij}(\mathbf{K},q,q',k^0). \tag{9.1.9}$$

Upon substituting this in Eq. (9.1.7) gives

$$(\mathbf{k}^2-k^{02})H^{ij}-k^ik^\ell H^{\ell j} = 0, \tag{9.1.10}$$

which upon multiplying by $k^i$ and summing over $i$, gives

$$-k^{02}k^iH^{ij} = 0,\ \Rightarrow\ k^iH^{ij} = 0, \tag{9.1.11}$$

for a non-zero energy $k^0$. The latter equation and Eq. (9.1.10), lead to

$$(k^{02}-\mathbf{k}^2)H^{ij} = 0,\ \Rightarrow\ (k^0-|\mathbf{k}|)(k^0+|\mathbf{k}|)H^{ij} = 0, \\ (k^0-|\mathbf{k}|)H^{ij} = 0, \tag{9.1.12}$$

for $k^0>0$. Upon using the last relation in Eq. (3.6.11), involving the delta distribution in Chap. 3, this allows one to write

$$H^{ij} = \delta(k^0-|\mathbf{k}|)\,G^{ij}. \tag{9.1.13}$$

where $G^{ij}$ is to be determined.

On the other hand, upon substituting $H^{>ij}(x,x')$, given in Eq. (9.1.9), in Eq. (9.1.8) gives

$$(\mathbf{k}'^2-k^{02})H^{ij}-k'^\ell k'^jH^{i\ell} = 0,\ \Rightarrow\ -k^{02}k'^\ell H^{i\ell} = 0,\ \ k'^jH^{ij} = 0, \tag{9.1.14}$$

as before, where now $\mathbf{k}' = (\mathbf{K}, q')$, leading, in particular, to

$$(\mathbf{k}'^2 - k^{02})H^{ij} = 0, \quad (q'^2 - q^2)H^{ij} = 0. \tag{9.1.15}$$

thus providing two solutions $q = \pm q'$. But if we take $q'$ to be negative $q' < 0$, in order to consider incidence on the reflecting surface, we must choose $q = |q'|$ to describe reflection falling in the region of the positive part of the $z-$ axis. By referring to Eq. (9.1.13), we may, conveniently, write

$$H^{ij} = \frac{\pi\,\delta(q' + |q|)}{|\mathbf{k}|}\,\delta(k^0 - |\mathbf{k}|)G_+^{ij}, \tag{9.1.16}$$

and $G_+^{ij}$ is to be determined.

Thus Eq. (9.1.9) takes the general form

$$H^{>ij}(x, x') = \mathrm{i}\int \frac{\mathrm{d}^2\mathbf{K}}{(2\pi)^2}\int \frac{\mathrm{d}q\,\mathrm{d}q'}{2\pi}\frac{\delta(q' + |q|)}{2|\mathbf{k}|}\mathrm{e}^{\mathbf{K}\cdot(\mathbf{x}_\parallel - \mathbf{x}'_\parallel)}\mathrm{e}^{iqz}\mathrm{e}^{-iq'z'}$$
$$\times\ \mathrm{e}^{-i|\mathbf{k}|(t-t')}\,G_+^{ij}(\mathbf{K}, q, q', |\mathbf{k}|). \tag{9.1.17}$$

Referring to Eqs. (9.1.11), (9.1.14), $G_+^{ij}$ must satisfy the equations

$$k^i\,G_+^{ij} = 0, \quad G_+^{ij}\,k'^j = 0. \tag{9.1.18}$$

Accordingly, with $q' = -q, q > 0$, considered subsequently, $G_+^{ij}$ may be expressed as follows:

$$G_+^{ab} = \left(\delta^{ab}A_+ - \frac{K^aK^b}{|\mathbf{k}|^2}B_+\right), \qquad a, b = 1, 2, \qquad |\mathbf{k}| = \sqrt{\mathbf{K}^2 + q^2}, \tag{9.1.19}$$

$$G_+^{3b} = -\frac{K^a}{|q|}G_+^{ab}, \quad G_+^{a3} = G_+^{ab}\frac{K^b}{|q|}, \quad G_+^{33} = -\frac{K^aG_+^{ab}K^b}{|q|^2}, \tag{9.1.20}$$

which satisfy the equations in Eq. (9.1.18), as is easily verified. The coefficients $A_+$ and $B_+$ will determined by the application of appropriate boundary conditions.

The components of $G_+^{ij}$ may be re-rewritten in more detail as

$$G_+^{a3} = \frac{K^a}{|q|}\left(A_+ - \frac{|\mathbf{K}|^2}{|\mathbf{k}|^2}B_+\right), \quad G_+^{3b} = -\frac{K^b}{|q|}\left(A_+ - \frac{|\mathbf{K}|^2}{|\mathbf{k}|^2}B_+\right),$$
$$G_+^{33} = -\frac{|\mathbf{K}|^2}{|q|^2}\left(A_+ - \frac{|\mathbf{K}|^2}{|\mathbf{k}|^2}B_+\right). \tag{9.1.21}$$

The propagator for $z > 0$ is then given by

$$D_+^{ij} = D^{>ij}(x, x') = D_0(x, x') + H^{>ij}(x, x'), \tag{9.1.22}$$

with $D_0(x,x')$ given in Eq. (9.1.5), $H^{>ij}(x,x')$ given in Eq. (9.1.17), where $G_+^{ij}$ is defined in Eq. (9.1.21), and $A_+$, $B_+$ are to be determined.

**The Refraction Region**: $z < 0$, $z' > 0$
The expression for the propagator in this region, quite generally, may be expressed as

$$D_+^{<ij}(x,x') = \mathrm{i}\int \frac{\mathrm{d}^2\mathbf{K}}{(2\pi)^2}\int \frac{\mathrm{d}q''\,\mathrm{d}q'}{2\pi}\int \frac{\mathrm{d}k^0}{2\pi}\,\mathrm{e}^{\mathbf{K}\cdot(\mathbf{x}_\parallel-\mathbf{x}'_\parallel)}\mathrm{e}^{iq''z''}\mathrm{e}^{-iq'z'}$$
$$\times\ \mathrm{e}^{-ik^0(t-t')}H_-^{ij}(\mathbf{K},q'',q',k^0). \tag{9.1.23}$$

which, in reference to Eqs. (5.4.52), (5.4.53) satisfies the homogeneous equations

$$\left[(-\boldsymbol{\partial}^2 + \varepsilon\partial^{02})\delta^{i\ell} + \partial^i\partial^\ell\right] D_+^{<\ell j}(x,x') = 0, \tag{9.1.24}$$
$$\left[(-\boldsymbol{\partial}'^2 + \partial'^{02})\delta^{\ell i} + \partial'^\ell\partial'^i\right] D_+^{<\ell j}(x,x') = 0. \tag{9.1.25}$$

recalling that for $z' > 0$, $\varepsilon(z') = 1$, and $z \neq z'$.

Note that with such a semi-infinite region, extending all the way down to $-\infty$, there is no reflection to be considered, i.e., of a particular solution as we have done before for the case $z > 0$.

Referring to Eqs. (9.1.15), and (9.1.18), the above two equations lead to

$$(\mathbf{k}''^2 - \varepsilon k^{02})H_-^{ij} = 0, \qquad (\mathbf{k}'^2 - k^{02})H_-^{ij} = 0, \tag{9.1.26}$$
$$k''^i H_-^{ij} = 0, \quad H_-^{ij}k'^j = 0, \quad \mathbf{k}' = (\mathbf{K},-q), \ \mathbf{k}'' = (\mathbf{K},q''). \tag{9.1.27}$$

where $q''$ is to be determined.

Equation (9.1.26), gives rise to the following equations

$$(k^0 - |\mathbf{k}'|)H_-^{ij} = 0, \qquad |\mathbf{k}'| = \sqrt{\mathbf{K}^2 + q^2}, \tag{9.1.28}$$
$$\left(q''^2 - q^2 - (\varepsilon-1)|\mathbf{k}'|^2\right)H_-^{ij} = 0. \tag{9.1.29}$$

Equation (9.1.29) provides two solutions. In the medium, however, falling in the region of the negative part of the $z-$ axis, we must choose $q'$, subsequently, to be negative, corresponding to refraction, i.e. from Eqs. (9.1.28), (9.1.29), we may conveniently write,

$$H_-^{ij} = \frac{\pi}{|\mathbf{k}'|}\,\delta\big(k^0 - |\mathbf{k}'|\big)\,\delta\big(q'' + Q\big)G_-^{ij}, \qquad Q = \sqrt{\varepsilon|\mathbf{k}|'^2 - |\mathbf{K}|^2} \tag{9.1.30}$$

where $G_-^{ij}$ will be determined by the application of appropriate boundary conditions.

The expression in Eq. (9.1.23) takes the form

$$D_{+}^{<ij}(x,x') = \mathrm{i}\int \frac{\mathrm{d}^2\mathbf{K}}{(2\pi)^2}\int \frac{\mathrm{d}q''\,\mathrm{d}q'}{2\pi} \mathrm{e}^{\mathbf{K}\cdot(\mathbf{x}_{\|}-\mathbf{x}'_{\|})}\mathrm{e}^{-iQz}\mathrm{e}^{-iq'z'}\frac{\delta(q''+Q)}{2|\mathbf{k}'|} \times \mathrm{e}^{-i|\mathbf{k}'|(t-t')} G_{-}^{ij}(\mathbf{K}, Q, q', |\mathbf{k}'|). \tag{9.1.31}$$

According to Eq. (9.1.27), $k''^{i} G_{-}^{ij} = 0$, and $G_{-}^{ij} k'^{j} = 0$, and hence, $G_{-}^{ij}$ may be expressed as follows:

$$G_{-}^{ab} = \Big(\delta^{ab} A_{-} - \frac{K^a K^b}{|\mathbf{k}|^2} B_{-}\Big), \quad a, b = 1, 2, \tag{9.1.32}$$

$$G_{-}^{3b} = \frac{K^a}{Q} G_{-}^{ab}, \qquad G_{-}^{a3} = G_{-}^{ab}\frac{K^b}{q}, \qquad G_{-}^{33} = \frac{K^a G_{-}^{ab} K^b}{qQ}, \tag{9.1.33}$$

$$\mathbf{k}'' = (\mathbf{K}, -Q), \;\; \mathbf{k}' = (\mathbf{K}, -q), \;\; Q = \sqrt{\varepsilon|\mathbf{k}|'^2 - |\mathbf{K}|^2}, \;\; |\mathbf{k}''| = \sqrt{\varepsilon}|\mathbf{k}'|. \tag{9.1.34}$$

We note that the energy of the photon in free space is given by $|\mathbf{k}'|$, in units of the speed of light c in it set to one. While in the medium the speed of light is given by $\mathrm{c}/\sqrt{\varepsilon} = 1/\sqrt{\varepsilon}$ as established below Eq. (5.5.17) in Chapter 5. Hence the energy of the photon *in* the medium is given by $\mathrm{c}|\mathbf{k}''|/\sqrt{\varepsilon} = |\mathbf{k}'|$ and energy is *conserved*, as expected.

From Eq. (9.1.33), the components of $G_{-}^{ij}$, together with Eq. (9.1.32), may be explicitly rewritten as

$$G_{-}^{a3} = \frac{K^a}{|q|}\Big(A_{-} - B_{-}\frac{|\mathbf{K}|^2}{|\mathbf{k}|^2}\Big), \tag{9.1.35}$$

$$G_{-}^{3b} = \frac{K^b}{Q}\Big(A_{-} - B_{-}\frac{|\mathbf{K}|^2}{|\mathbf{k}|^2}\Big), \tag{9.1.36}$$

$$G_{-}^{33} = \frac{|\mathbf{K}|^2}{|q|Q}\Big(A_{-} - B_{-}\frac{|\mathbf{K}|^2}{|\mathbf{k}|^2}\Big), \tag{9.1.37}$$

The propagator *for* $z < 0$ is now given by

$$D_{+}^{ij}(x,x') \equiv D_{+}^{<ij}(x,x'), \tag{9.1.38}$$

with the latter given in Eq. (9.1.31), and $G_{-}^{ij}$ defined in Eqs. (9.1.32), (9.1.35)–(9.1.37), $a, b$ taking values in $\{1, 2\}$.

It remains to apply the appropriate boundary conditions (B.C.). Here we note that $z$ in Eq. (9.1.31) will approach zero from negative values. The B.C. are given by

$$D_{+}^{aj}(x,x'), \quad \varepsilon(z) D_{+}^{3j}(x,x'), \quad [\partial^a D_{+}^{3j}(x,x') - \partial^3 D_{+}^{aj}(x,x')], \tag{9.1.39}$$

are continuous at $z$ taken at the limiting values of $z = +0, z = -0$. With the B.C. defined in the temporal gauge as given, e.g. in Eq. (5.3.23), in which $E^i = \partial^0 A^i$, with $A^i$ given in Eq. (5.4.51). Also $\mathbf{B} = \nabla \times \mathbf{A}$, and general boundary conditions are spelled out in Eqs. (5.4.27)–(5.4.30). There $\mathbf{n}$ denotes a unit vector *perpendicular* to an intervening surface.

The first B.C. for $j = b$ above gives from Eqs. (9.1.5), (9.1.19), (9.1.20), (9.1.31), and (9.1.32)

$$\Big(\delta^{ab} - \frac{K^a K^b}{|\mathbf{k}|^2}\Big) + G_+^{ab} = G_-^{ab}, \tag{9.1.40}$$

$$\text{i.e.}\ \ 1 + A_+ = A_-, \quad 1 + B_+ = B_-. \tag{9.1.41}$$

The second B.C. for $j = 3$ above gives

$$\Big(1 - \frac{q^2}{|\mathbf{k}|^2}\Big) + G_+^{33} = \varepsilon\, G_-^{33} \tag{9.1.42}$$

$$\Big(1 - \frac{|q|^2}{|\mathbf{k}|^2}\Big) - \frac{|\mathbf{K}|^2}{|q|^2}\Big(A_+ - \frac{|\mathbf{K}|^2}{|\mathbf{k}|^2} B_+\Big) = \frac{\varepsilon\, |\mathbf{K}|^2}{|q|\, Q}\Big(A_- - \frac{|\mathbf{K}|^2}{|\mathbf{k}|^2} B_-\Big), \tag{9.1.43}$$

and from Eq. (9.1.39), this gives

$$\Big(A_+ - \frac{|\mathbf{K}|^2}{|\mathbf{k}|^2} B_+\Big) = \frac{q^2}{|\mathbf{k}|^2}\Big(\frac{Q - \varepsilon |q|}{Q + \varepsilon |q|}\Big). \tag{9.1.44}$$

By using, in turn, Eq. (9.1.41) in the above equation also gives,

$$\Big(A_- - \frac{|\mathbf{K}|^2}{|\mathbf{k}|^2} B_-\Big) = \frac{2|q|^2 Q}{|\mathbf{k}|^2 (Q + \varepsilon |q|)}. \tag{9.1.45}$$

Now it is straightforward to find the explicit structure of $G_-^{ij}$ relevant to describe the intricacies of refraction. To this end, referring to Eqs. (9.1.32), (9.1.35)–(9.1.37), we expect that it has the general structure

$$G_-^{ij} = \Big(A_- \delta^{ij} - B_- \frac{k''^i k''^j}{\mathbf{k}^2}\Big), \tag{9.1.46}$$

$$\mathbf{k}' = (\mathbf{K}, -|q|), \ \ \mathbf{k}'' = (\mathbf{K}, -Q), \ \ \mathbf{k} = (\mathbf{k}, |q|), \ \ \mathbf{k}' \cdot \mathbf{k}'' = |\mathbf{K}|^2 + |q| Q. \tag{9.1.47}$$

Here $\mathbf{k}'$, $\mathbf{k}$, $\mathbf{k}''$ denote, respectively, vectors of incidence, reflection, and refraction. The above relation is certainly true, for $i = a$, $j = b$. For $i = a$, $j = 3$, we have from Eqs. (9.1.35), (9.1.45) *and* (9.1.46)

$$G_-^{a3} = \frac{K^a}{|q|}\Big(A_- - B_-\frac{|\mathbf{K}|^2}{|\mathbf{k}|^2}\Big) = \frac{2|q|K^a Q}{|\mathbf{k}|^2(Q+\varepsilon|q|)} \equiv -B_-\frac{k'^a k''^3}{|\mathbf{k}|^2}. \tag{9.1.48}$$

Similarly, from Eqs. (9.1.36), (9.1.45) *and* (9.1.46)

$$G_-^{3b} = \frac{K^b}{Q}\Big(A_- - B_-\frac{|\mathbf{K}|^2}{|\mathbf{k}|^2}\Big) = \frac{2|q|^2 K^b}{|\mathbf{k}|^2(Q+\varepsilon|q|)} \equiv -B_-\frac{k'^3 k''^b}{|\mathbf{k}|^2}. \tag{9.1.49}$$

Finally, from Eqs. (9.1.37), (9.1.45) *and* (9.1.46)

$$G_-^{33} = \frac{|\mathbf{K}|^2}{|q|Q}\Big(A_- - B_-\frac{|\mathbf{K}|^2}{|\mathbf{k}|^2}\Big) = \frac{2|q||\mathbf{K}|^2}{|\mathbf{k}|^2(Q+\varepsilon|q|)} \equiv A_- - B_-\frac{k'^3 k''^3}{|\mathbf{k}|^2}. \tag{9.1.50}$$

Equations (9.1.48)–(9.1.50) lead to the solutions:

$$B_- = \frac{2|q|}{Q+\varepsilon|q|}. \tag{9.1.51}$$

$$A_- = \frac{2|q||\mathbf{K}|^2}{|\mathbf{k}|^2(Q+\varepsilon|q|)} + \frac{2|q|}{Q+\varepsilon|q|}\frac{|q|Q}{|\mathbf{k}|^2} = \frac{2|q|\big(|\mathbf{K}|^2+|q|Q\big)}{|\mathbf{k}|^2\big(Q+\varepsilon|q|\big)}. \tag{9.1.52}$$

The latter two expressions may be simplified further by noting that

$$\begin{aligned}|\mathbf{k}|^2\,(Q+\varepsilon|q|) &= |\mathbf{k}|^2 Q + (\varepsilon|\mathbf{k}|^2)|q| = \big(|\mathbf{K}|^2+|q|^2\big)Q + |\mathbf{k}''|^2|q| \\ &= \big(|\mathbf{K}|^2+|q|^2\big)Q + \big(|\mathbf{K}|^2+Q^2\big)|q| \\ &= \big(|\mathbf{K}|^2+|q|Q\big)\big(Q+|q|\big).\end{aligned} \tag{9.1.53}$$

That is

$$\frac{1}{Q+\varepsilon|q|} = \frac{|\mathbf{k}|^2}{\big(|\mathbf{K}|^2+|q|Q\big)\,\big(Q+|q|\big)} \equiv \frac{|\mathbf{k}|^2}{(Q+|q|)}\frac{1}{\mathbf{k}'\cdot\mathbf{k}''}. \tag{9.1.54}$$

Hence we may rewrite

$$A_- = \frac{2|q|}{(Q+|q|)}, \qquad B_- = \frac{2|q|}{(Q+|q|)}\frac{|\mathbf{k}|^2}{\mathbf{k}''\cdot\mathbf{k}'} \equiv \frac{2|q|}{Q+\varepsilon|q|}. \tag{9.1.55}$$

Accordingly, $G_-^{ij}$ is explicitly given by

$$G_-^{ij} = \frac{2|q|}{(Q+|q|)}\Big(\delta^{ij} - \frac{k'^i k''^j}{\mathbf{k}''\cdot\mathbf{k}'}\Big). \tag{9.1.56}$$

Most importantly, we verify the transversality properties in Eq. (9.1.27), as applied to $G_-^{ij}$:

$$k''^{i}G_{-}^{ij} = \frac{2|q|}{Q+|q|}\Big(k''^{j} - \frac{\mathbf{k}''\cdot\mathbf{k}'}{\mathbf{k}''\cdot\mathbf{k}'}k''^{j}\Big) = 0, \tag{9.1.57}$$

$$G_{-}^{ij}k'^{j} = \frac{2|q|}{Q+|q|}\Big(k'^{i} - k'^{i}\frac{\mathbf{k}''\cdot\mathbf{k}'}{\mathbf{k}''\cdot\mathbf{k}'}\Big) = 0. \tag{9.1.58}$$

To find $A_+$, $B_+$, in Eqs. (9.1.19), (9.1.21) relevant to reflection, we use Eq. (9.1.41), and the expressions of $A_-$ and $B_-$ in Eq. (9.1.55), to obtain

$$A_+ = -\frac{Q-|q|}{Q+|q|}, \quad B_+ = -\frac{Q+\varepsilon|q|-2|q|}{Q+\varepsilon|q|} \equiv \frac{(|\mathbf{k}|^2+|q|^2)(1-\varepsilon)+qQ(1-\varepsilon)}{(Q+\varepsilon|q|)(Q+|q|)}. \tag{9.1.59}$$

Here we note that if we set $\varepsilon = 1$, $Q = |q|$, then both $A_+ = 0$, $B_+ = 0$, as expected, since the question of reflection doesn't even arise.

Now that the coefficients $A_+$, $B_+$ are known, and the components of $G_+^{ij}$, relevant to reflection, given in Eqs. (9.1.19), (9.1.21), are expressed in terms of these coefficients, the explicit expression of $G_+^{ij}$ may be readily found. We may first express it as follows:

$$G_+^{ij} = \delta^{ia}\delta^{bj}G_+^{ab} + \delta^{i3}\delta^{bj}G_+^{3b} + \delta^{ia}\delta^{3j}G_+^{a3} + \delta^{i3}\delta^{3j}G_+^{33}. \tag{9.1.60}$$

Using the explicit expressions of its components in Eqs. (9.1.19), (9.1.21), with $A_+$, $B_+$ defined in Eq. (9.1.59), give,

$$G_+^{ij} = \delta^{ia}\Big(\delta^{ab}A_+ - \frac{K^aK^b}{|\mathbf{k}|^2}B_+\Big)\delta^{bj}$$
$$+ \frac{1}{|\mathbf{k}|^2}\Big(\frac{Q-\varepsilon|q|}{Q+\varepsilon|q|}\Big)\Big( - \delta^{i3}\delta^{bj}K^b|q| + \delta^{ia}\delta^{3j}K^a|q| - \delta^{i3}\delta^{3j}|\mathbf{K}|^2\Big). \tag{9.1.61}$$

Most importantly, we readily verify the transversality properties in Eqs. (9.1.11), and (9.1.14) as applied to $G_+^{ij}$, by using, in the process, the identity in Eq. (9.1.44),

$$k^iG_+^{ij} = 0, \quad G_+^{ij}k'^j = 0, \quad \mathbf{k} = (\mathbf{K}, |q|), \ \mathbf{k}' = (\mathbf{K}, -|q|). \tag{9.1.62}$$

**Emergence of the Physical Processes**:

Let us consider first the reflection process. *Given* directions of propagations, one can readily introduce associated polarization vectors satisfying a set of properties as given below. To this end, we may introduce polarization vectors, as done in Eqs. (8.2.33)–(8.2.36), in reference to the direction of propagations $\mathbf{n}$, $\mathbf{n}'$ obtained above, associated with reflection and incidence denoted by $\mathbf{e}_\lambda$, $\mathbf{e}'_\lambda$, respectively, satisfying:

$$\mathbf{e}_\lambda \cdot \mathbf{e}_\alpha = \delta_{\lambda\alpha}, \quad \mathbf{n} \cdot \mathbf{e}_\lambda = 0, \quad \mathbf{e}_1 \times \mathbf{e}_2 = \mathbf{n}, \quad \mathbf{n} = \frac{\mathbf{k}}{|\mathbf{k}|}, \quad \mathbf{k} = (\mathbf{K}, |q|), \tag{9.1.63}$$

$$\mathbf{e}'_\lambda \cdot \mathbf{e}'_\alpha = \delta_{\lambda\alpha}, \quad \mathbf{n}' \cdot \mathbf{e}'_\lambda = 0, \quad \mathbf{e}'_1 \times \mathbf{e}'_2 = \mathbf{n}', \quad \mathbf{n}' = \frac{\mathbf{k}'}{|\mathbf{k}|}, \quad \mathbf{k}' = (\mathbf{K}, -|q|), \tag{9.1.64}$$

for $\lambda, \alpha = 1, 2$. With the transversality conditions given in Eq. (9.1.62),[4] the amplitude of transition of reflection is simply given by a familiar expression, by now,

$$(\mathbf{k}, \lambda \mid \mathbf{k}'\alpha) = \mathrm{e}^i_\lambda\, G^{ij}_+\, \mathrm{e}'^j{}_\alpha. \tag{9.1.65}$$

With polarization vectors as given in Eqs. (8.2.42), (8.2.43),

$$\mathbf{e}'_1 = \frac{(\mathbf{n} \times \mathbf{n}') \times \mathbf{n}'}{|(\mathbf{n} \times \mathbf{n}') \times \mathbf{n}'|} = \frac{\mathbf{n}'(\mathbf{n} \cdot \mathbf{n}') - \mathbf{n}}{|\mathbf{n} \times \mathbf{n}'|}, \qquad \mathbf{e}'_2 = \frac{\mathbf{n} \times \mathbf{n}'}{|\mathbf{n} \times \mathbf{n}'|}, \tag{9.1.66}$$

$$\mathbf{e}_1 = \frac{\big(\mathbf{n} \times \mathbf{n}'\big) \times \mathbf{n}}{|\big(\mathbf{n} \times \mathbf{n}'\big) \times \mathbf{n}|} = \frac{\mathbf{n}' - \mathbf{n}(\mathbf{n} \cdot \mathbf{n}')}{|\mathbf{n} \times \mathbf{n}'|}, \qquad \mathbf{e}_2 = \frac{\mathbf{n} \times \mathbf{n}'}{|\mathbf{n} \times \mathbf{n}'|}, \tag{9.1.67}$$

we have

$$\mathbf{e}_1 = \frac{\big(\mathbf{K}|q|, -|\mathbf{K}|^2\big)}{|\mathbf{k}|\,|\mathbf{K}|}, \quad \mathbf{e}'_1 = \frac{\big(-\mathbf{K}|q|, -|\mathbf{K}|^2\big)}{|\mathbf{k}|\,|\mathbf{K}|}, \quad \mathbf{e}_2 = \frac{\big(-K^2, K^1, 0\big)}{|\mathbf{K}|} = \mathbf{e}'_2. \tag{9.1.68}$$

They satisfy the properties spelled out in Eqs. (9.1.63), (9.1.64).

**Computing the amplitudes**

By using the explicit expression of $G^{ij}_+$ given in Eq. (9.1.61), and the polarization vectors computed in Eq. (9.1.68), some elementary algebra gives the relative amplitudes of reflection as

$$\big(\mathbf{k}, \mathbf{e}_\lambda \mid \mathbf{k}', \mathbf{e}'_\alpha\big) = \mathrm{e}^i_\lambda G^{ij}_+ \mathrm{e}'^j{}_\alpha = \Big(\frac{\varepsilon|q| - Q}{Q + \varepsilon|q|}\Big)\delta^{\lambda 1}\delta^{\alpha 1} + \Big(\frac{|q| - Q}{|q| + Q}\Big)\delta^{\lambda 2}\delta^{\alpha 2}, \tag{9.1.69}$$

$$\text{where} \quad \mathbf{k}' = (\mathbf{K}, -|q|), \quad \mathbf{k} = (\mathbf{K}, |q|), \quad Q = \sqrt{\varepsilon|\mathbf{k}|^2 - |\mathbf{K}|^2}, \tag{9.1.70}$$

given, respectively, for polarizations parallel and normal to the surface of incidence. The vectors $\mathbf{k}'$ and $\mathbf{k}$, in particular, lie in this surface.

For $\varepsilon \to \infty$, the relative |amplitudes| go to one, as we have encountered before in Sect. 8.2. For finite $\varepsilon$, the photon cannot make up its mind, it might plunge into the medium. This brings us to investigate the refraction process.

[4] Compare with the $B^{ij}$ term in Eq. (8.2.32).

To the above end, we may introduce polarization vectors $\mathbf{e}''_\lambda$ to investigate the nature of refraction such that

$$\mathbf{e}_1'' \times \mathbf{e}_2'' = \mathbf{n}'', \quad \mathbf{n}'' = \frac{\mathbf{k}''}{|\mathbf{k}''|}, \quad \mathbf{k}'' = (\mathbf{K}, -Q), \quad |\mathbf{k}''| = \sqrt{\varepsilon}\,|\mathbf{k}'|. \tag{9.1.71}$$

Given the direction of propagation specified by the unit vector $\mathbf{n}''$, polarization vectors may be readily introduced associated with it

$$\mathbf{e}''_1 = \frac{\mathbf{n}'' \times (\mathbf{n}'' \times \mathbf{n}')}{|\mathbf{n}'' \times (\mathbf{n}'' \times \mathbf{n}')|}, \quad \mathbf{e}''_2 = \frac{\mathbf{n}' \times \mathbf{n}''}{|(\mathbf{n}' \times \mathbf{n}'')|}, \tag{9.1.72}$$

and given explicitly by

$$\mathbf{e}''_1 = -\frac{(\mathbf{K}\,Q, |\mathbf{K}|^2)}{\sqrt{\varepsilon}|\mathbf{k}|\,|\mathbf{K}|}, \quad \mathbf{e}''_2 = \frac{(-K^2, K^1, 0)}{|\mathbf{K}|}, \tag{9.1.73}$$

satisfying the properties

$$\mathrm{e}''^{i}_{\lambda} \cdot \mathrm{e}''^{i}_{\alpha} = \delta_{\lambda\alpha}, \quad \mathbf{k}'' \cdot \mathbf{e}''_\lambda = 0, \quad \mathbf{e}''_1 \times \mathbf{e}''_2 = \mathbf{n}''. \tag{9.1.74}$$

The transition amplitude of refraction is given by

$$(\mathbf{k}'', \mathbf{e}''_\lambda \mid \mathbf{k}', \mathbf{e}'_\alpha) = \mathrm{e}''^{i}_{\lambda}\, G^{ij}_{-}\, e'^{j}_{\alpha}, \quad \mathbf{k}' = (\mathbf{K}, -|q|), \;\; \mathbf{k}'' = (\mathbf{K}, -Q), \tag{9.1.75}$$

where $G^{ij}_{-}$ is given in Eq. (9.1.56), and $\mathbf{e}'_\lambda$ in Eq. (9.1.68). We use the following identities:

$$\mathbf{e}''_\lambda \cdot \mathbf{e}'_\alpha = \begin{cases} \mathbf{k}' \cdot \mathbf{k}''/\sqrt{\varepsilon}\,|\mathbf{k}|, & \text{for } \lambda = \alpha = 1 \\ 1 & \text{for } \lambda = \alpha = 2 \\ 0 & \text{for } \lambda \neq \alpha \end{cases}, \tag{9.1.76}$$

$$\mathbf{e}''_\lambda \cdot \mathbf{k}' = \begin{cases} -(Q - |q|)/\sqrt{\varepsilon}|\mathbf{k}|, & \text{for } \lambda = 1 \\ 0 & \text{for } \lambda = 2 \end{cases}. \tag{9.1.77}$$

Using the above two equations, together with the expression of $G^{ij}_{-}$ in Eq. (9.1.56), we have for the amplitude in question,

$$\mathrm{e}''^{i}_{\lambda}\, \frac{2|q|}{(Q + |q|)} \Big( \delta^{ij} - \frac{k'^{i}\, k''^{j}}{\mathbf{k}' \cdot \mathbf{k}''} \Big)\, \mathrm{e}'^{j}_{\alpha}. \tag{9.1.78}$$

In particular, this gives for $\lambda = \alpha = 1$,

$$\begin{aligned} &{\mathrm{e}''}^{i}_{1}\,\frac{2|q|}{(Q+|q|)}\Big(\delta^{ij}-\frac{k'^{i}\,k''^{j}}{\mathbf{k}'\cdot\mathbf{k}''}\Big)\,{\mathrm{e}'}^{j}_{1} \\ &=\frac{2|q|}{(Q+|q|)}\,\frac{1}{\mathbf{k}'\cdot\mathbf{k}''\sqrt{\varepsilon}|\mathbf{k}|^2}\Big(\big(\mathbf{k}'\cdot\mathbf{k}''\big)^2+(Q-q)^2|\mathbf{K}|^2\Big), \end{aligned} \tag{9.1.79}$$

which by using the identity in Eq. (9.1.54), and

$$\big(|\mathbf{K}|^2+|q|Q)^2+(Q-q)^2|\mathbf{K}|^2\big)=\big(|\mathbf{K}|^2+q^2\big)\big(|\mathbf{K}|^2+Q^2\big)\equiv\big(|\mathbf{k}|^2\varepsilon|\mathbf{k}|^2\big). \tag{9.1.80}$$

That is

$${\mathrm{e}''}^{i}_{1}\,G^{ij}_{-}\,{\mathrm{e}'}^{j}_{1}\;=\;\frac{2\sqrt{\varepsilon}|q|}{Q+\varepsilon|q|}. \tag{9.1.81}$$

On the other hand, from Eqs. (9.1.76), (9.1.77),

$${\mathrm{e}''}^{i}_{2}\,\frac{2|q|}{(Q+|q|)}\Big(\delta^{ij}-\frac{k'^{i}k''^{j}}{\mathbf{k}'\cdot\mathbf{k}''}\Big)\,{\mathrm{e}'}^{j}_{2}=\frac{2|q|}{Q+|q|}\Big({\mathbf{e}''}_{2}\cdot\mathbf{e}'_{2}+0\Big)=\frac{2|q|}{Q+|q|}, \tag{9.1.82}$$

$$\begin{aligned} &{\mathrm{e}''}^{i}_{\lambda}\,\frac{2|q|}{Q+|q|}\Big(\delta^{ij}-\frac{k'^{i}k''^{j}}{\mathbf{k}'\cdot\mathbf{k}''}\Big)\,{\mathrm{e}'}^{j}_{\alpha} \\ &=\frac{2|q|}{Q+|q|}\Big(\mathbf{e}''_{\lambda}\cdot{\mathbf{e}'}^{j}_{\alpha}-\frac{(\mathbf{e}''_{\lambda}\cdot\mathbf{k}')\,(\mathbf{k}''\cdot{\mathbf{e}'}^{j}_{\alpha})}{\mathbf{k}'\cdot\mathbf{k}''}\Big)=0,\quad\text{for }\lambda\neq\alpha. \end{aligned} \tag{9.1.83}$$

All told, the relative amplitudes are, respectively, given by

$$\big(\mathbf{k}'',\mathbf{e}''_{\lambda}\mid\mathbf{k}',\mathbf{e}'_{\alpha}\big)\;=\;\frac{2\sqrt{\varepsilon}|q|}{Q+\varepsilon|q|}\delta_{\lambda\,1}\,\delta_{\alpha\,1}\;+\;\frac{2|q|}{Q+|q|}\delta_{\lambda\,2}\,\delta_{\alpha\,2}, \tag{9.1.84}$$

given, respectively, for polarizations parallel and normal to the surface of incidence. Recall that, $Q=\sqrt{\varepsilon|\mathbf{k}|^2-|\mathbf{K}|^2}$.

We note that the sines of the angles of incidence and refraction $\vartheta_{\text{inc}}$, $\vartheta_{\text{refrac}}$ are given, respectively, by $|\mathbf{K}/|\mathbf{k}|$, $|\mathbf{K}/\sqrt{\varepsilon}|\mathbf{k}|$, using, for the latter, that $|\mathbf{k}''|=\sqrt{\varepsilon}|\mathbf{k}|$. From this follows Snell's law $\sin\vartheta_{\text{inc}}=\sqrt{\varepsilon}\,\sin\vartheta_{\text{refrac}}$.

From Eq. (9.1.69), the relative amplitude of reflection of a photon with polarization $\mathbf{e}'_1$ parallel to the plane of incidence vanishes for $Q=\varepsilon|q|$. That is, by dividing the first equation below by $\sqrt{\varepsilon}|\mathbf{k}|$, and squaring the result, this implies that

$$\begin{aligned} \varepsilon|q|=\sqrt{\varepsilon|\mathbf{k}|^2-\mathbf{K}|^2},\;\Rightarrow\;\varepsilon\cos^2\vartheta_{\text{inc}}&=1-\sin^2\vartheta_{\text{refrac}}, \\ \sqrt{\varepsilon}\cos\vartheta_{\text{inc}}&=\cos\vartheta_{\text{refrac}}. \end{aligned} \tag{9.1.85}$$

Upon using this result together with Snell's law, one has

$$\sin\vartheta_{\text{inc}}\cos\vartheta_{\text{inc}} = \sin\vartheta_{\text{refrac}}\cos\vartheta_{\text{refrac}}, \Rightarrow \sin 2\vartheta_{\text{inc}} = \sin 2\vartheta_{\text{refrac}}, \tag{9.1.86}$$

$$\text{or} \quad \vartheta_{\text{inc}} = \frac{\pi}{2} - \vartheta_{\text{refrac}}, \qquad \cos\vartheta_{\text{inc}} = \sin\vartheta_{\text{refrac}}. \tag{9.1.87}$$

An angle incidence $\vartheta_{\text{inc}}$ such that $\vartheta_{\text{inc}} + \vartheta_{\text{refrac}} = \pi/2$ is referred to as Brewster's angle for which there is *no* reflection for incidence polarized in the plane of incidence.

From Snell's law $\sin\vartheta_{\text{inc}} = \sqrt{\varepsilon}\sin\vartheta_{\text{refrac}}$ and the condition $\cos\vartheta_{\text{inc}} = \sin\vartheta_{\text{refrac}}$, in Eq. (9.1.87), the Brewster's angle may be computed from the value of the index of refraction $\sqrt{\varepsilon}$, through the equation $\tan\vartheta_{\text{inc}} = \sqrt{\varepsilon}$.

It is interesting to note that an initially unpolarized photon becomes polarized (in a direction perpendicular to the incident plane) upon reflection at the Brewster's angle. Accordingly, a polarizer may be set up, in turn, to remove this polarization, thus eliminating reflection altogether.

In terms of the angle of incidence $\vartheta_{\text{inc}}$ defined through $\sin\vartheta_{\text{inc}} = |\mathbf{K}|/|\mathbf{k}|$, $\cos\vartheta_{\text{inc}} = |q|/|\mathbf{k}|$, $Q = |\mathbf{k}|\sqrt{\varepsilon - \sin^2\vartheta_{\text{inc}}}$, then according to Eqs. (9.1.69), (9.1.84), the amplitudes of reflections, may be equivalently rewritten as:

$$\begin{aligned}\big(\mathbf{k}, \mathbf{e}_\lambda \mid \mathbf{k}', \mathbf{e}'_\alpha\big) = &\left(\frac{\varepsilon\cos\vartheta_{\text{inc}} - \sqrt{\varepsilon - \sin^2\vartheta_{\text{inc}}}}{\varepsilon\cos\vartheta_{\text{inc}} + \sqrt{\varepsilon - \sin^2\vartheta_{\text{inc}}}}\right)\delta_{\lambda 1}\delta_{\alpha 1} + \\ &\left(\frac{\cos\vartheta_{\text{inc}} - \sqrt{\varepsilon - \sin^2\vartheta_{\text{inc}}}}{\cos\vartheta_{\text{inc}} + \sqrt{\varepsilon - \sin^2\vartheta_{\text{inc}}}}\right)\delta_{\lambda 2}\delta_{\alpha 2}.\end{aligned} \tag{9.1.88}$$

While from Eq. (9.1.84), the amplitudes of refractions, may be equivalently rewritten as:

$$\begin{aligned}\big(\mathbf{k}'', \mathbf{e}''_\lambda \mid \mathbf{k}', \mathbf{e}'_\alpha\big) = &\frac{2\sqrt{\varepsilon}\cos\vartheta_{\text{inc}}}{\varepsilon\cos\vartheta_{\text{inc}} + \sqrt{\varepsilon - \sin^2\vartheta_{\text{inc}}}}\,\delta_{\lambda\,1}\,\delta_{\alpha\,1} + \\ &\frac{2\cos\vartheta_{\text{inc}}}{\cos\vartheta_{\text{inc}} + \sqrt{\varepsilon - \sin^2\vartheta_{\text{inc}}}}\,\delta_{\lambda\,2}\,\delta_{\alpha\,2},\end{aligned} \tag{9.1.89}$$

both, given, respectively, for polarizations parallel and normal to the surface of incidence.

Accordingly, the relative amplitudes of reflection and refraction with polarization $\mathbf{e}'_1$ of incidence *parallel* to the plane of incidence are, respectively, given by

$$\frac{\varepsilon\cos\vartheta_{\text{inc}} - \sqrt{\varepsilon - \sin^2\vartheta_{\text{inc}}}}{\varepsilon\cos\vartheta_{\text{inc}} + \sqrt{\varepsilon - \sin^2\vartheta_{\text{inc}}}}, \tag{9.1.90}$$

and

$$\frac{2\sqrt{\varepsilon}\cos\vartheta_{\text{inc}}}{\varepsilon\cos\vartheta_{\text{inc}} + \sqrt{\varepsilon - \sin^2\vartheta_{\text{inc}}}}, \tag{9.1.91}$$

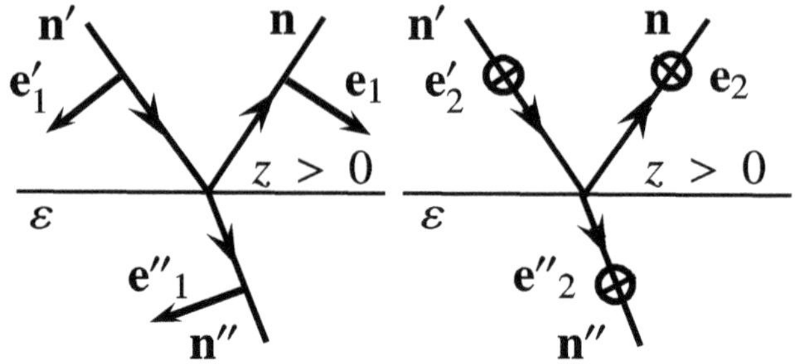


**Fig. 9.1** From left to right, depicting the reflection processes $(\mathbf{K}, -q, \mathbf{e}'_1) \to (\mathbf{K}, q, \mathbf{e}_1)$, with $\mathbf{e}'_1$ and $\mathbf{e}_1$ in the plane of incidence, and $(\mathbf{K}, -q, \mathbf{e}'_2) \to (\mathbf{K}, q, \mathbf{e}_2)$, with $\mathbf{e}'_2$ and $\mathbf{e}_2$ normal to it, respectively. The crosses indicate vectors that point into the page. Remember that if light of a given momentum, say, $\mathbf{p}$, is polarized along a direction specified by a polarization vector, say, $\mathbf{e}$, then the direction of the magnetic field $\mathbf{B}$ is given by the relation $\mathbf{e} \times \mathbf{B} \propto \mathbf{p}$

where all the polarization of incidence, reflection and refraction are parallel to the plane of incidence corresponding to the ones on the l.h.s of Fig. 9.1.

On the other hand, the relative amplitudes of reflection and refraction with polarization $\mathbf{e}'_2$ of incidence *normal* to the plane of incidence are, respectively, given by

$$\frac{\cos\vartheta_{\text{inc}} - \sqrt{\varepsilon - \sin^2\vartheta_{\text{inc}}}}{\cos\vartheta_{\text{inc}} + \sqrt{\varepsilon - \sin^2\vartheta_{\text{inc}}}}, \tag{9.1.92}$$

and

$$\frac{2\cos\vartheta_{\text{inc}}}{\cos\vartheta_{\text{inc}} + \sqrt{\varepsilon - \sin^2\vartheta_{\text{inc}}}}, \tag{9.1.93}$$

where all the polarization of incidence, reflection and refraction are normal to the plane of incidence corresponding to the ones on the r.h.s of Fig. 9.1.

For a given angle on incidence $\vartheta_{\text{inc}}$, with incidence from a region of index of refraction $\varepsilon_1$ to a region of index of refraction $\varepsilon_2$, the relative amplitudes may be obtained directly from the above ones by merely replacing $\varepsilon$ by the relative indices of refractions $\varepsilon_2/\varepsilon_1$.

## 9.2 Refraction and Reflection for $\varepsilon \neq 1$, $\mu \neq 1$ in Propagator Theory

We consider the problem of reflection and refraction for incidence from free space into a semi-infinite region of parameters $(\varepsilon, \mu)$, and conveniently set

$$\varepsilon(z),\ \mu(z) = \begin{cases} 1,\ 1, & \text{respectively, for } z > 0 \\ \varepsilon,\ \mu, & \text{respectively, for } z < 0 \end{cases}. \tag{9.2.1}$$

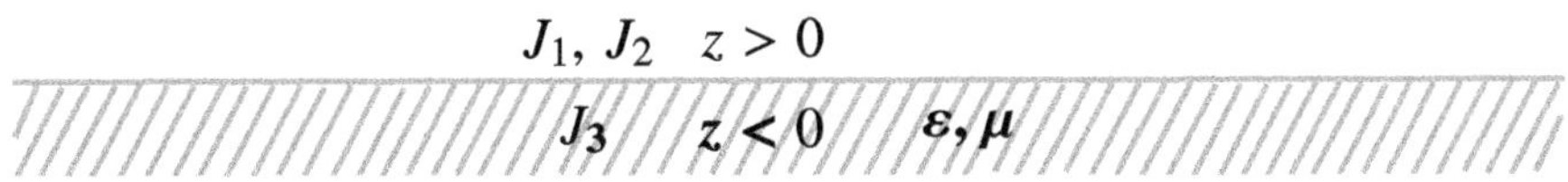


$$J(x) = J_1(x) + J_2(x) + J_3(x), \tag{9.2.2}$$

where the sources $J_2(x)$, $J_3(x)$ are switched on after $J_1(x)$, as an emitter, is switched off.

**The Reflection Region**: $z > 0$, $z' > 0$

Here $\varepsilon(z) = 1$, $\mu(z) = 1$. According, to Eqs. (5.4.52), (5.4.53), the propagator, in question, denoted by $D_+^{>ij}$ satisfies the equations:

$$\big[(-\boldsymbol{\partial}^2 + \partial^{02})\delta^{i\ell} + \partial^i\partial^\ell\big]D_+^{>\ell j}(x,x') = \delta^{ij}\,\delta^{(4)}(x-x'), \tag{9.2.3}$$

$$\big[(-\boldsymbol{\partial}'^2 + \partial'^{02})\delta^{\ell i} + \partial'^\ell\partial'^i\big]D_+^{>j\ell}(x,x') = \delta^{ij}\,\delta^{(4)}(x-x') \tag{9.2.4}$$

Directly,from Eqs. (9.1.22), (9.1.5), (9.1.17), the general solution in this region,for $t > t'$, is given by

$$D_+^{>ij}(x,x') = D_0(x,x') + H^{>ij}(x,x'). \tag{9.2.5}$$

and

$$D_0^{ij}(x,x') = \mathrm{i}\int\frac{\mathrm{d}^2\mathbf{K}}{(2\pi)^2}\int\frac{\mathrm{d}q}{2\pi}\,\frac{\mathrm{e}^{\mathbf{K}\cdot(\mathbf{x}_\parallel - \mathbf{x}'_\parallel)}\mathrm{e}^{iq(z-z')}}{|\mathbf{k}|}\,\mathrm{e}^{-\mathrm{i}|\mathbf{k}|(t-t')}\Big(\delta^{ij} - \frac{k^i k^j}{|\mathbf{k}|^2}\Big), \tag{9.2.6}$$

$$\text{where}\quad \mathbf{k} = (\mathbf{K}, q),\quad |\mathbf{k}| = \sqrt{\mathbf{K}^2 + q^2}, \tag{9.2.7}$$

$$H^{>ij}(x,x') = \mathrm{i}\int\frac{\mathrm{d}^2\mathbf{K}}{(2\pi)^2}\int\frac{\mathrm{d}q\,\mathrm{d}q'}{2\pi}\,\frac{\delta(q' + |q|)}{2|\mathbf{k}|}\mathrm{e}^{\mathbf{K}\cdot(\mathbf{x}_\parallel - \mathbf{x}'_\parallel)}\mathrm{e}^{iqz}\mathrm{e}^{-iq'z'} \times \mathrm{e}^{-i|\mathbf{k}|(t-t')}\,G_+^{ij}(\mathbf{K}, q, q', |\mathbf{k}|). \tag{9.2.8}$$

and from Eqs. (9.1.19), (9.1.21),

$$G_+^{ab} = \Big(\delta^{ab} A_+ - \frac{K^a k^b}{|\mathbf{k}|^2} B_+\Big), \qquad a, b = 1, 2, \qquad |\mathbf{k}| = \sqrt{\mathbf{K}^2 + q^2}, \tag{9.2.9}$$

$$G_+^{a3} = \frac{K^a}{|q|}\Big(A_+ - \frac{|\mathbf{K}|^2}{|\mathbf{k}|^2} B_+\Big), \quad G_+^{3b} = -\frac{K^b}{|q|}\Big(A_+ - \frac{|\mathbf{K}|^2}{|\mathbf{k}|^2} B_+\Big),$$
$$G_+^{33} = -\frac{|\mathbf{K}|^2}{|q|^2}\Big(A_+ - \frac{|\mathbf{K}|^2}{|\mathbf{k}|^2} B_+\Big), \tag{9.2.10}$$

where $A_+$ and $B_+$, now, are some new functions to be determined.

**The Refraction Region**: $z < 0, z' > 0$

The expression for the propagator whet it reaches this region, quite generally, may be expressed as

$$D_+^{<ij}(x, x') = \mathrm{i} \int \frac{\mathrm{d}^2\mathbf{K}}{(2\pi)^2} \int \frac{\mathrm{d}q'' \,\mathrm{d}q'}{2\pi}\, \mathrm{e}^{\mathbf{K}\cdot(\mathbf{x}_\| - \mathbf{x}'_\|)} \mathrm{e}^{iq''z''} \mathrm{e}^{-iq'z'}$$
$$\times\, \mathrm{e}^{-i(|\mathbf{k}|(t-t')} G_-^{ij}(\mathbf{K}, q'', q', k^0). \tag{9.2.11}$$

which, in reference to Eqs. (5.4.52), (5.4.53), now, satisfies the homogeneous equations

$$\big[(-\,\boldsymbol{\partial}^2 + \varepsilon\mu \partial^{02})\delta^{i\ell} + \partial^i \partial^\ell\big]\, D_+^{<\ell j}(x, x') = 0, \tag{9.2.12}$$
$$\big[(-\,\boldsymbol{\partial}'^2 + \partial'^{02})\delta^{\ell i} + \partial'^\ell \partial'^i\big]\, D_+^{<j\ell}(x, x') = 0, \tag{9.2.13}$$

where recall that since $z' > 0$, $\varepsilon(z') = 1$, $\mu(z') = 1$, *and* note that $z \neq z'$.

As in Eqs. (9.1.26), (9.1.27), the above two equations now lead, respectively, to

$$\big(|\mathbf{k}''|^2 - \varepsilon\mu k^{02}\big) G_-^{ij} = 0, \quad k''^i G_-^{ij} = 0, \tag{9.2.14}$$
$$\big(|\mathbf{k}'|^2 - k^{02}\big) G_-^{ij} = 0, \quad G_-^{ij} k'^j = 0, \tag{9.2.15}$$
$$\mathbf{k}'' = \big(\mathbf{K}, q''\big), \quad \mathbf{k}' = \big(\mathbf{K}, -q\big). \tag{9.2.16}$$

The first relations in Eqs. (9.2.15, (9.2.16) lead to

$$|\mathbf{k}''|^2 = \varepsilon\mu |\mathbf{k}'|^2, \quad \Rightarrow \quad \mathbf{k}'' = \big(\mathbf{K}, -Q\big), \quad Q = \sqrt{\varepsilon\mu |\mathbf{k}|^2 - |\mathbf{K}|^2}. \tag{9.2.17}$$

The second relations in Eqs. (9.2.15), (9.2.16) together with an identical analysis leads to Eqs. (9.1.32), (9.1.35)–(9.1.37) give

$$G_-^{ab} = \left(\delta^{ab} A_- - \frac{K^a K^b}{|\mathbf{k}|^2} B_-\right), \qquad a, b = 1, 2, \qquad |\mathbf{k}| = \sqrt{\mathbf{K}^2 + q^2}, \tag{9.2.18}$$

$$G_-^{a3} = \frac{K^a}{|q|}\left(A_- - \frac{|\mathbf{K}|^2}{|\mathbf{k}|^2} B_-\right), \quad G_-^{3b} = \frac{K^b}{Q}\left(A_- - \frac{|\mathbf{K}|^2}{|\mathbf{k}|^2} B_-\right),$$
$$G_-^{33} = \frac{|\mathbf{K}|^2}{|q|\,Q}\left(A_- - \frac{|\mathbf{K}|^2}{|\mathbf{k}|^2} B_-\right), \tag{9.2.19}$$

where $A_-$ and $B_-$, now, are some new functions to be determined, $|\mathbf{k}'| = |\mathbf{k}|$.

The energy of the photon in free space is given by $|\mathbf{k}|$ in units of the speed of light c in it set to one. While in the medium the speed of light is given by $\mathrm{c}/\sqrt{\varepsilon\mu} = 1/\sqrt{\varepsilon\mu}$ as established below Eq. (5.5.17) in Chap. 5. Hence the energy of the photon *in* the medium is given by $|\mathbf{k}''|\mathrm{c}/\sqrt{\varepsilon\mu} = |\mathbf{k}|$ and energy is *conserved*, as before.

The B.C. as a function[5] of $z$ are taken to be

$$D_+^{aj}(x, x'), \quad \varepsilon(z) D_+^{3j}(x, x'), \quad \frac{1}{\mu(z)}[\partial^a D_+^{3j}(x, x') - \partial^3 D_+^{aj}(x, x')], \tag{9.2.20}$$

are continuous at $z$ taken at the limiting values of $z = +0$, $z = -0$, working in the temporal gauge as given, e.g. in Eq. (5.3.23).

In the momentum description, these read as

$$G_0^{aj} + G_+^{aj} = G_-^{aj}, \quad \left[G_0^{3j} + G_+^{3j}\right], = \varepsilon G_-^{3j}, \quad G_0^{ij} \equiv \left(\delta^{ij} - \frac{k^i k^j}{|\mathbf{k}|^2}\right), \tag{9.2.21}$$

$$K^a\left(G_0^{3j} + G_+^{3j}\right) + |q|\left(G_0^{aj} - G_+^{aj}\right) = \frac{1}{\mu}\left[K^a G_-^{aj} + Q\, G_-^{aj}\right]. \tag{9.2.22}$$

For $j = b$, the first B.C. gives,

$$A_+ + 1 = A_-, \qquad , B_+ + 1 = B_-. \tag{9.2.23}$$

For $j = b$, the second B.C., gives

$$\left(A_+ - \frac{|\mathbf{K}|^2}{|\mathbf{k}|^2} B_+\right) = \frac{|q|^2}{|\mathbf{k}|^2}\left(\frac{Q - \varepsilon|q|}{Q + \varepsilon|q|}\right). \tag{9.2.24}$$

From the coefficient of $\delta^{ab}$ of the last B.C., together Eq. (9.2.23) give,

$$A_+ = \frac{\mu|q| - Q}{\mu|q| + Q}, \qquad A_- = \frac{2\mu|q|}{\mu|q| + Q}. \tag{9.2.25}$$

Finally, Eqs. (9.2.24), (9.2.25), together Eq. (9.2.23), lead to

[5] See, in particular Eqs. (5.4.27)–(5.4.30).

$$B_+ = \frac{\big(|\mathbf{k}|^2 + |q|^2\big)\big(1 - \varepsilon\mu\big) + Q|q|\big(\mu - \varepsilon\big)}{\big(Q + \mu|q|\big)\big(Q + \varepsilon|q|\big)}, \tag{9.2.26}$$

$$B_- = \frac{2|q|\big(\mu Q + |q|\big)}{\big(Q + \mu|q|\big)\big(Q + \varepsilon|q|\big)}. \tag{9.2.27}$$

Compare the expressions derived for $A_-$, $B_-$, $A_+$, $B_+$ in Eqs. (9.2.25)–(9.2.27) with those derived for $\mu = 1$ in Eqs. (9.1.55)/(9.1.59).

Now that all the components of $G_+^{ij}$ and $G_-^{ij}$ have been found, we may use the identity in Eq. (9.1.60) to express a given tensor $T^{ij}$ in the form,

$$T^{ij} = \delta^{ia}\delta^{bj}T^{ab} + \delta^{i3}\delta^{bj}T^{3b} + \delta^{ia}\delta^{3j}T^{a3} + \delta^{i3}\delta^{3j}T^{33}, \tag{9.2.28}$$

for $i, j = 1, 2, 3$, $a, b = 1, 2$, to express $G_+^{ij}$, with components given Eqs. (9.2.9), (9.2.10), and $G_-^{ij}$, with components given Eqs. (9.2.18), (9.2.19), as

$$G_+^{ij} = \delta^{ia}\delta^{bj}\Big(\delta^{ab}A_+ - \frac{K^aK^b}{|\mathbf{k}|^2}B_+\Big)$$
$$+ \Big(-\delta^{i3}\delta^{bj}\frac{K^b}{|q|} + \delta^{ia}\delta^{3j}\frac{K^a}{|q|} - \delta^{i3}\delta^{j3}\frac{|\mathbf{K}|^2}{|q|^2}\Big)\Big(A_+ - \frac{|\mathbf{K}|^2}{|\mathbf{k}|^2}B_+\Big), \tag{9.2.29}$$

$$\mathbf{k}' = (\mathbf{K}, -|q|), \quad \mathbf{k} = (\mathbf{K}, |q|), \quad k^iG_+^{ij} = 0, \quad G_+^{ij}k'^j = 0, \tag{9.2.30}$$

and

$$G_-^{ij} = \Big(\delta^{ij}A_- - \frac{k''^i k''^j}{|\mathbf{k}|^2}B_-\Big) +$$
$$\frac{1}{|q|Q}\big[\,\mathbf{k}''\cdot\mathbf{k}'\,\delta^{i3}\delta^{j3} + k''^j|q|\,\delta^{i3} + k''^iQ\,\delta^{j3}\,\big]\Big(A_- - \frac{\mathbf{k}''\cdot\mathbf{k}'}{|\mathbf{k}|^2}B_-\Big), \tag{9.2.31}$$

$$\mathbf{k}'' = (\mathbf{K}, -Q), \quad \mathbf{k}' = (\mathbf{K}, -|q|), \qquad k''^iG_-^{ij} = 0, \quad G_-^{ij}k'^j = 0. \tag{9.2.32}$$

and recall that $Q = \sqrt{\varepsilon\mu|\mathbf{k}|^2 - |\mathbf{K}|^2}$.

Given, the three momenta $\mathbf{k}'$, $\mathbf{k}$, $\mathbf{k}''$, of incidence, reflection and refraction, respectively, we may introduce, respective, pairs of polarization vectors

$$\mathbf{e}'_\lambda, \quad \mathbf{e}'_\lambda\cdot\mathbf{e}'_\alpha = \delta_{\lambda\alpha}, \quad \mathbf{e}'_1\times\mathbf{e}'_2 = \frac{\mathbf{k}'}{|\mathbf{k}'|} \equiv \mathbf{n}', \quad \mathbf{n}'\cdot\mathbf{e}'_\lambda = 0, \tag{9.2.33}$$

$$\mathbf{e}_\lambda, \quad \mathbf{e}_\lambda\cdot\mathbf{e}_\alpha = \delta_{\lambda\alpha}, \quad \mathbf{e}_1\times\mathbf{e}_2 = \frac{\mathbf{k}}{|\mathbf{k}|} \equiv \mathbf{n}, \quad \mathbf{n}\cdot\mathbf{e}_\lambda = 0, \tag{9.2.34}$$

$$\mathbf{e}''_\lambda, \quad \mathbf{e}''_\lambda\cdot\mathbf{e}''_\alpha = \delta_{\lambda\alpha}, \quad \mathbf{e}''_1\times\mathbf{e}''_2 = \frac{\mathbf{k}''}{|\mathbf{k}''|} \equiv \mathbf{n}'', \quad \mathbf{n}''\cdot\mathbf{e}''_\lambda = 0, \tag{9.2.35}$$

with $\lambda, \alpha = 1, 2$.

From Eqs. (9.1.68) and (9.1.73), the polarization vectors may be spelled out as

$$\mathbf{e}'_1 = \frac{\big(-\mathbf{K}|q|, -|\mathbf{K}|^2\big)}{|\mathbf{k}|\,|\mathbf{K}|}, \quad \mathbf{e}'_2 = \frac{\big(-K^2, K^1, 0\big)}{|\mathbf{K}|} = \mathbf{e}_2, \quad \mathbf{e}_1 = \frac{\big(\mathbf{K}|q|, -|\mathbf{K}|^2\big)}{|\mathbf{k}|\,|\mathbf{K}|}, \tag{9.2.36}$$

$$\mathbf{e}''_1 = -\frac{\big(\mathbf{K}\,Q, |\mathbf{K}|^2\big)}{\sqrt{\varepsilon\mu}|\mathbf{k}|\,|\mathbf{K}|}, \quad \mathbf{e}''_2 = \frac{\big(-K^2, K^1, 0\big)}{|\mathbf{K}|}, \tag{9.2.37}$$

where $|\mathbf{k}''|^2 = (|\mathbf{k}|^2 + Q^2) = \varepsilon\mu|\mathbf{k}|^2$.

The amplitudes of reflections and refractions are given, respectively, by, now, the familiar expressions

$$\big(\mathbf{k}, \mathbf{e}_\lambda \,|\, \mathbf{k}', \mathbf{e}'_\alpha\big) = \mathrm{e}^i_\lambda\, G^{ij}_+ \mathbf{e}'^j_\alpha, \quad \big(\mathbf{k}'', \mathbf{e}_\lambda \,|\, \mathbf{k}', \mathbf{e}'_\alpha\big) = \mathrm{e}''^i_\lambda\, G^{ij}_- \mathbf{e}'^j_\alpha, \quad \lambda, \alpha = 1, 2. \tag{9.2.38}$$

From the explicit expressions of $G^{ij}_+$ in Eq. (9.2.29), and $G^{ij}_-$ in Eq. (9.2.31), and of the functions $A_+$, $B_+$, $A_-$, $B_-$ in Eqs. (9.2.25)–(9.2.27), straightforward algebra gives

$$\big(\mathbf{k}, \mathbf{e}_\lambda \,|\, \mathbf{k}', \mathbf{e}'_\alpha\big) = \frac{\varepsilon|q| - Q}{\varepsilon|q| + Q}\,\delta_{\lambda 1}\delta_{\alpha 1} + \frac{\mu|q| - Q}{\mu|q| + Q}\,\delta_{\lambda 2}\delta_{\alpha 2}, \tag{9.2.39}$$

for amplitudes of reflections, and

$$\big(\mathbf{k}'', \mathbf{e}_\lambda \,|\, \mathbf{k}', \mathbf{e}'_\alpha\big) = \frac{2\sqrt{\varepsilon\mu}\,|q|}{\varepsilon|q| + Q}\,\delta_{\lambda 1}\delta_{\alpha 1} + \frac{2\mu|q|}{\mu|q| + Q}\,\delta_{\lambda 2}\delta_{\alpha 2}, \tag{9.2.40}$$

for amplitudes of refractions, both with polarizations parallel and normal to the plane of incidence, respectively.

We may equivalently, rewrite these amplitudes in terms of angle of incidence $\vartheta_{\text{inc}}$ and use the following identity

$$\sin\vartheta_{\text{inc}} = \frac{|\mathbf{K}|}{|\mathbf{k}|}, \quad \cos\vartheta_{\text{inc}} = \frac{|q|}{|\mathbf{k}|}, \quad Q = |\mathbf{k}|\sqrt{\varepsilon\mu - \sin^2\vartheta_{\text{inc}}}. \tag{9.2.41}$$

According to the first relations in Eqs. (9.2.39), (9.2.40), the relative amplitudes of reflection and refraction with polarization $\mathbf{e}'_1$ of incidence *parallel* to the plane of incidence, respectively, given by

$$\frac{\varepsilon\cos\vartheta_{\text{inc}} - \sqrt{\varepsilon\mu - \sin^2\vartheta_{\text{inc}}}}{\varepsilon\cos\vartheta_{\text{inc}} + \sqrt{\varepsilon\mu - \sin^2\vartheta_{\text{inc}}}}, \tag{9.2.42}$$

and

$$\frac{2\sqrt{\varepsilon\mu}\cos\vartheta_{\text{inc}}}{\varepsilon\cos\vartheta_{\text{inc}}+\sqrt{\varepsilon\mu-\sin^2\vartheta_{\text{inc}}}}, \tag{9.2.43}$$

where all the polarization of incidence, reflection and refraction are parallel to the plane of incidence corresponding to the ones on the l.h.s of Fig. 9.1 in the present general case.

On the other hand, according to second relations Eqs. (9.2.39), (9.2.40), the relative amplitudes of reflection and refraction with polarization $\mathbf{e}_2'$ of incidence *normal* to the plane of incidence are, respectively, given by

$$\frac{\mu\cos\vartheta_{\text{inc}}-\sqrt{\varepsilon\mu-\sin^2\vartheta_{\text{inc}}}}{\mu\cos\vartheta_{\text{inc}}+\sqrt{\varepsilon\mu-\sin^2\vartheta_{\text{inc}}}}, \tag{9.2.44}$$

and

$$\frac{2\mu\cos\vartheta_{\text{inc}}}{\mu\cos\vartheta_{\text{inc}}+\sqrt{\varepsilon\mu-\sin^2\vartheta_{\text{inc}}}}, \tag{9.2.45}$$

where all the polarization of incidence, reflection and refraction are normal to the plane of incidence corresponding to the ones on the r.h.s of Fig. 9.1 in the present case.

## References


1. Manoukian, E. B. (1987a). Reflection-off a reflecting surface in quantum mechanics. Where do the reflections actuall occur ?. *Il Nuovo Cimento B, 99 (2)*, 133–143.
2. Manoukian, E. B. (1987b). Reflection-off a reflecting surface in quantum mechanics. Where do the reflections actuall occur ? II. The law of reflection. *Il Nuovo Cimento B, 100 (2)*, 185–193.
3. Manoukian, E. B. (1992). Quantum field theory of reflection of light. *Hadronic Journal,20*, 253–265.
4. Manoukian, E. B. (1996). Quamntum field theory of refraction and reflection of light. I. *Hadronic Journal, 19*, 581–595.
5. Manoukian, E. B. (1997). Quamntum field theory of refraction and reflection of light. II. *Hadronic Journal, 20*, 251–265.
6. Feynman R. P. (1985). *QED: The strange theory of light and matter*. Princeton, New Jersey: Princeton University Press.

# Chapter 10
# Red Light, Blue Light, Feynman and a Layer of Glass

*Follow red and blue photons as they encounter a layer of glass and meet their fates upon detection; which one is more likely to be reflected, and which is more likely to be transmitted*

In the present chapter, a detailed study is carried out of the fascinating problem involving the fate of photons impinging on a layer of glass of flat parallel surfaces. We explicitly derive the propagator for transmission through, and reflection off, such a layer of glass. The corresponding transition amplitudes of a polarized or unpolarized photon are then obtained. The analysis is physically appealing done here through the study of the propagator, as a time evolution process, from the time of emission to the time of detection avoiding plane waves. The underlying interpretation that emerges from the theory is applied to study the fate of red and blue photons. We were inspired by Feynman's intuitive and highly non-technical study of red and blue photons in such a situation.[1]

The propagator was first derived by the author in the reference cited below.[2] The analysis already carried out in Sects. 9.1, 9.2 of Chap. 9 will be useful to obtain the solutions encountered in the present chapter.

---

[1] Feynman (1985), [1].

[2] Manoukian (2013) [2].



E. B. Manoukian, *Radiation*,
https://doi.org/10.1007/978-3-032-21506-2_10

## 10.1 Setting-Up the Scattering Problem

We consider a slab of dielectric constant $\varepsilon$ "embedded" in free space:

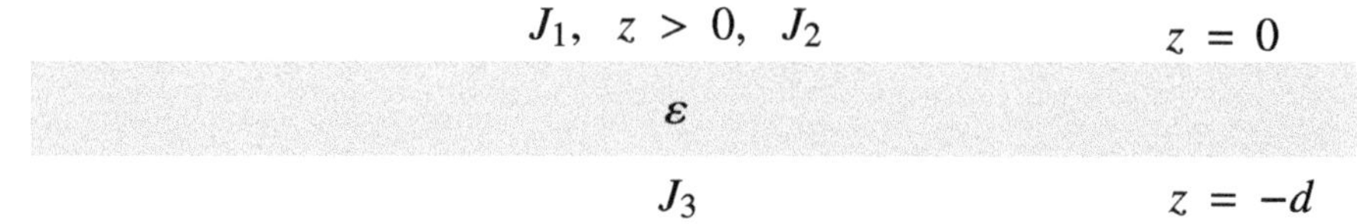


We set $\mu = 1$. Here we have three sources $J_1$, $J_2$, $J_3$, for a photon emission, of detection of a reflected photon, and a detection of a transmitted one, respectively.

Corresponding to the photon propagator $D(x, x')$, with $x = (t, \mathbf{x}_\parallel, z)$, we always have $z' > 0$, $t > t'$. The possible values of $z$ encountered below are: $z > 0$, as $z$ changes from the free region in the reflection region above the slab, to $0 > z > -d$ within the slab, and $-d > z$ in the transmission region. One then applies proper boundary conditions at these crossing points of the $z-$ values. Here $\mathrm{x}_\parallel$ is parallel to the surfaces of the slab.

As in Chap. 9, Sect. 9.1, in the reflection, free dpace, region $z > 0$, $z' > 0$, the propagator, in question, satisfies the following equations

$$\big[(-\boldsymbol{\partial}^2 + \partial^{02})\delta^{i\ell} + \partial^i\partial^\ell\big]D_+^{\ell j}(x, x') = \delta^{ij}\,\delta^{(4)}(x - x'), \tag{10.1.1}$$
$$\big[(-\boldsymbol{\partial}'^2 + \partial'^{02})\delta^{\ell i} + \partial'^\ell\partial'^i\big]D^{j\ell}(x, x') = \delta^{ij}\,\delta^{(4)}(x - x') \tag{10.1.2}$$

while for the $z$ variable within the slab ( $z \neq z'$, $z' > 0$), the propagator satisfies the equations

$$\big[(-\boldsymbol{\partial}^2 + \varepsilon\partial^{02})\delta^{i\ell} + \partial^i\partial^\ell\big]D_+^{\ell j}(x, x') = 0, \tag{10.1.3}$$
$$\big[(-\boldsymbol{\partial}'^2 + \partial'^{02})\delta^{\ell i} + \partial'^\ell\partial'^i\big]D_+^{j\ell}(x, x') = 0, \tag{10.1.4}$$

and for the $z$ variable in the transmission region ( $z \neq z'$, $z' > 0$), it satisfies the equations

$$\big[(-\boldsymbol{\partial}^2 + \partial^{02})\delta^{i\ell} + \partial^i\partial^\ell\big]D_+^{\ell j}(x, x') = 0, \tag{10.1.5}$$
$$\big[(-\boldsymbol{\partial}'^2 + \partial'^{02})\delta^{\ell i} + \partial'^\ell\partial'^i\big]D^{j\ell}(x, x') = 0, \tag{10.1.6}$$

where $x = (t, \mathrm{x}_\parallel, z)$, $x' = (t', \mathrm{x}'_\parallel, z')$, and $\mathrm{x}_\parallel$, $\mathrm{x}'_\parallel$ are parallel to the surfaces of the slab, as mentioned earlier.

The boundary conditions for the propagator as a function of $z$ are:

$$D_+^{aj}(x, x'),\quad \varepsilon(z)D_+^{3j}(x, x'),\quad \text{and}\quad [\partial^3 D_+^{aj}(x, x') - \partial^a D_+^{3j}(x, x')]$$
$$\text{are continuous at } z = 0 \text{ and at } z = -d,\quad a = 1, 2;\quad j = 1, 2, 3, \tag{10.1.7}$$

where $\varepsilon(z)$, corresponds to $\varepsilon$ within the slab, i.e. for $-d < z < 0$. We consider the causal relation $t > t'$.

## 10.2 General Expression of the Propagator

**The Reflection Region**: $z > 0$:

As in Sects. 9.1, 9.2, in Chap. 9, for $t \neq t'$, the general expression of the propagator in this region, for $t > t'$, may, from Eqs. (9.2.5)–(9.2.8), may, in a convenient notation, written as

$$D_{+}^{>ij}(x, x') = \mathrm{i}\int \frac{\mathrm{d}^2\mathbf{K}}{(2\pi)^2}\int \frac{\mathrm{d}q}{2\pi\ 2|\mathbf{k}|}\mathrm{e}^{\mathbf{K}\cdot(\mathbf{x}_{\parallel}-\mathbf{x}'_{\parallel})}\mathrm{e}^{-\mathrm{i}|\mathbf{k}|(t-t')}$$
$$\times\ \Big[\mathrm{e}^{\mathrm{i}(qz-qz')}\Big(\delta^{ij} - \frac{k^i k^j}{|\mathbf{k}|^2}\Big) + \mathrm{e}^{-\mathrm{i}(qz+qz')}G_{+}^{ij}\Big], \qquad (10.2.1)$$

where $\mathbf{k} = (\mathbf{K}, q)$, and we set $\mathbf{k}' = (\mathbf{K}, -q)$, $|\mathbf{k}'| = |\mathbf{k}|$, where from Eqs. (9.2.9) (9.2.10), we may write

$$G_{+}^{ab} = \Big(\delta^{ab}A_{+} - \frac{K^a K^b}{|\mathbf{k}|^2}B_{+}\Big), \qquad a, b = 1, 2, \qquad (10.2.2)$$

and

$$G_{+}^{a3} = \frac{K^a}{|q|}\Big(A_{+} - \frac{|\mathbf{K}|^2}{|\mathbf{k}|^2}B_{+}\Big), \qquad (10.2.3)$$

$$G_{+}^{3b} = -\frac{K^b}{|q|}\Big(A_{+} - \frac{|\mathbf{K}|^2}{|\mathbf{k}|^2}B_{+}\Big), \qquad (10.2.4)$$

$$G_{+}^{33} = -\frac{|\mathbf{K}|^2}{|q|^2}\Big(A_{+} - \frac{|\mathbf{K}|^2}{|\mathbf{k}|^2}B_{+}\Big), \qquad (10.2.5)$$

to use when applying the boundary conditions. As before, the above expressions follow from the tranversality conditions:

$$k^i G_{+}^{ij} = 0, \quad G_{+}^{ij}k'^j = 0, \qquad \mathbf{k} = (\mathbf{K}, q), \ \mathbf{k}' = (\mathbf{K}, -|q|), \ |\mathbf{k}'| = |\mathbf{k}|. \quad (10.2.6)$$

with, now, some new coefficients in Eqs. (10.2.4), (10.2.4) are to be determined. The vectors $\mathbf{k}', \mathbf{k}$ denote momenta of incidence and reflection respectively, as we have seen in Sect. 9.2.

**The Region Within the Slab**: $-d < z < 0$ :

Here the momentum component $q$ as in Sect. 9.1, changes to $Q = \sqrt{\varepsilon|\mathbf{k}|^2 - |\mathbf{K}|^2|}$. From Eq. (10.2.1), we may infer that we may, quite generally, introduce two structure functions $M_1^{ij}$, $M_2^{ij}$, and, in a convenient notation, write the propagator in this region as

$$\begin{aligned} \underline{D}_+^{<ij}(x,x') = \mathrm{i}\int \frac{\mathrm{d}^2\mathbf{K}}{(2\pi)^2}\int\frac{\mathrm{d}q}{2\pi\, 2\mathbf{k}|}\, \mathrm{e}^{\mathbf{K}\cdot(\mathbf{x}_\parallel - \mathbf{x}'_\parallel)}\mathrm{e}^{-\mathrm{i}|\mathbf{k}|(t-t')} \\ \times\ \big[\mathrm{e}^{\mathrm{i}(Qz-qz')}M_1^{ij} + \mathrm{e}^{-\mathrm{i}(Qz+qz')}M_2^{ij}\big]. \end{aligned} \tag{10.2.7}$$

Also

$$\partial^i \underline{D}_+^{<ij} = 0, \tag{10.2.8}$$

which lead to the relation

$$K^a\big(\mathrm{e}^{-\mathrm{i}Qd}M_1^{aj} + \mathrm{e}^{\mathrm{i}Qd}M_2^{aj}\big) = -Q\big(\mathrm{e}^{-\mathrm{i}Qd}M_1^{3j} - \mathrm{e}^{\mathrm{i}Qd}M_2^{3j}\big). \tag{10.2.9}$$

**The Transmission Region**: $z < -d$:

Quite generally, when the $z$ value of the propagator "leaves" the slab, i.e., $z < -d$, we may introduce general coefficients $A_-$, $B_-$, to be determined, and write it in the form

$$D_+^{<ij}(x,x') = \mathrm{i}\int\frac{\mathrm{d}^2\mathbf{K}}{(2\pi)^2}\int\frac{\mathrm{d}q}{2\pi\, 2|\mathbf{k}|}\mathrm{e}^{\mathbf{K}\cdot(\mathbf{x}_\parallel - \mathbf{x}'_\parallel)}\mathrm{e}^{-\mathrm{i}|\mathbf{k}|(t-t')}\mathrm{e}^{\mathrm{i}(qz-qz')}G_-^{ij}. \tag{10.2.10}$$

with

$$G_-^{ab} = \Big(\delta^{ab}A_- - \frac{K^aK^b}{|\mathbf{k}|^2}B_-\Big), \qquad a,b = 1,2, \qquad |\mathbf{k}| = \sqrt{\mathbf{K}^2 + q^2}, \tag{10.2.11}$$

$$\begin{aligned} G_-^{a3} = \frac{K^a}{|q|}\Big(A_- - \frac{|\mathbf{K}|^2}{|\mathbf{k}|^2}B_-\Big),\quad G_-^{3b} = \frac{K^b}{|q|}\Big(A_- - \frac{|\mathbf{K}|^2}{|\mathbf{k}|^2}B_-\Big), \\ G_-^{33} = \frac{|\mathbf{K}|^2}{|q|^2}\Big(A_- - \frac{|\mathbf{K}|^2}{|\mathbf{k}|^2}B_-\Big). \end{aligned} \tag{10.2.12}$$

We may then apply the boundary condition of the propagator when it is within the slab, at $z = -d + 0$, given in Eq. (10.2.7), with the expression of the propagator for $z = -d - 0$, given in Eq. (10.2.10).

It is readily verified that

$$k''^i G_-^{ij} = 0, \qquad G_-^{ij}k'^j = 0, \quad \mathbf{k}' = (\mathbf{K}, -|q|) = \mathbf{k}'', \tag{10.2.13}$$

where $\mathbf{k}' = \mathbf{k}''$ denote the momenta of incidence and transmission.

## 10.3 Satisfying the Boundary Conditions

**B.C. at** $z = 0$:

The first, second, and third B.C. for $j = b$ in Eq. (10.1.7), respectively, at $z = 0$ now correspond to the following conditions to be satisfied:

$$\left(\delta^{ab} - \frac{K^a K^b}{|\mathbf{k}|^2}\right) + G_+^{ab} = M_1^{ab} + M_2^{ab}, \tag{10.3.1}$$

$$\frac{|q|K^b}{|\mathbf{k}|^2} - \frac{K^c}{|q|}G_+^{cb} = \varepsilon\left(-\frac{K^c}{Q}M_1^{cb} + \frac{K^c}{Q}M_2^{cb}\right), \tag{10.3.2}$$

$$\begin{aligned} -|q|\left(\delta^{ab} - \frac{K^a K^b}{|\mathbf{k}|^2}\right) + |q|G_+^{ab} - \frac{|q|K^a K^b}{|\mathbf{k}|^2} + \frac{K^a K^c}{|q|}G_+^{cb} \\ = Q(M_1^{ab} - M_2^{ab}) - K^a\left(-\frac{K^c}{Q}M_1^{cb} + \frac{K^c}{Q}M_2^{cb}\right). \end{aligned} \tag{10.3.3}$$

These allow us to solve for $M_1^{ab}$, $M_2^{ab}$:

$$\begin{aligned} 2M_1^{ab} &= \left(\frac{Q - |q|}{Q}\right)\delta^{ab} + \left(\frac{Q + |q|}{Q}\right)G_+^{ab} \\ &+ \left(1 - \frac{1}{\varepsilon}\right)\frac{K^a K^c}{|q|Q}G_+^{cb} + \left(\frac{|q|}{\varepsilon Q} - 1\right)\frac{K^a K^b}{|\mathbf{k}|^2}, \end{aligned} \tag{10.3.4}$$

and

$$\begin{aligned} 2M_2^{ab} &= \left(\frac{Q + |q|}{Q}\right)\delta^{ab} + \left(\frac{Q - |q|}{Q}\right)G_+^{ab} \\ &- \left(1 - \frac{1}{\varepsilon}\right)\frac{K^a K^c}{|q|Q}G_+^{cb} - \left(\frac{|q|}{\varepsilon Q} + 1\right)\frac{K^a K^b}{|\mathbf{k}|^2}. \end{aligned} \tag{10.3.5}$$

**B.C. at** $z = -d$:

$$\mathrm{e}^{-\mathrm{i}Qd}M_1^{ab} + \mathrm{e}^{\mathrm{i}Qd}M_2^{ab} = \mathrm{e}^{-\mathrm{i}qd}G_-^{ab}, \tag{10.3.6}$$

$$\varepsilon\left[-\mathrm{e}^{-\mathrm{i}Qd}\left(\frac{K^c}{Q}M_1^{cb}\right) + \mathrm{e}^{\mathrm{i}Qd}\left(\frac{K^c}{Q}M_2^{cb}\right)\right] = \mathrm{e}^{-\mathrm{i}qd}\left(\frac{K^c}{|q|}G_-^{cb}\right), \tag{10.3.7}$$

$$\begin{aligned} Q[\mathrm{e}^{-\mathrm{i}Qd}M_1^{ab} - \mathrm{e}^{\mathrm{i}Qd}M_2^{ab}] - K^a\left[-\mathrm{e}^{-\mathrm{i}Qd}\left(\frac{K^c}{Q}M_1^{cb}\right) + \mathrm{e}^{\mathrm{i}Qd}\left(\frac{K^c}{Q}M_2^{cb}\right)\right] \\ = \mathrm{e}^{-\mathrm{i}qd}\left[|q|G_-^{ab} - \frac{K^a K^c}{|q|}G_-^{cb}\right]. \end{aligned} \tag{10.3.8}$$

From Eqs. (10.3.4)–(10.3.6), it readily follows that

$$2M_1^{ab} = e^{+i(Q+|q|)d}\Big[\frac{(Q-|q|)}{Q}G_-^{ab} - \Big(1-\frac{1}{\varepsilon}\Big)\frac{K^aK^c}{|q|Q}G_-^{cb}\Big], \tag{10.3.9}$$

$$2M_2^{ab} = e^{-i(Q-|q|)d}\Big[\frac{(Q+|q|)}{Q}G_-^{ab} + \Big(1-\frac{1}{\varepsilon}\Big)\frac{K^aK^c}{|q|Q}G_-^{cb}\Big]. \tag{10.3.10}$$

The coefficients $(A_+, B_+)$; $(A_-, B_-)$ of the structures $G_+^{ij}$ and $G_-^{ij}$ in Eqs. (10.2.2)–(10.2.5), and Eqs. (10.2.11), (10.2.12), respectively, relevant to the reflection and transmission processes, may be directly obtained upon comparisons of the expression of $M_1^{ab}$ in Eq. (10.3.4) with that one given for it in Eq. (10.3.9), and by comparing the expressions of $M_2^{ab}$ in Eq. (10.3.5) with that one given for it Eq. (10.3.10).

In particular upon these comparisons of the terms having $\delta^{ab}$ in them, involve, as follow from the given structures of $G_\pm^{ab}$, *only* the $A_\pm$ terms, i.e., no $B_\pm$ terms. The terms involving only $\delta^{ab}$ in $M_1^{ab}$, $M_2^{ab}$ give, respectively, the following two equations

$$\Big(\frac{Q-|q|}{Q}\Big) + \Big(\frac{Q+|q|}{Q}\Big)A_+ = e^{+i(Q+|q|)d}\,\frac{(Q-|q|)}{Q}A_-, \tag{10.3.11}$$

$$\Big(\frac{Q+|q|}{Q}\Big) + \Big(\frac{Q-|q|}{Q}\Big)A_+ = e^{-i(Q-|q|)d}\,\frac{(Q+|q|)}{Q}A_-, \tag{10.3.12}$$

involving only the unknown coefficients $A_\pm$. The solutions of these equation are easily obtained and are given by

$$A_+ = \frac{(Q^2-|q|^2)[e^{iQd}-e^{-iQd}]}{(Q+|q|)^2e^{-iQd}-(Q-|q|)^2e^{iQd}}, \tag{10.3.13}$$

$$A_- = \frac{-4|q|Qe^{-i|q|d}}{(Q+|q|)^2e^{-iQd}-(Q-|q|)^2e^{iQd}}. \tag{10.3.14}$$

With $A_\pm^{ab}$ explicitly known, we may now compare the coefficients *of* $K^aK^b$ in Eq. (10.3.4) with the one given in Eq. (10.3.9), and compare the coefficients $K^aK^b$ in $M_2^{ab}$ in Eq. (10.3.5) with the one given in Eq. (10.3.10). Some straightforward, though long algebraic steps, give

$$-\frac{|\mathbf{k}|^2}{|\mathbf{K}|^2}B_+ = \Big[\frac{|q|^2}{|\mathbf{k}|^2}\frac{(Q^2-\varepsilon^2|q|^2)}{(Q-\varepsilon|q|)^2\,e^{iQd}-(Q+\varepsilon|q|)^2\,e^{-iQd}} + \frac{(Q^2-|q|^2)}{(Q-|q|)^2\,e^{iQd}-(Q+|q|)^2\,e^{-iQd}}\Big]\big[e^{iQd}-e^{-iQd}\big], \tag{10.3.15}$$

and

$$-\frac{|\mathbf{k}|^2}{|\mathbf{K}|^2}B_- = -4|q|Q\Big[\frac{|q|^2}{|\mathbf{k}|^2}\frac{\varepsilon}{(Q-\varepsilon|q|)^2\,\mathrm{e}^{\mathrm{i}Qd}-(Q+\varepsilon|q|)^2\,\mathrm{e}^{-\mathrm{i}Qd}} - \frac{1}{(Q-|q|)^2\,\mathrm{e}^{\mathrm{i}Qd}-(Q+|q|)^2\,\mathrm{e}^{-\mathrm{i}Qd}}\Big]\mathrm{e}^{-\mathrm{i}|q|d}. \tag{10.3.16}$$

## 10.4 Reflections and Transmissions: Unpolarized Cases

With the explicit solutions of the components of $G_+^{ij}$, expressed in terms of $A_+$, $B_+$ in Eqs. (10.3.13), (10.3.15), relevant to the reflection process, one may readily express the latter, from Eqs. (10.2.2)–(10.2.5), in the convenient equivalent form as

$$G_+^{ij} = \Big(\delta^{ij}A_+ - \frac{k^i k'^j}{|\mathbf{k}|^2}B_+\Big) - \frac{1}{|q|^2}\big[\big((|\mathbf{k}|^2+2|q|^2)A_+ - |\mathbf{k}|^2 B_+\big]\delta^{i3}\delta^{3j} + \\ + \frac{1}{|q|}\big(k^i\delta^{3j}-\delta^{i3}k'^j\big)\Big(A_+ - B_+\Big), \qquad \mathbf{k}=(\mathbf{K},|q|),\ \mathbf{k}'=(\mathbf{K},-|q|), \tag{10.4.1}$$

as is easily verified by choosing $i, j = 1, 2, 3$ and by comparing the resulting expressions with the components spelled out in Eqs. (10.2.2)–(10.2.5) mentioned above. Recall that $\mathbf{k}'$, $\mathbf{k}$ represent the incident and reflected momenta, respectively. The important transversality properties in Eq. (10.2.6) are also satisfied by the expression of $G_+^{ij}$ above,

$$k^i G_+^{ij} = 0, \qquad G_+^{ij}k'^j = 0. \tag{10.4.2}$$

Similarly, we may conveniently write $G_-^{ij}$, relevant for the study of transmission, as

$$G_-^{ij} = \Big(\delta^{ij}A_- - \frac{k''^i k'^j}{|\mathbf{k}|^2}B_-\Big) + \\ + \Big(A_- - B_-\Big)\Big[\frac{|\mathbf{k}|^2}{|q|^2}\delta^{i3}\delta^{3j} + \frac{k''^i\delta^{3j}+\delta^{i3}k'^j}{|q|}\Big], \qquad \mathbf{k}'=(\mathbf{K},-|q|)=\mathbf{k}''. \tag{10.4.3}$$

Again by choosing $i, j = 1, 2, 3$ in the latter, it is readily verified that the resulting expressions coincide with one given in Eqs. (10.2.11), (10.2.12), and that the transversality conditions in Eq. (10.2.13) are satisfied as well

$$k''^i G_-^{ij} = 0, \qquad G_-^{ij}k'^j = 0. \tag{10.4.4}$$

Given the three momenta of incidence, reflection and transmission, $\mathbf{k}'$, $\mathbf{k}$ and transmission $\mathbf{k}'' = \mathbf{k}'$, respectively, one may, consistently introduce pairs of polarization vectors $(\mathbf{e}'_1, \mathbf{e}'_2)$, $(\mathbf{e}_1, \mathbf{e}_2)$, $(\mathbf{e}''_1, \mathbf{e}''_2) = (\mathbf{e}'_1, \mathbf{e}'_2)$, respectively, such that

$$\mathbf{e}'_\alpha \cdot \mathbf{e}'_\lambda = \delta_{\alpha\lambda}, \qquad \mathbf{k}' \cdot \mathbf{e}'_\lambda = 0, \qquad \mathbf{e}'_1 \times \mathbf{e}'_2 = \frac{\mathbf{k}'}{|\mathbf{k}'|} \equiv \mathbf{n}', \tag{10.4.5}$$

$$\delta^{ij} = \mathbf{n}'^i \mathbf{n}'^j + \sum_{\lambda=1,2} \mathbf{e}'^i_\lambda \mathbf{e}'^j_\lambda, \tag{10.4.6}$$

and

$$\mathbf{e}_\alpha \cdot \mathbf{e}_\lambda = \delta_{\alpha\lambda}, \qquad \mathbf{k} \cdot \mathbf{e}_\lambda = 0, \qquad \mathbf{e}_1 \times \mathbf{e}_2 = \frac{\mathbf{k}}{|\mathbf{k}|} \equiv \mathbf{n}, \tag{10.4.7}$$

$$\delta^{ij} = \mathbf{n}^i \mathbf{n}^j + \sum_{\lambda=1,2} \mathbf{e}^i_\lambda \mathbf{e}^j_\lambda, \tag{10.4.8}$$

where $\alpha, \lambda = 1, 2$, and note that $|\mathbf{k}'| = |\mathbf{k}| = |\mathbf{k}''| = \sqrt{\mathbf{K}^2 + q^2}$, of vectors lying *outside* the slab.

Given the above polarization vectors, the amplitude of reflection is given by

$$\big(\mathbf{k}, \mathbf{e}_\alpha \mid \mathbf{k}', \mathbf{e}'_\lambda\big) = \mathbf{e}^i_\alpha G^{ij}_+ \mathbf{e}'^j_\lambda, \tag{10.4.9}$$

By taking into account the transversality conditions in Eq. (10.4.2), we may eliminate the $k^i$, $k'^j$ terms, in the completeness relations Eqs. (10.4.6), (10.4.8) when applied to the polarization vectors on the right-hand side of the above equation.

For unpolarized photons, we have to average over the initial polarizations probabilities, i.e., over $\lambda = 1, 2$ in Eq. (10.4.9), and sum over the final ones, i.e., over $\alpha = 1, 2$, leading to the probability of reflection given by

$$\mathrm{Prob}|_{\mathrm{refl.}}[\mathbf{k}' \to \mathbf{k}] = \sum_{\alpha=1,2} \Big(\frac{1}{2} \sum_{\lambda=1,2} |\big(\mathbf{k}, \mathbf{e}_\alpha \mid \mathbf{k}', \mathbf{e}'_\lambda\big)|^2\Big) \tag{10.4.10}$$

$$= \frac{1}{2} \sum_{\alpha=1,2} \sum_{\lambda=1,2} \mathbf{e}^i_\alpha \, G^{ij}_+ \, \mathbf{e}'^j_\lambda \, \mathbf{e}^\ell_\alpha \, G^{\ell m}_+ \, \mathbf{e}'^m_\lambda$$

$$= \frac{1}{2} \delta^{i\ell} \delta^{jm} G^{ij}_+ G^{\ell m}_+ = \frac{1}{2} G^{ij}_+ G^{ij}_+. \tag{10.4.11}$$

From the explicit expression of $G^{ij}_+$ in Eq. (10.4.1), this, after some straightforward, though long algebraic steps, lead to

$$\mathrm{Prob}|_{\mathrm{refl.}}[\mathbf{k}' \to \mathbf{k}] = \frac{1}{2}\Big[ |A_+|^2 + \frac{|\mathbf{k}|^4}{|q|^4} \Big| A_+ - \frac{|\mathbf{K}|^2}{|\mathbf{k}|^2} B_+ \Big|^2 \Big]. \tag{10.4.12}$$

Both terms in the probability, within the square brackets, may be obtained from the explicit expressions of $A_+$ and $B_+$ in Eqs. (10.3.13) and (10.3.15), respectively, and are given by

$$|A_+|^2 = \frac{(Q^2 - |q|^2)^2 \sin^2 Qd}{4|q|^2 Q^2 + (Q^2 - |q|^2)^2 \sin^2 Qd}, \tag{10.4.13}$$

$$\frac{|\mathbf{k}|^4}{|q|^4}\left|A_+ - \frac{|\mathbf{K}|^2}{|\mathbf{k}|^2} B_+\right|^2 = \frac{(Q^2 - \varepsilon^2|q|^2)^2 \sin^2 Qd}{4\varepsilon^2|q|^2 Q^2 + (Q^2 - \varepsilon^2|q|^2)^2 \sin^2 Qd}, \tag{10.4.14}$$

where recall that $Q = \sqrt{\varepsilon|\mathbf{k}|^2 - |\mathbf{K}|^2}$.

Similarly, to derive the explicit expression of transmission, we note the amplitude of transmission is given by

$$\left(\mathbf{k}'', \mathbf{e}''_\alpha \mid \mathbf{k}', \mathbf{e}'_\lambda\right) = \mathrm{e}^i_\alpha G^{ij}_- \mathrm{e}'^j_\lambda, \qquad \mathbf{k}'' = \mathbf{k}', \ \mathbf{e}''_\alpha = \mathbf{e}'_\alpha, \tag{10.4.15}$$

where the polarizations vectors $\mathbf{e}'_\lambda$, $\lambda = 1, 2$, are introduced in Eqs. (10.4.5), (10.4.6).

By taking into account the transversality conditions in Eq. (10.4.4), we may eliminate the $k^i$, $k'^j$ terms, in the completeness relations Eq. (10.4.6), when applied to the polarization vectors on the right-hand side of the above equation.

For unpolarized photons, we have to average over the initial polarizations probabilities, i.e., over $\lambda = 1, 2$ in Eq. (10.4.15), and sum over the final ones, i.e., over $\alpha = 1, 2$, leading to the probability of transmission given by

$$\mathrm{Prob}|_{\text{trans.}}[\mathbf{k}' \to \mathbf{k}''] = \sum_{\alpha=1,2} \Big(\frac{1}{2} \sum_{\lambda=1,2} |(\mathbf{k}'', \mathbf{e}''_\alpha \mid \mathbf{k}', \mathbf{e}'_\lambda)|^2\Big) \tag{10.4.16}$$

$$= \frac{1}{2} \sum_{\alpha=1,2} \sum_{\lambda=1,2} \mathrm{e}''^i_\alpha \, G^{ij}_- \, \mathrm{e}'^j_\lambda \, \mathrm{e}''^\ell_\alpha \, G^{\ell m}_- \, \mathrm{e}'^m_\lambda$$

$$= \frac{1}{2} \delta^{i\ell} \delta^{jm} \, G^{ij}_- \, G^{\ell m}_- = \frac{1}{2} \, G^{ij}_- \, G^{ij}_- . \tag{10.4.17}$$

From the explicit expression of $G^{ij}_-$ in Eq. (10.4.3), this, after some straightforward, though long algebraic steps, lead to

$$\mathrm{Prob}|_{\text{trans.}}[\mathbf{k}' \to \mathbf{k}''] = \frac{1}{2}\left[|A_-|^2 + \frac{|\mathbf{k}|^4}{|q|^4}\left|A_- - \frac{|\mathbf{K}|^2}{|\mathbf{k}|^2} B_-\right|^2\right]. \tag{10.4.18}$$

Both terms in the probability, within the square brackets, may be obtained from the explicit expressions of $A_-$ and $B_-$ in Eqs. (10.3.14) and (10.3.16), respectively, and are given by

$$|A_-|^2 = \frac{4|q|^2 Q^2}{4|q|^2 Q^2 + (Q^2 - |q|^2)^2 \sin^2 Qd}, \tag{10.4.19}$$

$$\frac{|\mathbf{k}|^4}{|q|^4}\left|A_- - \frac{|\mathbf{K}|^2}{|\mathbf{k}|^2}B_-\right|^2 = \frac{4\varepsilon^2|q|^2Q^2}{4\varepsilon^2|q|^2Q^2 + (Q^2 - \varepsilon^2|q|^2)^2\sin^2 Qd}. \tag{10.4.20}$$

I hope you have noticed that the first terms in the probabilities of reflection and transmission $|A_+|^2$, $|A_-|^2$ as given, respectively, in Eqs. (10.4.13), (10.4.19) add up to 1, while the second terms, as given, respectively, in Eqs. (10.4.14), (10.4.20), also add up to 1. With the 1/2 factors multiplying both probabilities in Eqs. (10.4.12), (10.4.18), the conservation law of probabilities

$$\text{Prob[reflection]} + \text{Prob[transmission]} = 1, \tag{10.4.21}$$

is established with flying colors.

## 10.5 Reflections and Transmissions: Polarized Processes

In the present section, we work out the probabilities of reflection and transmission for a polarized photon. For definiteness, the momenta of incidence, reflection and transmission are chosen to be

$$\mathbf{k}' = (0, K, -|q|), \qquad \mathbf{k} = (0, K, |q|), \qquad \mathbf{k}'' = (0, K, -|q|). \tag{10.5.1}$$

Given these momenta, respective pairs of polarization vectors, consistent with the conditions spelled out in Eqs. (10.4.5), (10.4.7) are given by

$$\mathbf{e}'_1 = (1, 0, 0), \quad \mathbf{e}'_2 = \frac{(0, -|q|, -K)}{|\mathbf{k}|}, \qquad \mathbf{k}' \cdot \mathbf{e}'_\lambda = 0, \qquad \mathbf{e}'_1 \times \mathbf{e}'_2 = \frac{\mathbf{k}'}{|\mathbf{k}|}, \tag{10.5.2}$$

$$\mathbf{e}_1 = (1, 0, 0), \quad \mathbf{e}_2 = \frac{(0, |q|, -K)}{|\mathbf{k}|}, \qquad \mathbf{k} \cdot \mathbf{e}_\lambda = 0, \qquad \mathbf{e}_1 \times \mathbf{e}_2 = \frac{\mathbf{k}}{|\mathbf{k}|}, \tag{10.5.3}$$

$$\mathbf{e}''_1 = (1, 0, 0), \quad \mathbf{e}''_2 = \frac{(0, -|q|, -K)}{|\mathbf{k}|}, \qquad \mathbf{k}'' \cdot \mathbf{e}''_\lambda = 0, \quad \mathbf{e}''_1 \times \mathbf{e}''_2 = \frac{\mathbf{k}''}{|\mathbf{k}|}. \tag{10.5.4}$$

For the subsequent analysis, we record some of the basic properties encountered with the polarization vectors:

$$\mathbf{e}_\alpha \cdot \mathbf{e}'_\lambda = \delta_{\alpha 1}\delta_{\lambda 1} + \frac{|K|^2 - |q|^2}{|\mathbf{k}|^2}\delta_{\alpha 2}\delta_{\lambda 2}, \qquad e^3_\alpha e'^3_\lambda = \frac{|K|^2}{|\mathbf{k}|^2}\delta_{\alpha 2}\delta_{\lambda 2}, \tag{10.5.5}$$

$$\mathbf{e}''_\alpha \cdot \mathbf{e}'_\lambda = \delta_{\alpha 1}\delta_{\lambda 1} + \delta_{\alpha 2}\delta_{\lambda 2}, \qquad e''^3_\alpha e'^3_\lambda = \frac{|K|^2}{|\mathbf{k}|^2}\delta_{\alpha 2}\delta_{\lambda 2}. \tag{10.5.6}$$

The transition amplitude for reflection is given in Eq. (10.4.9). Due to the conditions $\mathbf{k}' \cdot \mathbf{e}' = 0, \mathbf{k} \cdot \mathbf{e} = 0$ in Eqs. (10.5.2), (10.5.3) above, we may omit the vector component $k'^j, k^i$ in the expression of $G_+^{ij}$ given in Eq. (10.4.1). Thus the reflection amplitude is from Eqs. (10.4.9), (10.4.1) given by

$$\begin{aligned}\big(\mathbf{k}, \mathbf{e}_\alpha \mid \mathbf{k}', \mathbf{e}'_\lambda\big) &= \mathrm{e}^i_\alpha G_+^{ij} \mathrm{e}'^j_\lambda \\ &= A_+\, \mathbf{e}_\alpha \cdot \mathbf{e}'_\lambda - \frac{1}{|q|^2}\Big[\big(|\mathbf{k}|^2 + 2|q|^2\big)A_+ - |\mathbf{k}|^2 B_+\Big]\mathrm{e}^3_\alpha \mathrm{e}'^3_\lambda \quad (10.5.7)\\ &= A_+\, \delta_{\alpha 1}\delta_{\lambda 1} - \frac{|\mathbf{k}|^2}{|q|^2}\Big[A_+ - \frac{|\mathbf{K}|^2}{|\mathbf{k}|^2} B_+\Big]\,\delta_{\alpha 2}\delta_{\lambda 2}, \quad (10.5.8)\end{aligned}$$

using both equalities in Eq. (10.5.5), in deriving the expression in the last line above, and note that $\mathbf{K} = (0, K)$, $|\mathbf{K}|^2 + q^2 = |\mathbf{k}|^2$.

Similarly for the transition amplitude for transmission from Eqs. (10.4.3), (10.4.15), and Eqs. (10.5.3), (10.5.4),

$$\begin{aligned}\big(\mathbf{k}'', \mathbf{e}''_\alpha \mid \mathbf{k}', \mathbf{e}'_\lambda\big) &= \mathrm{e}^i_\alpha G_-^{ij} \mathrm{e}'^j_\lambda, \qquad \mathbf{k}'' = \mathbf{k}',\ \mathbf{e}''_\alpha = \mathbf{e}'_\alpha \\ &= \mathbf{e}''_\alpha \cdot \mathbf{e}'_\lambda\, A_- + \frac{|\mathbf{k}|^2}{|q|^2}(A_- - B_-)\mathrm{e}''^3_\alpha \mathrm{e}'^3_\lambda \quad (10.5.9)\\ &= A_-\, \delta_{\alpha 1}\delta_{\lambda 1} + \frac{|\mathbf{k}|^2}{|q|^2}\Big[A_- - \frac{|\mathbf{K}|^2}{|\mathbf{k}|^2} B_-\Big]\,\delta_{\alpha 2}\delta_{\lambda 2}, \quad (10.5.10)\end{aligned}$$

where in the last line we have used both equalities in Eq. (10.5.6).

Therefore the probabilities of reflection and transmission for a photon with initial polarizations $\mathbf{e}'_1$, $\mathbf{e}'_2$ are from Eqs. (10.5.8), (10.5.10), respectively, given by

$$\text{Init. Pol. } \mathbf{e}'_1 \rightarrow \begin{cases} |A_+|^2, & \text{reflection} \\ |A_-|^2, & \text{transmission} \end{cases}, \quad (10.5.11)$$

and

$$\text{Init. Pol. } \mathbf{e}'_2 \rightarrow \begin{cases} \big(|\mathbf{k}|^4/|q|^4\big)\,\Big|A_+ - \big(|\mathbf{K}|^2/|\mathbf{k}|^2\big)\, B_+\Big|^2, & \text{reflection} \\ \big(|\mathbf{k}|^4/|q|^4\big)\,\Big|A_- - \big(|\mathbf{K}|^2/|\mathbf{k}|^2\big)\, B_-\Big|^2, & \text{transmission} \end{cases}. \quad (10.5.12)$$

From Eqs. (10.4.13)/(10.4.19), and Eqs. (10.4.14)/(10.4.20), we note that for each of the initial polarizations, the upper and lower entries in (10.5.11), as well in the upper and lower entries in Eq. (10.5.12), separately, add up to one. Thus the conservation of probabilities is again established for both initial polarizations, as for the unpolarized case.

## 10.6 Red Light, Blue Light and a Layer of Glass

For unpolarized light and normal incidence, the respective probabilities in Eqs. (10.4.12)–(10.4.14), and Eqs. (10.4.18)–(10.4.20), for reflection and transmission, are given by

$$|A_+|^2 = \frac{(\varepsilon-1)^2\sin^2(Qd)}{4\varepsilon+(\varepsilon-1)^2\sin^2(Qd)}, \tag{10.6.1}$$

$$|A_-|^2 = \frac{4\varepsilon}{4\varepsilon+(\varepsilon-1)^2\sin^2(Qd)}, \tag{10.6.2}$$

since $Q=\sqrt{\varepsilon}|q|$ in this case. Also note that $|A_+|^2+|A_-|^2=1$, as expected.

The reflection probability has a $\sin^2(\sqrt{\epsilon}|q|d)$ factor, while the transmission does not. This means, depending on the wavelength of a photon considered, for a specific medium the reflection amplitude may vanish.

For example consider glass of thickness $d=3\sqrt{2}/2$ cm, and $\varepsilon=(3+\sqrt{3})/2$. For red and blue light, e.g., we may take for their wavelengths

$$\lambda_{\text{red}} = 3[3+\sqrt{3}]^{1/2}10^{-5}\,\text{cm}, \qquad \text{and } \lambda_{\text{blue}} = 3[(3+\sqrt{3})/2]^{1/2}10^{-5}\,\text{cm}.$$

Recall that $|q|=2\pi/\lambda$. The above data lead to

$$\sqrt{\varepsilon}q_{\text{red}}\,d = 2\pi(50000), \quad \sqrt{\varepsilon}q_{\text{blue}}\,d = 2\pi(70710+0.6781). \tag{10.6.3}$$

That is, this unpolarized red photon has just the correct wavelength to make the reflection coefficient vanish, it is transmitted, and has no other option where to go. While this blue photon has a non-zero probability of being reflected off a glass with such properties. Does the blue photon know how thick the glass is before it gets reflected instead of being transmitted?.

For another illustration, consider initially polarized light with polarization vector $\mathbf{e}_2'$ as given in Eq. (10.5.2). For a $45^\circ$ degree of incidence, for example, $|\mathbf{K}|=|q|$, i.e., $|\mathbf{k}|=\sqrt{2}|q|$, $Q=|q|\sqrt{2\varepsilon-1}$. We note that with $\varepsilon$, as given above, $\varepsilon^2=(3/2)(2\varepsilon-1)$, for which $Q^2=2\varepsilon^2|q|^2/3$, and

$$(Q^2-\varepsilon^2|q|^2)^2 = \frac{1}{9}\varepsilon^4|q|^4, \quad 4\varepsilon^2q^2Q^2 = \frac{8}{3}\varepsilon^4|q|^4.$$

The probabilities of reflection and transmission are, from Eq. (10.5.12) and Eqs. (10.4.14), (10.4.20), given by

$$\begin{aligned}\text{Prob[reflec.]} &= \frac{(Q^2-\varepsilon^2|q|^2)^2\sin^2 Qd}{4\varepsilon^2|q|^2Q^2+(Q^2-\varepsilon^2|q|^2)^2\sin^2 Qd}\\ &= \frac{\sin^2\left(\sqrt{2/3}\varepsilon|q|d\right)}{24+\sin^2\left(\sqrt{2/3}\varepsilon|q|d\right)},\end{aligned} \tag{10.6.4}$$

and

$$\begin{aligned}\text{Prob[trans.]} &= \frac{4\varepsilon^2|q|^2Q^2}{4\varepsilon^2|q|^2Q^2 + (Q^2 - \varepsilon^2|q|^2)^2\sin^2 Qd} \\ &= \frac{24}{24 + \sin^2\left(\sqrt{2/3}\varepsilon|q|d\right)}.\end{aligned} \tag{10.6.5}$$

Here we see from the theory the true *character* letting more light through it than it is reflected by simply comparing the sine-square factor with the large number 24 in the numerators of the above two equations, without going into the structure of glass.

For definiteness, consider the following values for the wavelengths of red and blue light in the spectrum:

$$\lambda_{\text{red}} = (3+\sqrt{3})\frac{11}{10}\sqrt{\frac{\pi}{2}}\,10^{-5}\,\text{cm}, \quad \lambda_{\text{blue}} = (3+\sqrt{3})\frac{11}{10}\frac{\sqrt{\pi}}{2}\,10^{-5}\,\text{cm}.$$

For the case considered, $|\mathbf{K}| = |q|$, i.e. $|\mathbf{k}| = \sqrt{2}|q| = 2\pi/\lambda$, and hence $|q| = \pi\sqrt{2}/\lambda$.

Consider now glass of thickness $d = 11\sqrt{3\pi}/10$ cm. This gives for blue light $Q_{\text{blue}}d = (2\pi)10^5$. That is, for such blue light $\sin Qd = 0$, and such a blue photon gets transmitted. But the red photon, nevertheless has a small probability of .033 to be reflected.

On the other hand, for a glass of thickness $d = (11/10)\sqrt{3\pi/2}$ cm, the opposite happens, such a red photon gets transmitted, while such a blue one has a probability of .033 of being reflected. Needless to say, additional such applications may be given for both polarized and unpolarized light.

## References

1. Feynman, R. P. (1985). *QED: The strange theory of light and matter*. Princeton, New Jersey: Princeton University Press.
2. Manoukian, E.B. (2013). Red light, blue light, Feynman and a layer of glass. The causal propagator, *Journal of Electromagnetic Waves and Applications, 27. No.15*, 1986-1996.

# Chapter 11
# Radiation Seeping Through a Thin Metal

*Observe how radiation seeps through a thin metal and witness, in particular, how the propagator reveals, through transmission, that the exponential decay is, generally, modified in real conductors*

Accurate measurements of the complex dielectric functions of various metals such as silver and aluminum have been made over the years.[1] Accordingly, the problem of considering the transmission of electromagnetic radiation through a metal is important in applications. It also turns up to be quite challenging theoretically. This is the subject matter of the present chapter. The metal silver, in particular, has a high level of electric conductivity, and it is considered here as an application of the theory developed.

As before, the investigation of the underlying propagator, describing the time evolution, in particular, of such a process of transmission, turns up to be most interesting and certainly physically appealing. An exact expression of the propagator will be derived, allowing one to asses the accuracy of the simple exponential damping often quoted in the literature. The analysis carried out applies to an arbitrary thickness of a metal as well with emphasis, however, given here to a thin metal. This chapter is based on the author's work cited below.[2]

---

[1] See, e.g., Nash and Sanders (1996) [3], Tillin and Sanders (1988). See also Kovacs (1978) [1].
[2] Manoukian (2016) [2].



E. B. Manoukian, *Radiation*,
https://doi.org/10.1007/978-3-032-21506-2_11

## 11.1 Propagator Equations and Boundary Conditions

We consider a thin metal of thickness $d$, and parameters $\varepsilon$, $\mu$, $\sigma$ "embedded" in free space:

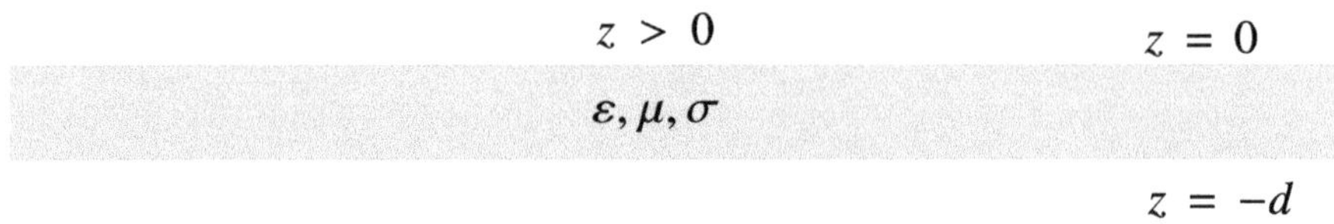


We recall that from Eqs. (5.4.53), (5.4.55), in Chap. 5, that for any point $\mathbf{x}' = (\mathbf{x}'_\parallel, z')$ with $z' > 0$, i.e., in free space, and any point $\mathbf{x} = (\mathbf{x}_\parallel, z)$ with $z < 0$, within the metal, the propagator $D_+^{ij}(x, x')$, denoted by $\underline{D}_+^{ij}(x, x')$ for such variables, satisfies the basic equations:

$$\big[\big(-\,\boldsymbol{\partial}^2 + \varepsilon\mu\partial^{02} - \mu\sigma\partial^0\big)\,\delta^{i\ell} + \partial^i\partial^\ell\big]\,\underline{D}_+^{\ell j}(x, x') = 0, \tag{11.1.1}$$

$$\big[\big(-\,\boldsymbol{\partial}'^2 + \partial'^{02}\big)\,\delta^{\ell i} + \partial'^\ell\partial'^i\big]\,\underline{D}_+^{j\ell}(x, x') = 0, \tag{11.1.2}$$

for $t > t'$, $z' > 0$, $0 > z > -d$, i.e., $z \neq z'$. Note that, in reference to the variable $z' > 0$, corresponding to a point outside the metal, i.e., in free space, $\varepsilon = 1$, $\mu = 1$, $\sigma = 0$, as seen in Eq. (11.1.2).

Upon multiplying the first equation by $\partial^i$ and the second by $\partial'^i$, we obtain two sets of, conveniently labeled, equations

$$\big(-\,\boldsymbol{\partial}^2 + \varepsilon\mu\partial^{02} - \mu\sigma\partial^0\big)\,\underline{D}_+^{ij}(x, x') = 0, \qquad \partial^i\,\underline{D}_+^{ij}(x, x') = 0, \tag{11.1.3}$$

$$\big(-\,\boldsymbol{\partial}'^2 + \partial'^{02}\big)\,\underline{D}_+^{ij}(x, x') = 0, \qquad \partial'^j\,\underline{D}_+^{ij}(x, x') = 0. \tag{11.1.4}$$

With translation invariance broken along the $z-$ axis, we may quite generally write in this region $-d < z < 0$,

$$\underline{D}_+^{ij}(x, x') = \mathrm{i}\int\frac{\mathrm{d}^2\mathbf{K}}{(2\pi)^2}\int\frac{\mathrm{d}q'\,\mathrm{d}\underline{q}}{(2\pi)\,2|\mathbf{k}|}\;\mathrm{e}^{\mathrm{i}\mathbf{K}\cdot(\mathbf{x}_\parallel - \mathbf{x}'_\parallel)}\,\mathrm{e}^{-\mathrm{i}|\mathbf{k}|(t-t')}\,\mathrm{e}^{\mathrm{i}(\underline{q}z - q'z')}\,\underline{G}^{ij}, \tag{11.1.5}$$

where from Eq. (11.1.4), $|\mathbf{k}| = \sqrt{|\mathbf{K}|^2 + q'^2}$, and $z' > 0$ always.

The first relation in Eq. (11.1.3) then leads to

$$\underline{q} = \pm\,Q, \quad \text{where} \quad Q = \sqrt{\big(\varepsilon\mu - 1\big)|\mathbf{K}|^2 + \varepsilon\mu|q'|^2 + \mathrm{i}\,\mu\sigma\,|\mathbf{k}|}. \tag{11.1.6}$$

For normal incidence, i.e., $\mathbf{K} = 0$, in particular, and optical properties, we may set $\mu = 1$, giving

$$Q = |q'|\sqrt{\varepsilon + \mathrm{i}\frac{\sigma}{|q'|}}. \tag{11.1.7}$$

We may, in general, introduce two structures $M_1^{ij}$, $M_2^{ij}$, to be determined, and express the general solution in this region as

$$\underline{D}_+^{ij}(x,x') = \mathrm{i}\int \frac{\mathrm{d}^2\mathbf{K}}{(2\pi)^2}\int \frac{\mathrm{d}q'}{(2\pi)\,2|\mathbf{k}|}\,\mathrm{e}^{\mathrm{i}\mathbf{K}\cdot(\mathbf{x}_\parallel-\mathbf{x}'_\parallel)}\,\mathrm{e}^{-\mathrm{i}|\mathbf{k}|(t-t')}$$
$$\times\left(\mathrm{e}^{\mathrm{i}(Qz-q'z')}M_1^{ij} + \mathrm{e}^{-\mathrm{i}(Qz+q'z')}M_2^{ij}\right),\qquad -d < z < 0. \tag{11.1.8}$$

The transversality condition in Eq. (11.1.3), gives for $z \to -d$

$$K^a\left(\mathrm{e}^{-\mathrm{i}Qd}M_1^{aj} + \mathrm{e}^{\mathrm{i}Qd}M_2^{aj}\right) = -\,Q\left(\mathrm{e}^{-\mathrm{i}Qd}M_1^{3j} - \mathrm{e}^{\mathrm{i}Qd}M_2^{3j}\right). \tag{11.1.9}$$

In the region $z > 0$, we may directly use the solution already worked out in Sect. 10.2 of Chap. 10, Eq. (10.2.1), with the propagator $D_+^{ij}$ in this region denoted by $D_+^{>ij}$, to be given by

$$D_+^{>ij}(x,x') = \mathrm{i}\int \frac{\mathrm{d}^2\mathbf{K}}{(2\pi)^2}\int \frac{\mathrm{d}q'}{2\pi\,2|\mathbf{k}|}\mathrm{e}^{\mathbf{K}\cdot(\mathbf{x}_\parallel-\mathbf{x}'_\parallel)}\mathrm{e}^{-\mathrm{i}|\mathbf{k}|(t-t')}$$
$$\times\ \left[\mathrm{e}^{\mathrm{i}((q'z-q'z')}\left(\delta^{ij} - \frac{k^ik^j}{|\mathbf{k}|^2}\right) + \mathrm{e}^{-\mathrm{i}(q'z+q'z')}G_+^{ij}\right],\qquad z > 0, \tag{11.1.10}$$

where we eventually take $\mathbf{k}' = (\mathbf{K}, -|q|)$ for incidence, $\mathbf{k} = (\mathbf{K}, |q|)$, for reflection, i.e., $q' = -|q|$, $|\mathbf{k}'| = |\mathbf{k}|$, i.e., from Eqs. (10.2.2)–(10.2.5),

$$G_+^{ab} = \left(\delta^{ab}A_+ - \frac{K^ak^b}{|\mathbf{k}|^2}B_+\right),\qquad a,b = 1,2, \tag{11.1.11}$$

and the various components of $G_+^{ij}$ may be spelled out as follows,

$$G_+^{a3} = \frac{K^a}{|q|}\left(A_+ - \frac{|\mathbf{K}|^2}{|\mathbf{k}|^2}B_+\right), \tag{11.1.12}$$
$$G_+^{3b} = -\frac{K^b}{|q|}\left(A_+ - \frac{|\mathbf{K}|^2}{|\mathbf{k}|^2}B_+\right), \tag{11.1.13}$$
$$G_+^{33} = -\frac{|\mathbf{K}|^2}{|q|^2}\left(A_+ - \frac{|\mathbf{K}|^2}{|\mathbf{k}|^2}B_+\right), \tag{11.1.14}$$

with $\mathbf{k}' = (\mathbf{K}, -|q|)$ representing the momentum of incidence. In this chapter, we are interested in the radiation seeping through the metal rather than reflections (described by $G_+^{ij}$), which will arise from the $z < -d$ region.

In the region $z < -d$, we may directly use the solution already worked out in Sect. 10.2 of Chap. 10, Eqs. (10.2.10)–(10.2.12), with the propagator $D_+^{ij}$ in this region denoted by $D_+^{<ij}$, given by

$$
\begin{aligned}
D_{+}^{<ij}(x,x') = \mathrm{i}\int \frac{\mathrm{d}^2\mathbf{K}}{(2\pi)^2}\int \frac{\mathrm{d}q}{2\pi\, 2|\mathbf{k}'|}\, \mathrm{e}^{\mathbf{K}\cdot(\mathbf{x}_\parallel-\mathbf{x}'_\parallel)}\, \mathrm{e}^{-\mathrm{i}|\mathbf{k}|(t-t')} \\
\times\, \mathrm{e}^{\mathrm{i}q(z-z')}\, G_{-}^{ij}, \qquad z<-d,
\end{aligned}
\tag{11.1.15}
$$

and in view of investigating the radiation seeping through the metal we spell out the general structure of $G_{-}^{ij}$:

$$
G_{-}^{ab} = \Big(\delta^{ab}A_{-} - \frac{K^a k^b}{|\mathbf{k}|^2}B_{-}\Big), \qquad a,b=1,2, \qquad |\mathbf{k}| = \sqrt{\mathbf{K}^2+q^2}, \tag{11.1.16}
$$

$$
\begin{aligned}
G_{-}^{a3} &= \frac{K^a}{|q|}\Big(A_{-} - \frac{|\mathbf{K}|^2}{|\mathbf{k}|^2}B_{-}\Big), \quad G_{-}^{3b} = \frac{K^b}{|q|}\Big(A_{-} - \frac{|\mathbf{K}|^2}{|\mathbf{k}|^2}B_{-}\Big), \\
G_{-}^{33} &= \frac{|\mathbf{K}|^2}{|q|^2}\Big(A_{-} - \frac{|\mathbf{K}|^2}{|\mathbf{k}|^2}B_{-}\Big),
\end{aligned}
\tag{11.1.17}
$$

where $A_{-}$, $B_{-}$, now, are some new functions to be determined from applying the appropriate boundary conditions.

The momenta involved in the analysis are given by

$$
\text{incident: } \mathbf{k}' = (\mathbf{K}, -|q|), \quad \text{reflected: } \mathbf{k} = (\mathbf{K}, |q|), \quad \text{transmitted: } \mathbf{k}'' = \mathbf{k}'. \tag{11.1.18}
$$

The boundary conditions to be taken are:

$$
\begin{aligned}
&D_{+}^{aj}(x,x'), \qquad \varepsilon(z)D_{+}^{3j}(x,x') \\
&\frac{1}{\mu(z)}\big[\partial^3 D_{+}^{aj}(x,x') - \partial^a D_{+}^{3j}(x,x')\big], \quad \big[\partial^1 D_{+}^{2j}(x,x') - \partial^2 D_{+}^{1j}(x,x')\big],
\end{aligned}
\tag{11.1.19}
$$

are continuous at $z=0$ and at $z=-d$, where $a=1,2$, $j=1,2,3$, $\varepsilon(z), \mu(z) = (\varepsilon,\mu)$ for $-d<z<0$, and $=(1,1)$, otherwise.

## 11.2 Satisfying the Boundary Conditions

The boundary conditions set up in Eq. (11.1.19) will be considered in turn at $z=0$ and at $z=-d$.

**Boundary Conditions at $z=0$:**

For $i=a$, $j=b$, we have, in turn,

$$
\Big(\delta^{ab} - \frac{K^a K^b}{|\mathbf{k}|^2}\Big) + G_{+}^{ab} = M_1^{ab} + M_2^{ab}, \tag{11.2.1}
$$

$$\frac{|q|K^b}{|\mathbf{k}|^2} - \frac{K^c}{|q|}G_+^{cb} = \varepsilon\Big(-\frac{K^c}{Q}M_1^{cb} + \frac{K^c}{Q}M_2^{cb}\Big), \tag{11.2.2}$$

$$\begin{aligned} -|q|\Big(\delta^{ab} - \frac{K^aK^b}{|\mathbf{k}|^2}\Big) + |q|G_+^{ab} - \frac{|q|K^aK^b}{|\mathbf{k}|^2} + \frac{K^aK^c}{|q|}G_+^{cb} \\ = \frac{1}{\mu}\Big[Q(M_1^{ab} - M_2^{ab}) - K^a\Big(-\frac{K^c}{Q}M_1^{cb} + \frac{K^c}{Q}M_2^{cb}\Big)\Big]. \end{aligned} \tag{11.2.3}$$

One may then express the structures $M_1^{ab}$, $M_2^{ab}$ as

$$\begin{aligned} 2M_1^{ab} &= \Big(\frac{Q-\mu|q|}{Q}\Big)\delta^{ab} + \Big(\frac{Q+\mu|q|}{Q}\Big)G_+^{ab} \\ &+ \mu\Big(1-\frac{1}{\varepsilon\mu}\Big)\frac{K^aK^c}{|q|Q}G_+^{cb} + \Big(\frac{|q|}{\varepsilon Q} - 1\Big)\frac{K^aK^b}{|\mathbf{k}|^2}, \end{aligned} \tag{11.2.4}$$

and

$$\begin{aligned} 2M_2^{ab} &= \Big(\frac{Q+\mu|q|}{Q}\Big)\delta^{ab} + \Big(\frac{Q-\mu|q|}{Q}\Big)G_+^{ab} \\ &- \mu\Big(1-\frac{1}{\varepsilon\mu}\Big)\frac{K^aK^c}{|q|Q}G_+^{cb} - \Big(\frac{|q|}{\varepsilon Q} + 1\Big)\frac{K^aK^b}{|\mathbf{k}|^2}. \end{aligned} \tag{11.2.5}$$

**Boundary Conditions at $z = -d$:**

$$\mathrm{e}^{-\mathrm{i}Qd}M_1^{ab} + \mathrm{e}^{\mathrm{i}Qd}M_2^{ab} = \mathrm{e}^{\mathrm{i}|q|d}G_-^{ab}, \tag{11.2.6}$$

$$\varepsilon\Big[-\mathrm{e}^{-\mathrm{i}Qd}\Big(\frac{K^c}{Q}M_1^{cb}\Big) + \mathrm{e}^{\mathrm{i}Qd}\Big(\frac{K^c}{Q}M_2^{cb}\Big)\Big] = \mathrm{e}^{\mathrm{i}|q|d}\Big(\frac{K^c}{|q|}G_-^{cb}\Big), \tag{11.2.7}$$

$$\begin{aligned} \frac{1}{\mu}\Big[Q(\mathrm{e}^{-\mathrm{i}Qd}M_1^{ab} - \mathrm{e}^{\mathrm{i}Qd}M_2^{ab}) - K^a\Big(-\mathrm{e}^{-\mathrm{i}Qd}\Big(\frac{K^c}{Q}M_1^{cb}\Big) + \mathrm{e}^{\mathrm{i}Qd}\Big(\frac{K^c}{Q}M_2^{cb}\Big)\Big)\Big] \\ = \mathrm{e}^{\mathrm{i}|q|d}\Big[|q|G_-^{ab} - \frac{K^aK^c}{|q|}G_-^{cb}\Big]. \end{aligned} \tag{11.2.8}$$

From Eqs. (11.2.6)–(11.2.8), we may also express $M_1^{ab}$, $M_2^{ab}$ as

$$2M_1^{ab} = \mathrm{e}^{+\mathrm{i}(Q+|q|)d}\Big[\frac{(Q-\mu|q|)}{Q}G_-^{ab} - \mu\Big(1-\frac{1}{\varepsilon\mu}\Big)\frac{K^aK^c}{|q|Q}G_-^{cb}\Big], \tag{11.2.9}$$

$$2M_2^{ab} = \mathrm{e}^{-\mathrm{i}(Q-|q|)d}\Big[\frac{(Q+\mu|q|)}{Q}G_-^{ab} + \mu\Big(1-\frac{1}{\varepsilon\mu}\Big)\frac{K^aK^c}{|q|Q}G_-^{cb}\Big]. \tag{11.2.10}$$

With the aim of investigating the radiation seeping through the thin metal we must obtain the general expression of the propagator as it emerges through the metal. To this end, we compare the pairs of Eqs. (11.2.4)/(11.2.5) with those in Eqs. (11.2.9)/(11.2.10) for the derived expressions of the pair of structures $M_1^{ab}/M_2^{ab}$, and using the decomposition of $G_-^{ab}$ in terms of the coefficients $A_-$, $B_-$ as given in Eq. (11.1.16), we obtain after some straightforward algebra, though a long one, its relevant coefficients

$$A_- = \frac{\Big[(Q - \mu|q|)^2 - (Q + \mu|q|)^2\Big]\mathrm{e}^{-\mathrm{i}|q|d}}{(Q - \mu|q|)^2\,\mathrm{e}^{\mathrm{i}Qd} - (Q + \mu|q|)^2\,\mathrm{e}^{-\mathrm{i}Qd}}, \tag{11.2.11}$$

where $Q = \sqrt{(\varepsilon\mu - 1)|\mathbf{K}|^2 + \varepsilon\mu|q|^2 + \mathrm{i}\,\mu\sigma\,|\mathbf{k}|}$, and

$$\frac{|\mathbf{k}|^2}{|q|^2}\Big(A_- - \frac{|\mathbf{K}|^2}{|\mathbf{k}|^2}B_-\Big) =$$

$$\frac{\Big[\big(\varepsilon\mu|\mathbf{k}|^2 - |\mathbf{K}|^2 - \varepsilon|q|Q\big)^2 - \big(\varepsilon\mu|\mathbf{k}|^2 - |\mathbf{K}|^2 + \varepsilon|q|Q\big)^2\Big]\mathrm{e}^{-\mathrm{i}|q|d}}{\big(\varepsilon\mu\mathbf{k}|^2 - |\mathbf{K}|^2 - \varepsilon|q|Q\big)^2\mathrm{e}^{\mathrm{i}Qd} - \big(\varepsilon\mu\mathbf{k}|^2 - |\mathbf{K}|^2 + \varepsilon|q|Q\big)^2\mathrm{e}^{-\mathrm{i}Qd}}. \tag{11.2.12}$$

## 11.3 The Emerging Propagator from the Metal

From the explicit expressions of the coefficients $A_-$ and $B_-$ of $G_-^{ij}$ arising in the integrand of the propagator $D_+^{<ij}$ in Eq. (11.1.15) as it emerges from the metal may be written explicitly as

$$G_-^{ij} = \Big(\delta^{ij}A_- - \frac{k'^i k'^j}{|\mathbf{k}|^2}B_-\Big) + \Big(A_- - B_-\Big)\Big[\frac{|\mathbf{k}|^2}{|q|^2}\delta^{i3}\delta^{3j} + \frac{k'^i\delta^{3j} + \delta^{i3}k'^j}{|q|}\Big], \tag{11.3.1}$$

where $\mathbf{k}' = (\mathbf{K}, -|q|)$. The transversality character of $G_-^{ij}$:

$$k'^i G_-^{ij} = 0, \qquad G_-^{ij}k'^j = 0, \tag{11.3.2}$$

is readily verified.

As in Eqs. (10.4.5), (10.4.6) in Sect. 10.4, we may introduce polarization vectors $\mathbf{e}'_\lambda = \mathbf{e}''_\lambda$, associated with incident and transmitted momenta $\mathbf{k}' = (\mathbf{K}, -|q|) = \mathbf{k}''$,

$$\mathbf{e}'_\alpha \cdot \mathbf{e}'_\lambda = \delta_{\alpha\lambda}, \qquad \mathbf{k}' \cdot \mathbf{e}'_\lambda = 0, \qquad \mathbf{e}'_1 \times \mathbf{e}'_2 = \frac{\mathbf{k}'}{|\mathbf{k}'|} \equiv \mathbf{n}', \tag{11.3.3}$$

$$\delta^{ij} = \frac{\mathbf{k}'^i}{|\mathbf{k}|}\frac{\mathbf{k}'^j}{|\mathbf{k}|} + \sum_{\lambda=1,2} \mathrm{e}'^i_\lambda \mathrm{e}'^j_\lambda, \tag{11.3.4}$$

and infer that the amplitude of radiation seeping is given by

$$e''^{i}{}_{\alpha} G^{ij}_{-} e'^{i}_{\lambda}. \tag{11.3.5}$$

By averaging over the initial polarizations, and summing over the final ones, exactly as we have done in deriving Eq. (10.4.18) in Sect. 10.4, the probability of transmission emerges as

$$\text{Prob}|_{\text{trans.}}[\mathbf{k}' \to \mathbf{k}''] = \frac{1}{2}\Big[|A_-|^2 + \frac{|\mathbf{k}|^4}{|q|^4}\Big|A_- - \frac{|\mathbf{K}|^2}{|\mathbf{k}|^2}B_-\Big|^2\Big]. \tag{11.3.6}$$

for unpolarized radiation, where now the expression on the right-hand side of the equation is obtained from Eqs. (11.2.11), (11.2.12).

In particular, for normal incidence $|\mathbf{K}| = 0$. Then, from Eqs. (11.2.11), (11.2.12), the second term in the square brackets in the above becomes simply $|A_-|^2$, and the probability simply reduces to $|A_-|^2$. For optical properties, we may set $\mu = 1$. The momentum $Q$, now, takes the simple form from given in Eq. (11.1.7), with $|q'| = |q|$,

$$Q = |q|\sqrt{\varepsilon + \mathrm{i}\frac{\sigma}{|q|}} \equiv |q|(\gamma + \mathrm{i}\beta), \tag{11.3.7}$$

thus, in turn, defining its real and imaginary parts $|q|\gamma$ and $|q|\beta$, respectively.

The complex index of refraction $n_c$ is given by

$$n_c = \sqrt{\varepsilon + \mathrm{i}\frac{\sigma}{|q|}}, \qquad \varepsilon \equiv \varepsilon_{\text{Re}}, \qquad \frac{\sigma}{|q|} \equiv \varepsilon_{\text{Im}}, \tag{11.3.8}$$

with the latter parameters $\varepsilon_{\text{Re}}$, $\varepsilon_{\text{Im}}$, thus simply denoted, are determined experimentally.

The probability of seepage through a metal of arbitrary thickness $d$ may be thus written from Eqs. (11.2.11), and (11.3.7), in detail, as

$$\text{Prob} = \frac{\mathcal{N}}{\mathcal{D}}, \tag{11.3.9}$$

where

$$\mathcal{N} = [(\gamma-1)^2+\beta^2]^2 + [(\gamma+1)^2+\beta^2]^2 - 2\Big([\gamma^2-1+\beta^2]^2 - 4\beta^2\Big), \tag{11.3.10}$$

and

$$\begin{aligned}\mathcal{D} &= [(\gamma-1)^2+\beta^2]^2 \mathrm{e}^{-2|q|\beta d} + [(\gamma+1)^2+\beta^2]^2 \mathrm{e}^{2|q|\beta d}\\ &- 2\Big([\gamma^2-1+\beta^2]^2 - 4\beta^2\Big)\cos(2|q|\gamma d) + 8\beta[\gamma^2-1+\beta^2]\sin(2|q|\gamma d).\end{aligned} \tag{11.3.11}$$

As an application, we consider silver metal of thickness $d = 50$ nm. The optical dielectric function of Silver was well investigated experimentally,[3] in which $\varepsilon_{\text{Re}}$ $\varepsilon_{\text{Im}}$ where obtained. We consider about the lower and about the upper ends of the visible spectrum.

For red light of wavelength $\lambda_{\text{red}} = 625$ nm, $\varepsilon_{\text{Re}} = -18.18$, $\varepsilon_{\text{Im}} = 0.43$. These give $\gamma = 0.0504$, $\beta = 4.264$, leading to a probability of 0.011. As the thickness increases above the latter value, the probability decreases, while it increases for smaller thicknesses, as expected. For blue light of wavelength $\lambda_{\text{blue}} = 450$ nm, $\varepsilon_{\text{Re}} = -7.16$, $\varepsilon_{\text{Im}} = 0.28$. These give $\gamma = 0.0523$, $\beta = 2.6763$, leading to a probability of 0.031 of seeping through.

An important application of the exact expression of the probability given in Eqs. (11.3.9)–(11.3.11), is assessing the accuracy of the simple exponential damping often quoted in the literature on which some considerations were given in Eqs. (5.4.44), (5.4.46) in Sect. 5.4.

To the above end, consider $d \gg 1/(2|q|\beta)$. Equations (11.3.9)–(11.3.11) give

$$\text{Prob}_{\text{exact}} \rightarrow \left[1 + \frac{\left[(\gamma-1)^2+\beta^2\right]^2}{\left[(\gamma+1)^2+\beta^2\right]^2} - 2\frac{\left([\gamma^2-1+\beta^2]^2-4\beta^2\right)}{\left[(\gamma+1)^2+\beta^2\right]^2}\right] \mathrm{e}^{-2|q|\beta d} \equiv \text{Prob}_{\infty}, \tag{11.3.12}$$

where now $\text{Prob}_{\infty}$ is the limiting expression involving a simple exponential damping.

To assess the accuracy of $\text{Prob}_{\infty}$ involving a simple exponential damping, we may, for any thickness $d$, define the absolute error $\mathcal{E}(d)$

$$\mathcal{E}(d) = \left|\frac{\text{Prob}_{\text{exact}} - \text{Prob}_{\infty}}{\text{Prob}_{\text{exact}}}\right|. \tag{11.3.13}$$

From Eq. (11.3.12), and Eqs. (11.3.9)–(11.3.11), this works out to read

$$\mathcal{E}(d) = \Bigg| 2\frac{\left([\gamma^2-1+\beta^2]^2-4\beta^2\right)}{\left[(\gamma+1)^2+\beta^2\right]^2}\cos(2|q|\gamma d)\,\mathrm{e}^{-2|q|\beta d} \tag{11.3.14}$$

$$- 8\,\beta\frac{\left[(\gamma^2-1+\beta^2\right]}{\left[(\gamma+1)^2+\beta^2\right]^2}\sin(2|q|\gamma d)\,\mathrm{e}^{-2|q|\beta d} \tag{11.3.15}$$

$$- \frac{\left[(\gamma-1)^2+\beta^2\right]^2}{\left[(\gamma+1)^2+\beta^2\right]^2}\mathrm{e}^{-4|q|\beta d}\Bigg|. \tag{11.3.16}$$

The absolute relative error of the departure from a simple exponential damping, for $d = 50$ nm, is about 1.1% for red light, while it is only 0.26% for blue light, and the latter decreases rapidly for blue light, having a smaller wavelength. Needless to say, the exact expression of the probability of transmission given through Eqs. (11.3.9)–(11.3.11) is quite general and applies for any thickness.

---

[3] Nash and Sambles (1996) [3].

## References

1. Kovacs G. J. (1978). Sulphide formation on evaporated Ag films. *Surface Science, 78*, L245-L249.
2. Manoukian, E. B. (2016). Electromagnetic propagation across a metal: The causal propagator. *Journal of Electromagnetic Waves and Applications, 30 (5)*, 677-688.
3. Nash, D. J., & Sanders, J. R. (1996). Surface plasmon-polariton study of the optical dielectric funcion of silver. *Journal of Modern Optics, 43*, 81-91.
4. Tillin, M.D., & Sanders, J. R. (1988). A surface-polariton study of the dielectric constants of metals: aluminum. *Thin Solid Films, 167*, 73-84.

# Chapter 12
# Čerenkov Radiation in Various Configurations: Beyond the Classic Theory

*Experience the elegance and simplicity in the derivation of power of, experimentally verified, radiation emitted by a charged particle moving in different media, and experienced even by astronauts, in flight, with their eyes closed*

In simplest terms, Čerenkov radiation is the radiation emitted by a charged particle in a medium when its speed exceeds the speed of light *in* the medium.The history of Čerenkov radiation began with its discovery in 1934–1936 by Pavel Čerenkov, and its theoretical explanation by Igor Tamm and Ilya Frank in 1937.[1] Experimentally, it has been verified repeatedly many times over the years.[2]

As mentioned in Sect. 7.3, it is interesting to know that Apollo Astronauts during Apollo Missions have reported of "seeing" flashes of light even with their eyes closed. An explanation of this was generally attributed to high energy cosmic charged particles, encountered freely in outer space, that would pass one's eyelids causing Čerenkov radiation within one's eye itself.[3]

The classic theory of radiation emission in a medium of infinite extension was treated in Sect. 7.3. In the present chapter, we consider modifications of the classic theory arising from two physically interesting situations discussed in the beginning of the following sections. For such applications and additional ones, the reader may consider the references to the author below.[4]

---

[1] See, e.g., Čerenkov, Frank and Tamm (1958) [2].

[2] See, e.g., Rich et al. (1953) [14]; Hartman et al. (1979) [4]; Wang et al. (1991) [15]. See also Protheroe and Patterson (1984) [13].

[3] See, e.g., Fazio et al. (1970) [3]; Pinsky et al. (1974) [12]; McNulty et al. (1976) [11].

[4] Manoukian (1991–1999) [5–10].

 
E. B. Manoukian, *Radiation*,
https://doi.org/10.1007/978-3-032-21506-2_12

## 12.1 Čerenkov Radiation in a Medium with a Conducting Plate

We consider the physically interesting case of a charged particle in an isotropic permeable medium, moving parallel to a perfectly conducting neutral plate. The interest in the Čerenkov of the spectrum of radiation observed does not only provide information on the presence of a charged particle in the conventional sense but would also indicate the presence of a reflecting surface deep within the medium.

The conducting plate is placed in the $x^1, x^2$−plane at $x^3 \equiv z = 0$. The charged particle is supposed to be moving parallel to the plate, at $z = a$, along the $x^1$−axis:

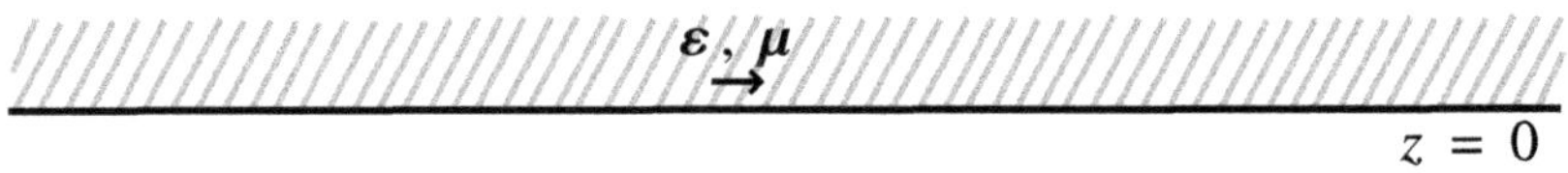


In particular, we investigate the enhancement of the power of radiation attainable as a function of the distance $a$ of the charged particle due to the presence of the conducting plate over the one of infinite extension derived in Sect. 7.3.

As before, we work in the temporal gauge. With the above description and such a trajectory, only the $D_+^{11}$ will be relevant in considering the power of radiation. It satisfies the boundary condition at the boundary

$$D_+^{11}(x, x') = 0, \quad \text{at } z = 0, \tag{12.1.1}$$

$x = (t, x^1, x^2, z)$, and we my, conveniently, use the Dirac delta representation given in Sect. 3.6, Eq. (3.6.41),

$$\delta(z, z') = \int_{-\infty}^{\infty} \frac{\mathrm{d}q}{\pi} \sin qz \, \sin qz' \tag{12.1.2}$$

The current along the $x^1$−axis, due to the charged particle of charge e is defined by

$$J^1(x) = \mathrm{e}\, v\delta(x^1 - vt)\, \delta(x^2)\, \delta(z - a). \tag{12.1.3}$$

Consider first the case where $\varepsilon = 1$, $\mu = 1$. The propagator in this case, in the temporal gauge satisfies, by now, the familiar equation

$$[(-\partial^2 + \partial^{02})\delta^{i\ell} + \partial^i \partial^\ell] D_+^{\ell j}(x, x') = \delta^{ij} \delta^{(4)}(x, x'). \tag{12.1.4}$$

Upon multiplying this equation by $\partial^i$, and then again by $\partial^i$, give in turn

$$\Rightarrow\ \partial^{02}\partial^{\ell} D_+^{\ell j}(x,x') = \partial^j \delta^{(4)}(x,x'), \tag{12.1.5}$$

$$\Rightarrow\ \partial^{02}\partial^i\partial^{\ell} D_+^{\ell j}(x,x') = \partial^i\partial^j \delta^{(4)}(x,x'). \tag{12.1.6}$$

$$\Rightarrow\ \partial^i\partial^{\ell} D_+^{\ell j}(x,x') = \frac{\partial^i\partial^j}{\partial^{02}}\,\delta^{(4)}(x,x'). \tag{12.1.7}$$

The last relation allows us to rewrite Eq. (12.1.4) as

$$(-\boldsymbol{\partial}^2 + \partial^{02}) D_+^{ij}(x,x') = \left(\delta^{ij} - \frac{\partial^i\partial^j}{\partial^{02}}\right)\delta^{(4)}(x,x'), \tag{12.1.8}$$

$$\Rightarrow\ D_+^{ij}(x,x') = \left(\delta^{ij} - \frac{\partial^i\partial^j}{\partial^{02}}\right)\left(\frac{1}{-\boldsymbol{\partial}^2 + \partial^{02}}\right)\delta^{(4)}(x,x'). \tag{12.1.9}$$

As before $x = (t, \mathbf{x}_{\|}, z)$, $\mathbf{x}_{\|} = (x^1, x^2)$.

With $\delta(z,z')$ given in Eq. (12.1.2), it follows from the above equation that

$$D_+^{11}(x,x') = \int \frac{\mathrm{d}^2\mathbf{K}}{(2\pi)^2}\int_{-\infty}^{\infty}\frac{\mathrm{d}q}{\pi}\int_{-\infty}^{\infty}\frac{\mathrm{d}Q^0}{2\pi}\,\frac{\sin qz\,\sin qz'}{\left[|\mathbf{K}|^2 + |q|^2 - Q^{02} - \mathrm{i}\delta\right]}$$
$$\times\ \mathrm{e}^{\mathrm{i}\mathbf{K}\cdot(\mathbf{x}_{\|}-\mathbf{x}'_{\|})}\mathrm{e}^{-\mathrm{i}Q^0(t-t')}\left[1 - \frac{(K^1)^2}{Q^{02}}\right], \qquad \delta \to +0. \tag{12.1.10}$$

In order to take into account of the parameters $\varepsilon$ and $\mu$, we simply have to consider the scalings $x \to \underline{x}$, $x' \to \underline{x}'$, i.e., $(t,t') \to (t,t')/\sqrt{\varepsilon\mu}$, spelled out on the first line in Eq. (7.1.7), in $D_+^{11}(x,x')$. On the other hand, recall, from Eq. (7.2.1), that the average number of photons $< N >$ emitted by the charged particle is given by

$$< N > = \sqrt{\frac{\mu}{\varepsilon}}\int (\mathrm{d}x)(\mathrm{d}x')J^1(x)\,\mathcal{I}m\,D_+^{11}(\underline{x},\underline{x}')\,J^1(x'). \tag{12.1.11}$$

Due to the reality of the currents, and the symmetry of the above integral in the exchange $\underline{x} \leftrightarrow \underline{x}'$, the imaginary part of $D_+^{11}(\underline{x},\underline{x}')$ may be taken by applying Eq. (3.6.16) directly to the denominator in the integrand in Eq. (12.1.10):

$$\mathcal{I}m\,D_+^{11}(x,x') \to \pi\int \frac{\mathrm{d}^2\mathbf{K}}{(2\pi)^2}\int_{-\infty}^{\infty}\frac{\mathrm{d}q}{\pi}\int_{-\infty}^{\infty}\frac{\mathrm{d}Q^0}{2\pi}\,\mathrm{e}^{\mathrm{i}\,\mathbf{K}\cdot(\mathbf{x}_{\|}-\mathbf{x}'_{\|})}\mathrm{e}^{-\mathrm{i}\,Q^0(t-t')/\sqrt{\varepsilon\mu}}$$
$$\times\ \sin(qz)\,\sin(qz')\,\delta\big(Q^{02} - (|\mathbf{K}|^2 + q^2)\big)\left[1 - \left(\frac{K^1}{Q^0}\right)^2\right]. \tag{12.1.12}$$

Upon setting

$$t = T + \frac{\tau}{2}, \quad t' = T - \frac{\tau}{2}, \quad t - t' = \tau, \tag{12.1.13}$$

the average number $< N >$ of photons emitted *per unit time*, immediately follows from Eqs. (12.1.3), (12.1.11), (12.1.12), upon integration over $\mathbf{x}$, $\mathbf{x}'$, to be

$$< N > = \pi e^2 \sqrt{\frac{\mu}{\varepsilon}} \int \frac{dK^1 dK^2}{(2\pi)^2} \int_{-\infty}^{\infty} \frac{dq}{\pi} \int_{-\infty}^{\infty} \frac{dQ^0}{2\pi} \int_{-\infty}^{\infty} d\tau \, e^{i\left(K^1 v - Q^0/\sqrt{\varepsilon\mu}\right)\tau}$$
$$\times \left[1 - \left(\frac{K^1}{Q^0}\right)^2\right] \sin^2(qa)\, \delta\big(Q^{02} - (|K^1|^2 + |K^2|^2 + q^2)\big). \tag{12.1.14}$$

By integrating first over $\tau$ giving $2\pi\delta\big(K^1 v - Q^0/\sqrt{\varepsilon\mu}\big)$, and then over $K^1$, lead to

$$< N > = \pi\, v\, e^2 \sqrt{\frac{\mu}{\varepsilon}} \Big[1 - \frac{1}{v^2\varepsilon\mu}\Big] \int \frac{dK^2}{2\pi} \int_{-\infty}^{\infty} \frac{dq}{\pi}\, 2 \int_0^{\infty} \frac{dQ^0}{2\pi}$$
$$\times\, \delta\Big( Q^{02}\Big(1 - \frac{1}{v^2\varepsilon\mu}\Big) - \big((K^2)^2 + q^2\big) \Big) \sin^2(qa), \tag{12.1.15}$$

where due to the fact that the integrand is an even function of $Q^0$, we have multiplied the corresponding integral by a factor of 2, and defined an integral from 0 to $\infty$.

Now we use the fact that

$$\int_0^{\infty} \frac{dq}{\pi} \delta\Big( Q^{02}\Big(1 - \frac{1}{v^2\varepsilon\mu}\Big) - \big((K^2)^2 + q^2\big) \Big) \sin^2(qa)$$
$$= \int_{-\infty}^{\infty} \frac{dq}{2\pi} \delta\Big( Q^{02}\Big(1 - \frac{1}{v^2\varepsilon\mu}\Big) - \big((K^2)^2 + q^2\big) \Big)\big[1 - e^{2iqa}\big],$$

to integrate Eq. (12.1.15) explicitly, by setting $q = R\cos\theta$, $K^2 = R\sin\theta$, and using the fact $dK^2 dq \to R dR\, d\theta$. Accordingly, we may, in turn, use the *definition* of the Bessel function $J_0^{\mathrm{B}}$,[5] of order zero

$$\int_0^{2\pi} d\theta \exp[i\, 2aR\cos\theta] = 2\pi\, J_0^{\mathrm{B}}(2aR), \tag{12.1.16}$$

to readily obtain upon integration over $\theta$, and then over $R^2$, upon writing, in the process, $R dR = dR^2/2$,

[5] In order not to confuse the Bessel function here with current, we have introduced the letter B.

$$< N >= v\,\mathrm{e}^2\sqrt{\frac{\mu}{\varepsilon}}\Big[1-\frac{1}{v^2\varepsilon\mu}\Big]\int_0^\infty \frac{R\mathrm{d}R}{(2\pi)^2}\int_0^\infty \mathrm{d}Q^0$$

$$\times\,\delta\Big(Q^{02}\Big(1-\frac{1}{v^2\varepsilon\mu}\Big)-R^2\Big)\int_0^{2\pi}\mathrm{d}\theta\Big[1-\mathrm{e}^{\mathrm{i}R2a\cos\theta}\Big]. \tag{12.1.17}$$

$$=\frac{v\,\mathrm{e}^2}{4\pi}\sqrt{\frac{\mu}{\varepsilon}}\Big[1-\frac{1}{v^2\varepsilon\mu}\Big]\int_0^\infty \mathrm{d}Q^0\Big[1-J_0^{\mathrm{B}}\Big(2Q^0a\sqrt{1-\frac{1}{v^2\varepsilon\mu}}\Big)\Big], \tag{12.1.18}$$

*for* $v^2 > 1/\varepsilon\mu$, and note that $Q^0 \geq 0$, and $R \geq 0$.

To derive the expression of power radiation, we insert the following identity, and finally multiply the resulting net integrand, obtained upon integrating over $Q^0$, by $\omega$. That is,

$$1=\int_0^\infty \mathrm{d}\omega\,\delta\Big(\omega-\frac{Q^0}{\sqrt{\varepsilon\mu}}\Big)=\sqrt{\varepsilon\mu}\int_0^\infty \mathrm{d}\omega\,\delta\big(\sqrt{\varepsilon\mu}\omega-Q^0\big), \tag{12.1.19}$$

leads to

$$\mathrm{P}(\omega)=\frac{\mu v\,\mathrm{e}^2\omega}{4\pi}\Big[1-\frac{1}{v^2\varepsilon\mu}\Big]\Big[1-J_0^{\mathrm{B}}\Big(2\omega a\sqrt{\varepsilon\mu}\sqrt{1-\frac{1}{v^2\varepsilon\mu}}\Big)\Big], \tag{12.1.20}$$

with $v^2 > 1/\varepsilon\mu$, where, in particular,

$$\mathrm{P}_0(\omega)\equiv\frac{\mu v\,\mathrm{e}^2\omega}{4\pi}\Big[1-\frac{1}{v^2\varepsilon\mu}\Big], \tag{12.1.21}$$

denotes the classic Čerenkov radiation formula for infinite extension in space, involving no obstacles, such as the conducting plate, derived in Eq. (7.3.8). For $a\to\infty$, when the charged particle is far away from the conducting plate, $|J_0^{\mathrm{B}}|\to 0$, and we recover the expected classic result, as the plate becomes irrelevant.

Due to the oscillatory nature of $J_0^{\mathrm{B}}(x)$, $\mathrm{P}(\omega)$ may take on values smaller or larger than $\mathrm{P}_0(\omega)$, where it should be noted that $|J_0^{\mathrm{B}}(x)|\leq 1$. Numerically, $[1-J_0^{\mathrm{B}}(x)]=1.403$, as a maximum,[6] for $x=3.83$, corresponding to the distance $a$ of the charged particle above the conducting plate, obtained from the argument of the Bessel function in Eq. (12.1.20), leading to the bound $P(\omega)/P_0(\omega)\leq 1.403$ as a maximum possible enhancement due to the conducting plate.

[6] Ambramowitz and Stegan (1972) [1], p. 390, Table 9.1.

## 12.2 Čerenkov Radiation Within a Cylinder

We consider Čerenkov radiation off a charged particle moving in an isotropic permeable medium within a cylinder with a perfectly conducting surface, along its center specified by the $z-$axis:

As before, we work in the temporal gauge. With the above description, the current associated with the charged particle of charge e in cylindrical coordinates is, by using the expression of the delta distribution in Eq. (3.6.17), given by

$$J^3(x) = \mathrm{e}\,\mathrm{v}\,\delta(z - vt)\frac{\delta(r)}{r}\delta(\theta), \tag{12.2.1}$$

with v denoting the speed of the charged particle along the $z-$axis. As before, we work in the temporal gauge. Moreover, only the component, $D_+^{33}$ of the propagator will be involved here. With a perfecting conducting surface, the boundary to be satisfied is simply

$$D_+^{33}(x, x')\Big|_{r=a} = 0. \tag{12.2.2}$$

where $a$ is the radius of the cylinder.

In this system of coordinates, the delta distribution, spelled out in Chap. 3, Sect. 3.6, Eqs. (3.6.17), (3.6.57), (3.6.61), is given by

$$\delta^{(4)}(x, x') = \frac{\delta(r - r')}{r}\delta(\theta - \theta')\delta(z - z')\delta(x^0 - x'^0). \tag{12.2.3}$$

In order to satisfy the above boundary condition, the different delta distributions may be written, in turn, as in Eqs. (3.6.57), (3.6.61)

$$\frac{\delta(r, r')}{r} = \frac{2}{a^2}\sum_{n=1}^{\infty}\frac{J_0^{\mathrm{B}}(\gamma_n r)J_0^{\mathrm{B}}(\gamma_n r')}{|J_1^{\mathrm{B}}(\gamma_n a)|^2}, \qquad J_1^{\mathrm{B}}(x) = -\frac{\mathrm{d}}{\mathrm{d}x}J_0^{\mathrm{B}}(x), \tag{12.2.4}$$

$$\delta(\theta - \theta') = \frac{1}{2\pi}\sum_{N=-\infty}^{\infty}\mathrm{e}^{\mathrm{i}N(\theta-\theta')}, \tag{12.2.5}$$

$$\delta(z - z') = \int_{-\infty}^{\infty}\frac{\mathrm{d}q}{2\pi}\,\mathrm{e}^{\mathrm{i}q(z-z')}, \tag{12.2.6}$$

$$\delta(x^0 - x'^0) = \int_{-\infty}^{\infty}\frac{\mathrm{d}Q^0}{2\pi}\mathrm{e}^{\mathrm{i}Q^0(x^0-x'^0)}. \tag{12.2.7}$$

where $J_0^{\mathrm{B}}(\gamma_n r)$ is the Bessel function of order zero,[7] and where $\gamma_n a$ define the positive roots of the Bessel function

$$J_0^{\mathrm{B}}(\gamma_n r)\Big|_{r=a} = 0, \quad n = 1, 2, \ldots, \tag{12.2.8}$$

The Bessel function $J_0^{\mathrm{B}}(\gamma_n r)$ satisfies the differential equation

$$\left[\frac{\mathrm{d}^2}{\mathrm{d}r^2} + \frac{1}{r}\frac{\mathrm{d}}{\mathrm{d}r} + \gamma_n^2\right] J_0^{\mathrm{B}}(\gamma_n r) = 0. \tag{12.2.9}$$

First recall from, e.g., Eq. (12.1.9), that in the absence of a medium,

$$D_+^{33}(x, x') = \frac{1}{(-\boldsymbol{\partial}^2 + \partial^{02})}\left[1 - \left(\frac{\partial^3}{\partial^0}\right)^2\right]\delta^{(4)}(x, x'), \tag{12.2.10}$$

where in cylindrical coordinates,

$$\boldsymbol{\partial}^2 = \left(\frac{\partial}{\partial r}\right)^2 + \frac{1}{r}\frac{\partial}{\partial r} + \frac{1}{r^2}\left(\frac{\partial}{\partial \theta}\right)^2 + \left(\frac{\partial}{\partial z}\right)^2. \tag{12.2.11}$$

With $\delta^{(4)}(x, x')$ given above, and the differential equation satisfied by the Bessel functions, we explicitly have

$$D_+^{33}(x, x') = \int_{-\infty}^{\infty}\frac{\mathrm{d}q}{2\pi}\mathrm{e}^{\mathrm{i}q(z-z')}\int_{-\infty}^{\infty}\frac{\mathrm{d}Q^0}{2\pi}\mathrm{e}^{\mathrm{i}Q^0(x^0-x'^0)}\left[1 - \left(\frac{q}{Q^0}\right)^2\right]\frac{1}{2\pi}\sum_{N=-\infty}^{\infty}\mathrm{e}^{\mathrm{i}N(\theta-\theta')}$$
$$\times\frac{2}{a^2}\sum_{n=1}^{\infty}\frac{J_0^{\mathrm{B}}(\gamma_n r)J_0^{\mathrm{B}}(\gamma_n r')}{\left[\frac{N^2}{r^2} + \left(\gamma_n^2 + q^2 - Q^{02}\right) - \mathrm{i}\,\delta\right]\left|J_1^{\mathrm{B}}(\gamma_n a)\right|^2}, \quad \delta \to +0. \tag{12.2.12}$$

In cylindrical coordinates, the following expressions should be noted

$$(\mathrm{d}x) \to \mathrm{d}t\, r\mathrm{d}r\, \mathrm{d}\theta\, \mathrm{d}z, \qquad (\mathrm{d}x') \to \mathrm{d}t'\, r'\mathrm{d}r'\, \mathrm{d}\theta'\, \mathrm{d}z', \tag{12.2.13}$$

$$J^3(x)J^3(x') = \mathrm{e}^2\mathrm{v}^2\frac{\delta(r)}{r}\frac{\delta(r')}{r'}\delta(\theta)\,\delta(\theta')\,\delta(z - vt)\,\delta(z' - vt'). \tag{12.2.14}$$

Using the fact that $J_0^{\mathrm{B}}(0) = 1$, and due to the presence of $\delta(r)\delta(r')$, the following expression simplifies, as indicated below

[7] Additional properties of the Bessel functions were spelled out in Sect. 6.7.

$$
\begin{aligned}
&\sum_{N=-\infty}^{\infty} \mathrm{e}^{\mathrm{i}N(\theta-\theta')} \sum_{n=1}^{\infty} \frac{J_0^{\mathrm{B}}(\gamma_n r) J_0^{\mathrm{B}}(\gamma_n r')\, \delta(r)\, \delta(r')}{\left[\frac{N^2}{r^2} + \left(\gamma_n^2 + q^2 - Q^{02}\right) - \mathrm{i}\,\delta\right] |J_1^{\mathrm{B}}(\gamma_n^2 a)|^2} \\
&\quad = \sum_{N=-\infty}^{\infty} \mathrm{e}^{\mathrm{i}N(\theta-\theta')} \sum_{n=1}^{\infty} \frac{\delta_{N0}\, \delta(r)\, \delta(r')}{\left[\left(\gamma_n^2 + q^2 - Q^{02}\right) - \mathrm{i}\,\delta\right] |J_1^{\mathrm{B}}(\gamma_n a)|^2} \\
&\quad = \sum_{n=1}^{\infty} \frac{\delta(r)\, \delta(r')}{\left[\gamma_n^2 + q^2 - Q^{02} - \mathrm{i}\,\delta\right] |J_1^{\mathrm{B}}(\gamma_n a)|^2},
\end{aligned}
\tag{12.2.15}
$$

where note that for $r \to 0$, only the $N = 0$ term can survive due to the presence of the term $N^2/r^2$ in the denominator, in the first line, which would give no contribution for $N \neq 0$, as it introduces a factor $1/\infty$.

Using the symmetry in the exchange $z \leftrightarrow z'$, $x^0 \leftrightarrow x'^0$, $\theta \leftrightarrow \theta'$, allows us to write

$$
\begin{aligned}
\mathcal{I}m\, D_+^{11}(\underline{x}, \underline{x}')\, \delta(r)\, \delta(r') &= \frac{1}{a^2} \int_{-\infty}^{\infty} \frac{\mathrm{d}q}{2\pi} \mathrm{e}^{\mathrm{i}q(z-z')} \int_{-\infty}^{\infty} \frac{\mathrm{d}Q^0}{2\pi} \mathrm{e}^{\mathrm{i}\, Q^0 (x^0 - x'^0)/\sqrt{\varepsilon\mu}} \\
&\times \left[1 - \left(\frac{q}{Q^0}\right)^2\right] \sum_{n=1}^{\infty} \frac{\delta(r)\, \delta(r') \delta\left(\gamma_n^2 + q^2 - Q^{02}\right)}{|J_1^{\mathrm{B}}(\gamma_n a)|^2},
\end{aligned}
\tag{12.2.16}
$$

where we have, in the process, used Eq. (3.6.16), with $\underline{x}$ defined in the first line of Eq. (7.1.7), as we have done in the previous section below Eq. (12.1.10).

As in Eqs. (12.1.12)–(12.1.14), the average number of photons emitted *per unit time* is then given by

$$
\begin{aligned}
< N > = \frac{\mathrm{e}^2 \mathrm{v}^2}{a^2} \sqrt{\frac{\mu}{\varepsilon}} \int_{-\infty}^{\infty} \frac{\mathrm{d}q}{2\pi} \int_{-\infty}^{\infty} \frac{\mathrm{d}Q^0}{2\pi} \int_{-\infty}^{\infty} \mathrm{d}\tau\, \mathrm{e}^{\mathrm{i}\,(q\mathrm{v} - Q^0/\sqrt{\varepsilon\mu})\tau} \\
\times \left[1 - \left(\frac{q}{Q^0}\right)^2\right] \sum_{n=1}^{\infty} \frac{\delta\left(Q^{02} - (\gamma_n^2 + q^2)\right)}{|J_1^{\mathrm{B}}(\gamma_n a)|^2},
\end{aligned}
\tag{12.2.17}
$$

where we have used Eq. (12.1.13) and set $t - t' = \tau$ as in the previous section.

Upon integration over $\tau$, and $q$, gives

$$
\begin{aligned}
< N > = \mathrm{v} \frac{\mathrm{e}^2}{2\pi\, a^2} \sqrt{\frac{\mu}{\varepsilon}} \left[1 - \frac{1}{\mathrm{v}^2 \varepsilon\mu}\right] \int_{-\infty}^{\infty} \mathrm{d}Q^0 \\
\times \sum_{n=1}^{\infty} \delta\left(Q^{02}\left(1 - \frac{1}{\mathrm{v}^2 \varepsilon\mu}\right) - \gamma_n^2\right) |J_1^{\mathrm{B}}(\gamma_n a)|^{-2}.
\end{aligned}
\tag{12.2.18}
$$

We may now insert the identity element

$$
1 = \int_0^{\infty} \mathrm{d}\omega\, \delta\left(\omega - \frac{Q^0}{\sqrt{\varepsilon\mu}}\right),
\tag{12.2.19}
$$

in the integrand in Eq. (12.2.18), and extract the power of radiation as a function of $\omega$ upon multiplying it by the latter, and using the symmetry in $Q^0 \leftrightarrow -Q^0$, to get

$$\mathrm{P}(\omega) = \frac{\mathrm{v}\,\mathrm{e}^2\,\omega}{2\pi\, a^2}\sqrt{\frac{\mu}{\varepsilon}}\Big[1-\frac{1}{\mathrm{v}^2\varepsilon\mu}\Big]2\int_0^\infty \mathrm{d}Q^0$$
$$\delta(\omega - Q^0/\sqrt{\varepsilon\mu})\sum_{n=1}^{\infty}\delta\Big(Q^{02}\Big(1-\frac{1}{\mathrm{v}^2\varepsilon\mu}\Big)-\gamma_n^2\Big)|J_1^{\mathrm{B}}(\gamma_n a)|^{-2}, \qquad \omega > 0. \tag{12.2.20}$$

To simplify the notation, set

$$\frac{1}{Z} \equiv \Big[1-\frac{1}{\mathrm{v}^2\varepsilon\mu}\Big], \tag{12.2.21}$$

to write

$$\delta\Big(Q^{02}\Big(1-\frac{1}{\mathrm{v}^2\varepsilon\mu}\Big)-\gamma_n^2\Big) = Z\delta\Big(Q^{02}-Z\gamma_n^2\Big) = \sqrt{Z}\frac{\delta(Q^0-\sqrt{Z}\gamma_n)}{2\gamma_n},$$
$$= Z\frac{\delta(Q^0-\sqrt{\varepsilon\mu}\omega_n)}{2\sqrt{\varepsilon\mu}\omega_n}, \tag{12.2.22}$$

for $Q^0 > 0$, where we have used Eq. (3.6.14) for $\delta(x^2 - a^2)$, $x > 0$, and introduced the notation

$$\omega_n \equiv \frac{\sqrt{Z}\gamma_n}{\sqrt{\varepsilon\mu}} = \frac{\gamma_n}{\sqrt{\varepsilon\mu}}\Big[1-\frac{1}{\mathrm{v}^2\varepsilon\mu}\Big]^{-1/2}. \tag{12.2.23}$$

Using the fact that $J_0^{\mathrm{B}}(\gamma_n a) = 0$, we may, conveniently, define

$$\lambda_n = \gamma_n a > 0, \quad \text{for } J_0^{\mathrm{B}}(\lambda_n) = 0, \tag{12.2.24}$$

as the positive roots of $J^{\mathrm{B}}(x) = 0$.

Upon integration over $Q^0$, the expression of the power finally emerges as

$$\mathrm{P}(\omega) = \frac{\mathrm{v}\mathrm{e}^2}{2\pi\varepsilon a^2}\sum_{n=1}^{\infty}\frac{\delta\big(\omega-\omega_n\big)}{|J_1^{\mathrm{B}}(\lambda_n)|^2}, \qquad \mathrm{v} > 1/\sqrt{\varepsilon\mu}. \tag{12.2.25}$$

The first[8] few positive zeros $\lambda_n$, in Eq. (12.2.24), are 2.41, 5.52, 8.65, 11.79, 14.93, .... For given parameters, the allowed angular frequencies may be computed from Eq. (12.2.23) for different speeds ($\mathrm{c} > \mathrm{v} > \mathrm{c}/\sqrt{\varepsilon\mu}$, with $\mathrm{c} = 1$).

---

[8] Abramowitz and Stegun (1972) [1], p. 409, Table 9.5.

## References


1. Abramowitz, M, & Stegun, I. A. (1972). *Handbokk of Mathematical Functions.* New York: Dover Publications.
2. Čerenkov, P. A., Frank, I. M., &, Tamm, I. E. (1958). Discovery and the interpretation of the Cerenkov effect. *NobelPrize.org. 1958.*
3. Fazio, G. G., et al (1970). Generation of Čerenkov light flashes by cosmic radiation within the eyes of the Apollo Astronauts. *Nature, 228*, 260-264.
4. Hartman, D., Fan, C. Y., Gough, M. P., Turver, K. E., & Weekes, T. C. (1979). The primary cosmic-radiation spectrum from a $10^{12}$ eV to $10^{15}$ eV from atmospheric Čerenkov-radiation measurements. *Il Nuovo Cimento, A 51*, 131-158.
5. Manoukian, E. B. (1991). Čerenkov radiation from a Nambu string. *Il Nuovo Cimento 104 A (10)*, 1459-1466.
6. Manoukian, E. B. (1993). Discrete Čerenkov power spectrum. *Physical Review B 48 (1)*, 543-547.
7. Manoukian, E. B. (1994). Čerenkov radiation in a medium with a conducting plate. *Il Nuovo Cimento, 107 A (3)*, 503-506.
8. Manoukian, E. B. (1997). Čerenkov radiation near a dielectric medium at finite temperature. *Physica B 240*,58-75.
9. Manoukian, E. B. (1997). Čerenkov photon emission near a conducting medium. *Progress of Theoretical Physics, 98 (1)*, 1-7.
10. Manoukian, E. B. (1997). Čerenkov radiation in discontinuous media: A quantum view-point. *Physica B 269*,115-122.
11. McNulty, P. J., et al. (1976). Role of Čerenkov radiation in the eye-flashes observed by Apollo Astronauts. *Life Science and Space Research, 14*, 205-217.
12. Pinsky, L. S., et al. (1974). Light flashes observed by Astronauts on Apollo 11 through Apollo 17. *Science,183*, 957-959.
13. Protheroe, R. J., & Patterson, J. R. (1984). Interpretation of Čerenkov density spectra. *Journal of PHysics, G: Nuclear Physics, 10 (6)*, 841-855.
14. Rich, J. A., Slovacek, R.E., & Studer, F. J. (1953). Čerenkov radiation from a $Co^{60}$ source in water. *Journal of the Optical Society of America, 43 (9)*, 750-752.
15. Wang, Q., *et al.* (1991). First operation of a multi-electron-beam Čerenkov free-electron laser oscillator. *Applied Physics Letters, 59*, 2378-2380.

# Chapter 13
# Rayleigh Radiation, Mie Radiation: Blue Skies, Red Skies, White Clouds

*Experience the theoretical developments, with all their technical subtelties, of Rayleigh and Mie radiations regarding visible light applicable, respectively, to molecules encountered in our atmosphere, and, more generally, applicable to larger ones, as water droplets, in clouds*

This chapter deals with radiation, as a scattering problem, off dielectric spheres of different sizes. For spheres of small radii, in comparison to the wavelengths of visible light, in particular, the underlying theory is referred to as Rayleigh theory of radiation, and, on the other hand, the underlying theory for radii comparable to the wavelengths of visible light or larger, the underlying theory has been referred to as Mie theory of radiation. The latter theory, however, is also applicable to dielectric spheres of small radii as well as a limiting case.

This subject matter has fascinated physicists for years as these theories may be successfully applied to explain why the sky is blue or red, and clouds appear white to us.

The wavelength of visible light varies from about the order $7 \times 10^{-7}$ m, for red light, down to the order about $4 \times 10^{-7}$ m, for blue (violet) light. Our atmosphere consists of about 78% Nitrogen molecules $N_2$, and about 21% of oxygen molecules $O_2$ (in addition to about 1% of other ones), and are of radii too small in comparison to the wavelengths of light, in the visible region, and the Rayleigh theory is applicable.

On the other hand, water droplets in clouds may have arbitrary sizes comparable to the wavelengths of light in visible spectrum, and may be even larger such as $\sim 10^{-4}$ m. The Rayleigh theory then breaks down, and the need arises to use the Mie theory.

Get ready, to embark on a "fascinating journey" involved in the description of these theories.

 
E. B. Manoukian, *Radiation*,
https://doi.org/10.1007/978-3-032-21506-2_13

## 13.1 Rayleigh Radiation

Our starting point is the averaged power of radiation, per unit frequency, per unit solid angle $\mathrm{dP}(\omega)/\mathrm{d}\Omega$, radiated by a given current distribution $J^{\mu}(x)$, which follows directly from Eq. (7.4.10) in Chap. 7, which in free space, between the molecules themselves, is given by

$$\frac{\mathrm{dP}(\omega)}{\mathrm{d}\Omega} = \frac{\omega^2}{32\pi^2}\int \mathrm{d}^3\mathbf{x}\, J^{*i}(\omega,\mathbf{x})\,\mathrm{e}^{\mathrm{i}\omega\,\mathbf{n}\cdot\mathbf{x}}\left[\delta^{ij}-\mathbf{n}^i\mathbf{n}^j\right]\int \mathrm{d}^3\mathbf{x}'\, J^{j}(\omega,\mathbf{x}')\,\mathrm{e}^{-\mathrm{i}\omega\,\mathbf{n}\cdot\mathbf{x}'}. \tag{13.1.1}$$

Here $\mathbf{n}$ is a unit vector which specifies the direction of emission of radiation off a particle in consideration, and due to the symmetry in the indices $i$ and $j$, we had the liberty in interchanging them, as well as $\mathbf{x} \leftrightarrow \mathbf{x}'$.

Rayleigh scattering is applicable if the wavelength $\lambda$ of radiation in question, is much larger than the size of a particle. This is indeed what is being treated here considering wavelengths in the visible range, and particles such as Oxygen $0_2$, and Nitrogen $N_2$ molecules in the atmosphere as discussed above.

The given currents are sources of scattered radiation off particles, of bounded supports in space, with $\mathbf{x}$, $\mathbf{x}'$ varying over distances of a given small order of the size of a molecule. Accordingly, one may use the fact that $\omega/\mathrm{c} = 2\pi/\lambda$, to neglect

$$\frac{\omega}{\mathrm{c}}(\mathbf{x}\cdot\mathbf{n}),\ \ \frac{\omega}{\mathrm{c}}(\mathbf{x}'\cdot\mathbf{n})\ \approx 0, \quad \text{in Eq. (13.1.1)}$$
$$\text{and set}\ \exp[\,\mathrm{i}\,\omega\,\mathbf{x}\cdot\mathbf{n}/\mathrm{c}\,],\ \ \exp[-\mathrm{i}\,\omega\,\mathbf{x}'\cdot\mathbf{n}/\mathrm{c}\,]\ \cong 1 \text{ in it.} \tag{13.1.2}$$

where $\lambda$ is the wavelength of light. This is referred to as the dipole approximation. Here we have, conveniently, re-inserted the speed of light.

We first consider the space integral of the total derivative of $x^i\mathbf{J}(x)$, integrated over all space. This, by Gauss's Theorem, it is zero for an integrand of bounded support and a surface integral over an infinite radius, i.e.,

$$\int \mathrm{d}^3\mathbf{x}\,\nabla\cdot\left[x^i\mathbf{J}(x)\right] = 0, \qquad i = 1,2,3. \tag{13.1.3}$$

By explicitly applying the operation of $\nabla$ in the latter, we get

$$\begin{aligned}\int \mathrm{d}^3\mathbf{x}\; J^i(x) &= -\int \mathrm{d}^3\mathbf{x}\; x^i\;\nabla\cdot\mathbf{J}(x)\\ &= +\frac{\partial}{\partial t}\int \mathrm{d}^3\mathbf{x}\; x^i\;\rho(x),\end{aligned} \tag{13.1.4}$$

where $\rho = J^0$ is the charge density, and we finally used current conservation.

Upon setting

$$\int d^3\mathbf{x}\,\mathbf{x}\,\rho(x) \;=\; \mathbf{p}(t). \tag{13.1.5}$$

We consider harmonic fields, as defined in Eq. (7.4.2), involving a time factor $\exp[-\mathrm{i}\omega t]$, and denote, in particular, $\mathbf{p}(t) = \exp[-\mathrm{i}\omega t]\mathbf{p}$. Equation (13.1.4) then leads to

$$\int d^3\mathbf{x}\,\mathbf{J}(\omega,\mathbf{x}) \;=\; -\mathrm{i}\,\omega\,\mathbf{p}, \tag{13.1.6}$$

which from Eq. (13.1.1) gives

$$\frac{dP(\omega)}{d\Omega} = \frac{\omega^4}{32\pi^2}\mathbf{p}^i\left[\delta^{ij} - \mathbf{n}^i\mathbf{n}^j\right]\mathbf{p}^j \;=\; \frac{|\mathbf{p}|^2\omega^4}{32\,\pi^2}\,\sin^2\beta, \tag{13.1.7}$$

where we have set

$$\mathbf{p}\cdot\mathbf{n} \;=\; |\mathbf{p}|\cos\beta. \tag{13.1.8}$$

Also, upon integrating over the solid angle $\Omega$, with $d\Omega = 2\pi\,\sin\beta\,d\beta$, and using the integral

$$2\pi\int_0^{\pi}\sin^3\beta\,d\beta \;=\; \frac{8\pi}{3}, \tag{13.1.9}$$

the power per unit frequency simply becomes

$$P(\omega) \;=\; \frac{|\mathbf{p}|^2\omega^4}{12\,\pi}. \tag{13.1.10}$$

Here $\mathbf{p}$ is the static electric dipole moment.

We consider particles as spherical bodies, of radii much smaller than the wavelength of radiation in question, that become polarized in an external constant electric field $\mathbf{E}_0$ (see Fig. 13.1). In the so-called static approximation, the dipole moment is assumed to be a static permanent property acquired by a dielectric sphere in the external electric field.

From the second Maxwell equation, in Eq. (5.5.2), $\nabla\times\mathbf{E} = 0$,

We may thus introduce a scalar field $\phi(\mathbf{x})$, such that $\mathbf{E}(\mathbf{x}) = -\nabla\phi(\mathbf{x})$. Moreover, in the absence of charges, i.e., $\nabla\cdot\mathbf{E} = 0$, $\nabla^2\phi(x) = 0$.

Due to the symmetry along rotations through the azimuthal angle $\varphi$, we may consider only part of the operator $\nabla^2$, in spherical coordinates, depending only on $r$ and $\theta$,

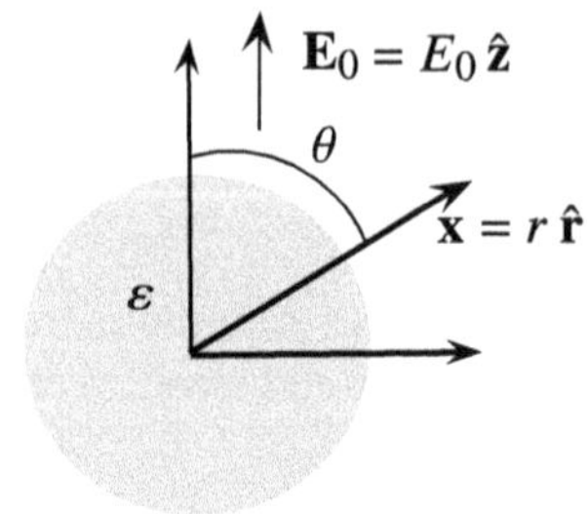


**Fig. 13.1** An example of a dielectric sphere in an external constant electric field $\mathbf{E}_0$. The total external field $\mathbf{E}_{\text{out}} \to \mathbf{E}_0$ for $r \to \infty$

$$\nabla^2\phi = \frac{1}{r^2}\Big[\frac{\partial}{\partial r}\Big(r^2\frac{\partial\phi}{\partial r}\Big) + \frac{1}{\sin\theta}\frac{\partial}{\partial\theta}\Big(\sin\theta\frac{\partial\phi}{\partial\theta}\Big)\Big], \tag{13.1.11}$$

and

$$\nabla^2\phi_{\text{in}}(r,\theta) = 0, \qquad \nabla^2\phi_{\text{out}}(r,\theta) = 0, \tag{13.1.12}$$

$$\text{with radial B.C.:} \quad -\frac{\partial\phi_{\text{out}}}{\partial r} = -\varepsilon\frac{\partial\phi_{\text{in}}}{\partial r}, \tag{13.1.13}$$

$$\text{and tangential B.C.:} \quad -\frac{1}{r}\frac{\partial\phi_{\text{out}}}{\partial\theta} = -\frac{1}{r}\frac{\partial\phi_{\text{in}}}{\partial\theta}, \tag{13.1.14}$$

at r=a, corresponding to the radial and tangential electric field components, where $a$ is the radius of a given sphere, and $\phi_{\text{in/out}}$ is the potential in/out the sphere.

These equations are solved in the Appendix in the next Sect. 13.2, and $\varphi_{\text{out}}$, in particular of interest, is given by

$$\phi_{\text{out}}(r,\theta) = -E_0\,r\cos\theta + \Big(\frac{\varepsilon-1}{\varepsilon+2}\Big)a^3\frac{E_0}{r^2}\cos\theta, \tag{13.1.15}$$

with the B.C. of the electric field $\mathbf{E}_{\text{out}} \to \mathbf{E}_0$ for $r \to \infty$, for which $-\partial\phi_{\text{out}}(r,\theta)/\partial r \to E_0\cos\theta$. We will learn in the Appendix, next Sect. 13.2, that the vector $\mathbf{p}$ in Eq. (13.1.6), is given by

$$\mathbf{p} = 4\pi\Big(\frac{\varepsilon-1}{\varepsilon+2}\Big)a^3\,\mathbf{E}_0. \tag{13.1.16}$$

which is what one calls the electric *dipole moment* associated with a dielectric sphere.

To see how the above expression may be read directly from Eq. (13.1.15), consider the problem involving of a scalar field $\tilde{\phi}(\mathbf{x})$ in the presence of a static charge distribution $\rho(\mathbf{x})$, of bounded support, satisfying

$$-\nabla^2\tilde{\phi}(\mathbf{x}) = \rho(\mathbf{x}). \tag{13.1.17}$$

Using the identity

$$\nabla^2 \frac{1}{|\mathbf{x} - \mathbf{x}'|} = -4\pi\, \delta^{(3)}(\mathbf{x} - \mathbf{x}'), \tag{13.1.18}$$

given in Eq. (3.6.46) in Sect. 3.6 of Chap. 3, the general solution of Eq. (13.1.17) is given by

$$\tilde{\phi}(\mathbf{x}) = \psi(\mathbf{x}) + \frac{1}{4\pi} \int_{|\mathbf{x}'| \leq R} \mathrm{d}^3\mathbf{x}' \, \frac{\rho(\mathbf{x}')}{|\mathbf{x} - \mathbf{x}'|}, \tag{13.1.19}$$

where $R$ is some finite number, and $\psi(\mathbf{x})$ is a function satisfying the homogeneous equation

$$\nabla^2 \psi(\mathbf{x}) = 0. \tag{13.1.20}$$

Considering a physical situation for which $|\mathbf{x}| \gg |\mathbf{x}'|$, and hence

$$\frac{1}{|\mathbf{x} - \mathbf{x}'|} = \frac{1}{r} + \frac{1}{r^3} \mathbf{x} \cdot \mathbf{x}' + \ldots, \tag{13.1.21}$$

gives

$$\tilde{\phi}(\mathbf{x}) = \psi(\mathbf{x}) + \frac{1}{4\pi} \frac{1}{r} \int \mathrm{d}^3\mathbf{x}' \, \rho(\mathbf{x}') + \frac{\mathbf{x}}{4\pi r^3} \cdot \int \mathrm{d}^3\mathbf{x}' \, \mathbf{x}' \rho(\mathbf{x}'). \tag{13.1.22}$$

Upon comparing the above equation with Eqs. (13.1.5), (13.1.15), we may infer that

$$\psi(\mathbf{x}) = -E_0\, r \cos\theta, \qquad 0 = \frac{1}{4\pi} \int \mathrm{d}^3\mathbf{x}' \, \rho(\mathbf{x}'), \tag{13.1.23}$$

$$\mathbf{p} = \int \mathrm{d}^3\mathbf{x}' \, \mathbf{x}' \rho(\mathbf{x}') = 4\pi \left(\frac{\varepsilon - 1}{\varepsilon + 2}\right) a^3 \, \mathbf{E}_0, \tag{13.1.24}$$

which coincides with the expression in Eq. (13.1.16). Also note that $(\mathbf{x}/r) \cdot \mathbf{E}_0 = E_0 \cos\theta$, as well as that $\nabla^2 (r \cos\theta) = 0$. Also remember that the integral of an odd function $f(\mathbf{x})$ vanishes, but $\mathbf{x} f(\mathbf{x})$, being an even function, may not vanish. This corresponds to the physical situation where the positive charges cancel out the negative ones.

Of particular interest is to derive the expression of the power of radiation per unit solid angle per unit frequency, given in Eq. (13.1.7) for polarized light, for unpolarized light as well, relevant to light emitted by the sun.

To the above end, it is most convenient, now, to set the polarization vector $\mathbf{p}$, i.e., also $\mathbf{E}_0$, in the $x - y$ plane, and readily average its direction in a plane. In the process, we set $\mathbf{n} \cdot \mathbf{p} = |\mathbf{p}| \cos\beta$, where $\beta$ is now to be determined, and

$$\mathbf{n} = (\sin\theta\cos\phi, \sin\theta\sin\phi, \cos\theta), \quad \mathbf{p} = |\mathbf{p}|(\cos\phi_0, \sin\phi_0, 0), \tag{13.1.25}$$

$$\mathbf{n}\cdot\mathbf{p} = |\mathbf{p}|\cos\beta, \tag{13.1.26}$$

defining the angle $\beta$. Recall that **n** specifies the direction of emission of radiation off a molecule. Thus we have

$$|\mathbf{p}|\cos\beta = \mathbf{n}\cdot\mathbf{p} = |\mathbf{p}|\sin\theta\cos(\phi_0 - \phi), \tag{13.1.27}$$

$$\Rightarrow \sin^2\beta = 1 - \sin^2\theta\cos^2(\phi_0 - \phi), \tag{13.1.28}$$

with the latter equation being relevant to the expression of the power per unit solid angle given in Eq. (13.1.7). Accordingly, by expanding $\cos(\phi_0 - \phi)$, and averaging over the angle $\phi_0$, from 0 to $2\pi$, in the $x - y$ plane, we obtain

$$< \sin^2\beta > = \frac{1}{2\pi}\int_0^{2\pi} \mathrm{d}\phi_0\,[1 - \sin^2\theta\cos^2(\phi_0 - \phi)] \tag{13.1.29}$$

$$= 1 - \sin^2\theta\Big[ < \cos^2\phi_0 > \cos^2\phi + < \sin^2\phi_0 > \sin^2\phi + < \sin\phi_0\cos\phi_0 > \sin(2\phi)$$

$$= 1 - \sin^2\theta\Big[\frac{1}{2}\cos^2\phi + \frac{1}{2}\sin^2\phi + 0\Big] = \frac{1}{2}[1 + \cos^2\theta], \tag{13.1.30}$$

and hence

$$\frac{\mathrm{d}P(\omega)}{\mathrm{d}\Omega} = \frac{|\mathbf{p}|^2\omega^4}{64\,\pi^2}[1 + \cos^2\theta], \tag{13.1.31}$$

for *unpolarized* incident light.

We may express, the power of radiation per unit frequency, per unit solid angle in terms of the volume $V$ of a particle in question, and the wavelength $\lambda$, of radiation, using,

$$V = \frac{4\pi}{3}a^3, \quad \text{and} \quad \omega = \frac{2\pi}{\lambda}, \quad |\mathbf{p}|^2 = 9\,V^2\Big(\frac{\varepsilon - 1}{\varepsilon + 2}\Big)^2 |\mathbf{E}_0|^2, \tag{13.1.32}$$

leading to

$$\frac{\mathrm{d}P(\omega)}{\mathrm{d}\Omega} = \frac{9}{64\,\pi^2}V^2\Big(\frac{\varepsilon - 1}{\varepsilon + 2}\Big)^2 \omega^4\,|\mathbf{E}_0|^2\big[1 + \cos^2\theta\big], \quad \mathrm{d}\Omega = \sin\theta\,\mathrm{d}\theta\,\mathrm{d}\phi, \tag{13.1.33}$$

$$= \frac{9\pi^2}{4\,\lambda^4}V^2\Big(\frac{\varepsilon - 1}{\varepsilon + 2}\Big)^2 |\mathbf{E}_0|^2\big[1 + \cos^2\theta\big]. \tag{13.1.34}$$

Before considering the many particles involved in the atmosphere, we already see that in the visible spectrum, the blue (violet) light of being of shortest wavelength, is scattered most, while red light of longest wavelength is scattered least.

For many particles involved in the system, one has to take into account the departure of the expression involving the dielectric constant in a dipole moment, of a given volume $V$, due to its departure from the average value, which accounts to consider the average of the square of the expression

$$V\delta\Big(\frac{\varepsilon-1}{\varepsilon+2}\Big) \simeq V\,\frac{\delta\varepsilon}{3}. \tag{13.1.35}$$

using the fact that for the particles involved in the medium $\varepsilon \approx 1$.

For many particles involved in the system, we ought to consider the realistic situation in which the medium consists of cells of volumes $v_i$ in each of which the dielectric constants differ from the average dielectric constant $\varepsilon$ in the medium. In this case, we make the replacement

$$V\,\frac{\delta\varepsilon}{3} \;\rightarrow\; \sum_i v_i\frac{\delta\varepsilon_i}{3}, \tag{13.1.36}$$

corresponding to the various dipole moments associated with different cells. By invoking the Clausius-Mossotti Equation, derived in the Appendix, §13.3, which states that[1]

$$\sum_i v_i\frac{\delta\varepsilon_i}{3} \simeq (\varepsilon-1)\sum_i \frac{v_i\,\delta N}{3\,N}, \tag{13.1.37}$$

where $N$ is the number of molecules per unit volume. Accordingly, $v_i N = \mathcal{N}_i$, by definition, denotes the number of particles in a cell of volume $v_i$. Let $\mathcal{N}$ denotes the average number of particles per cell. Accordingly, considering the average of the square of the expression in Eq. (13.1.37), in reference to Eq. (13.1.33), the power of radiation per solid angle, per unit frequency, now becomes replaced by

$$\frac{\mathrm{d}P(\omega)}{\mathrm{d}\Omega} = \frac{1}{64\,\pi^2}(\varepsilon-1)^2\frac{\sum_{i,j}\sigma_{i,j}}{N^2}\omega^4\,|\mathbf{E}_0|^2\big[1+\cos^2\theta\big], \tag{13.1.38}$$

where,

$$\sigma_{i,j} = \big\langle(\mathcal{N}_i-\mathcal{N})(\mathcal{N}_j-\mathcal{N})\big\rangle, \tag{13.1.39}$$

describes for $i \neq j$, the correlations between the number of particles in different cells, and $\varepsilon$ is the average dielectric constant of the medium.

In Sect. 13.5 we will develop the full theory avoiding the dipole approximation. As a rule of thumb the Rayleigh theory is reliable as long as the radius of a particle is not larger than 1/20 of the wavelength. From the full theory, one may access its

---

[1] See Eq. (13.3.13).

accuracy numerically more precisely. Much work has been done in this regard and some references are given below.[2]

We have seen that in Rayleigh scattering, the power of radiation is proportional to $1/\lambda^4$, and, as a consequence of which, blue light is radiated more than red, which makes the sky blue. On the other hand, during sunrises and sunsets, with the sun closer to the horizon, blue light is removed as it passes through a much thicker atmosphere,[3] grazing the Earth's surface, involving a higher average density of scatterers, leaving light near the red end of the spectrum and the sky looks red.

Some recommended technical references on Rayleigh radiation are given below.[4] For easy reading on the blue sky see. e.g., the two references cited below.[5]

## 13.2 Appendix: Dipole of a Dielectric Sphere

Refer to Fig. 13.1 in the previous section. Consider the function $\phi$ below, due to symmetry, to be independent of the azimuthal angle $\varphi$, as mentioned above Eq. (13.1.11),

$$\nabla^2\phi = \frac{1}{r^2}\left[\frac{\partial}{\partial r}\left(r^2\frac{\partial\phi}{\partial r}\right) + \frac{\cos\theta}{\sin\theta}\frac{\partial\phi}{\partial\theta} + \frac{\partial^2\phi}{\partial\theta^2}\right], \tag{13.2.1}$$

and the two equations, as well as the relevant boundary conditions below

$$\nabla^2\phi_{\text{out}}(r,\theta) = 0, \qquad \nabla^2\phi_{\text{in}}(r,\theta) = 0, \tag{13.2.2}$$

$$\text{with radial B.C.:} \qquad -\frac{\partial\phi_{\text{out}}}{\partial r} = -\varepsilon\frac{\partial\phi_{\text{in}}}{\partial r}, \tag{13.2.3}$$

$$\text{and tangential B.C.:} \qquad -\frac{1}{r}\frac{\partial\phi_{\text{out}}}{\partial\theta} = -\frac{1}{r}\frac{\partial\phi_{\text{in}}}{\partial\theta}, \tag{13.2.4}$$

at r=a, corresponding to the radial and tangential electric field components. Here, $a$ is the radius of a sphere, and $\phi_{\text{in/out}}$ is the potential in/out the sphere.

It is most illuminating to solve these equations by inspection. Regarding $\phi_{\text{out}}(r,\theta)$, we know that for $r$ large, $\phi_{\text{out}} \to -E_0\, r\, \cos\theta$, consistent with the presence of an incident electric field applied in the direction specified. It is easily verified that the just given expression satisfies Eq. (13.2.2) for $\phi_{\text{out}}(r,\theta)$. Now we have to add a term which vanishes for $r \to \infty$, consistent with the condition just stated. Consider a vanishing term like $1/r$. The derivatives w.r.t. $r$, as well the ones w.r.t. $\theta$, within the

[2] See, e.g., Kerker (1969) [13], Bohren and Huffman (1983) [2].

[3] For various aspects of our atmosphere, the following book by Wallace (2006) [20] is well recommended.

[4] Stratton (2007) [18], Bohren and Huffman (1983) [2], van Hulst (1981) [12], Kerker (1969) [13], Born and Wolf, E. (2004) [4], Strutton (aka Lord Rayleigh) (1899) [19], Kerker, Wang, and Chew (1980) [14], Mackowski (1991) [15], Eom (2004) [6].

[5] Hoeppe (2007) [11], Bohren and Alistair (1985) [3].

square brackets in Eq. (13.2.1), will give 0 for such a term. Accordingly, we may add a term $c_0/r$ to $-E_0\, r\cos\theta$. On the other hand, a vanishing term like $1/r^2$, by itself, will not work as the term involving derivatives with respect to $r$, within the square brackets in Eq. (13.2.1), will give $2/r^2$, while the angular part will give zero. This is remedied by considering the expression $\cos\theta/r^2$, instead, which is easily verified to satisfies Eq. (13.2.2) for $\phi_{\text{out}}(r,\theta)$. Thus, we may write

$$\phi_{\text{out}}(r,\theta) \;=\; -E_0\, r\,\cos\theta + \frac{c_0}{r} + \frac{c_1\,\cos\theta}{r^2}. \tag{13.2.5}$$

On the other hand, terms of the sort $1/r$, $1/r^2$ are not allowed in $\phi_{\text{in}}(r,\theta)$ as they will blow up at the origin. It is easily verified that $r\cos\theta$ will satisfy Eq. (13.2.2) for $\phi_{\text{in}}(r,\theta)$ as well. Accordingly, we write

$$\phi_{\text{in}}(r,\theta) \;=\; b_1\, r\,\cos\theta. \tag{13.2.6}$$

The B.C. in Eq. (13.2.3), from Eq. (13.2.5), and Eq. (13.2.6), gives

$$-\,E_0\,\cos\theta\; - 0\,\frac{c_0}{a^2} \;-\; \frac{2\,c_1}{a^3}\cos\theta \;=\; \varepsilon\, b_1\cos\theta. \tag{13.2.7}$$

That is,

$$\varepsilon\, b_1 \;=\; -\,E_0 \;-\; \frac{2\,c_1}{a^3}, \qquad c_0 \;=\; 0. \tag{13.2.8}$$

While the B.C. in Eq. (13.2.4), from Eq. (13.2.5), and Eq. (13.2.6), gives

$$E_0\sin\theta \;-\; c_1\frac{\sin\theta}{a^3} \;=\; -b_1\sin\theta. \tag{13.2.9}$$

That is

$$b_1 \;=\; -\,E_0 \;+\; \frac{c_1}{a^3}, \tag{13.2.10}$$

From Eqs. (13.2.8), (13.2.10), we have

$$c_1 \;=\; E_0\, a^3\Big(\frac{\varepsilon-1}{\varepsilon+2}\Big), \qquad c_0 \;=\; 0, \tag{13.2.11}$$

and from Eq. (13.2.5), the following expression emerges for $\phi_{\text{out}}(r,\theta)$,

$$\phi_{\text{out}}(r,\theta) \;=\; -\,E_0\, r\cos\theta \;+\; \Big(\frac{\varepsilon-1}{\varepsilon+2}\Big)\, a^3\,\frac{E_0}{r^2}\cos\theta, \tag{13.2.12}$$

relevant to the radiation problem. Through the problem of the analysis, we also learn, in the process, that there are no $1/r^3$ or higher order terms contributions to $\phi_{\text{out}}(r, \theta)$.

## 13.3 Appendix: Clausius-Mossotti Equation

In order to find the electric field at the point where a given molecule is situated in the medium, one may, as a working hypothesis, consider that it may be included within a spherical cavity in the dielectric medium of radius $R$ large enough to engulf the molecule in question, but small enough not to be large in comparison to a typical distance between the molecules. On the average, a molecule would be surrounded by other molecules and its extension may be expected to be well approximated by the spherical hole.

The dielectric outside the sphere may be treated as a continuous medium. Remember, the medium is a macroscopic one, while inside the sphere, including the molecule in question, is microscopic. As a working hypothesis, all other molecules are considered as part of the continuum.

The dielectric medium is uniformly polarized by a mean electric field $\mathbf{E}$, and directed along the $z$ -axis as shown in Fig. 13.2. The polarization vector is defined by

$$\mathbf{P} = (\varepsilon - 1)\mathbf{E}, \quad \mathbf{E} = E\,\hat{\mathbf{z}}. \tag{13.3.1}$$

It arises by considering the left-hand side of Maxwell's first equation, upon writing it as follows

$$\nabla \cdot \varepsilon\,\mathbf{E} = \nabla \cdot \mathbf{E} + (\varepsilon - 1)\nabla \cdot \mathbf{E}. \tag{13.3.2}$$

resulting a polarization charge density defined by $\nabla \cdot \mathbf{P}$. We may refer to Gauss' law, for which a given vector $\mathbf{V}$ arising in a volume integral may be replaced by a surface integral, in the process, by the substitution rule

$$\nabla \cdot \mathbf{V}\,\mathrm{d(volume)} \;\to\; \mathbf{V} \cdot \mathbf{n}\,\mathrm{d(surface)}. \tag{13.3.3}$$

This will generate a surface charge density

$$\sigma = \mathbf{P} \cdot \mathbf{n}, \quad \text{of magnitude} \quad P\cos\theta. \tag{13.3.4}$$

with unit vector directed *toward* the center (not shown) of the cavity. This, in turn, will give rise to an electric field, necessarily in a direction along the $z$−axis, as shown in the figure. Clearly, due to the charge distribution on the wall of the cavity only the horizontal components of the electric fields, due to infinitesimal charges of magnitudes $P\cos\theta\sin\theta\mathrm{d}\theta\mathrm{d}\phi$ directed along the $z$−axis will arise. Here note the important fact that the polarization vector $\mathbf{P}$ is *along the $z$−axis* as given in Eq.

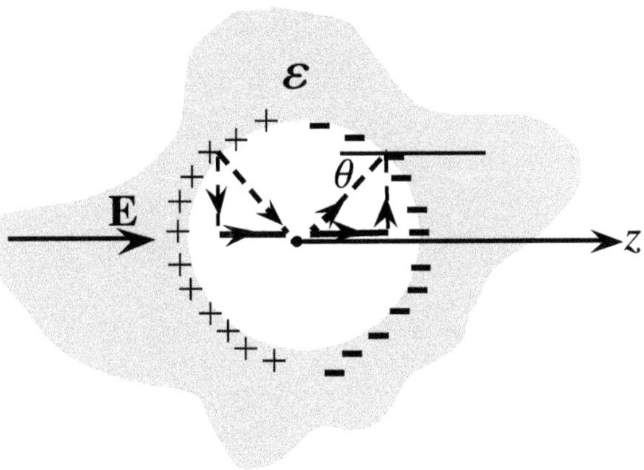


**Fig. 13.2** A given molecule is considered to be included within a spherical cavity in the dielectric medium uniformly polarized by a mean electric field **E**, as well as the polarization vector, directed along the $z-$axis. The unit vector **n**, not shown here, in Eq. (13.3.3) emanating from the surface of the cavity is directed towards the center of the cavity. The net electric field within the cavity is also directed along the $z-$axis, and by the symmetry of the surface charge density, no perpendicular components, to the $z-$axis, arising by dashed lines, parallel to the $x - y$ plane could contribute

(13.3.1). Perpendicular components, to the $z-$axis, arising by dashed lines, parallel to the $x - y$ plane could not contribute by symmetry.

Accordingly the net electric field $\mathbf{E}_s$, along the $z-$*axis* in the direction of **E**, within the cavity, due to this surface charge distribution, is given by

$$\mathbf{E}_s = \frac{1}{4\pi}\int_0^{\pi} d\theta \int_0^{2\pi} d\phi \, \frac{(P\cos\theta)(\cos\theta)(R^2\sin\theta)}{R^2}\hat{\mathbf{z}}$$
$$= \frac{P}{2}\int_{-1}^{+1} \cos^2\theta \, d(\cos\theta)\hat{\mathbf{z}} = \frac{\mathbf{P}}{3}, \tag{13.3.5}$$

*where* $\mathbf{P} = P\,\hat{\mathbf{z}}$, $dS = R^2 \sin\theta \, d\theta \, d\phi$.

Hence the effective electric field $\mathbf{E}_{\text{mol}}$ within the cavity, where the molecule at the center resides, is given by

$$\mathbf{E}_{\text{mol}} = \mathbf{E} + \frac{\mathbf{P}}{3}. \tag{13.3.6}$$

Thus introducing the phenomenological parameter *molecular polarizability* $\gamma_{\text{mol}}$, the polarization vector **P** due to $N$ molecules per unit volume is expressed as

$$\mathbf{P} = N\gamma_{\text{mol}}\,\mathbf{E}_{\text{mol}}$$
$$= N\,\gamma_{\text{mol}}\left(\mathbf{E} + \frac{\mathbf{P}}{3}\right), \tag{13.3.7}$$
$$\Rightarrow \mathbf{P} = \frac{N\gamma_{\text{mol}}}{1 - \frac{N\gamma_{\text{mol}}}{3}}\,\mathbf{E}. \tag{13.3.8}$$

On the other hand from

$$\mathbf{P} = (\varepsilon - 1)\mathbf{E}, \tag{13.3.9}$$
$$\Rightarrow \quad \varepsilon - 1 = \frac{N\gamma_{\text{mol}}}{1 - \frac{N\gamma_{\text{mol}}}{3}}, \qquad (\varepsilon + 2) = \frac{3}{1 - \frac{N\gamma_{\text{mol}}}{3}} \tag{13.3.10}$$

Equation (13.3.10), which relates the dielectric constant to the molecular number density $N$ and the polarizability $\gamma_{\text{mol}}$, is referred to as the Clausius-Mossotti equation.

In particular, for the expression involving $\varepsilon$ of interest in Sect. 13.1,

$$\frac{(\varepsilon - 1)}{(\varepsilon + 2)} = \frac{N\gamma_{\text{mol}}}{3}, \tag{13.3.11}$$
$$\frac{1}{(\varepsilon + 2)^2}\,\delta\varepsilon = \frac{\gamma_{\text{mol}}}{9}\,\delta N, \tag{13.3.12}$$
$$\frac{1}{(\varepsilon - 1)(\varepsilon + 2)}\,\delta\varepsilon = \frac{\delta N}{3\,N}. \tag{13.3.13}$$

The Clausius–Mossotti relation is named after Ottaviano Fabrizio Mossotti and Rudolf Julius Emanuel Clausius.[6]

## 13.4 Polarization Aspects of Rayleigh Radiation

Light from the sun is unpolarized and it is interesting to investigate the nature of the polarization of light we observe when it is scattered by the molecules of our atmosphere and reaches our eyes.

Our equipment (eyes) needed to observe light may be represented by a current source $J_2^\mu(x)$ as we now discuss. By varying $\delta J_2^\mu(x)$, we may re-orient the direction along which we observe light, we may turn our polarizers around and may get a different *response* from the source of light that is scattered each time. The source of light scattered before reaching us may be also represented by another current source $J_1^\mu(x')$, such that this source of emission is causally arranged with the source of detection $J_2^\mu(x)$ such that $t > t'$. We are free to choose how to orient our direction of observation and vary $J_2^\mu(x)$ as described. That is, we need to describe the interaction between these two current sources, one as a source of radiation emission, upon

[6] Clausius (1850) [17], Mossotti (1879) [5].

scattering, from the atmosphere, and one of detecting it with our eyes, polarizers and all that, and then see what *response* we get as we *vary* our equipment.

Our starting point is to apply the expression of *current-current interaction energy* between the two currents $J_2^{\mu}(x)$, $J_{1\mu}(x')$, causally arranged as described above, related to the following expression derived in Eq. (3.5.22) in Chap. 3. It reads as:

$$\mathcal{E} = \int (\mathrm{d}x) \int (\mathrm{d}x') J_2^{\mu}(x) \frac{1}{4\pi\,|\mathbf{x}-\mathbf{x}'|} \delta\Big((t-t') - |\mathbf{x}-\mathbf{x}'|\Big)\, J_{1\mu}(x'), \quad t > t'. \tag{13.4.1}$$

By using the conservation law: $\partial_{\mu} J^{\mu} = 0$ with $J^0(x) = \rho(x)$, the charge density, we may conveniently write

$$J_2^{\mu}(x) J_{1\mu}(x') = J_2^i(x) J_1^i(x') - \Big(\frac{\partial^i}{\partial_0} J_2^i(x)\Big)\Big(\frac{\partial'^j}{\partial_0'} J_1^j(x')\Big), \tag{13.4.2}$$

and upon partial integration with respect to $t, t'$, we may re-express $\mathcal{E}$ as

$$\mathcal{E} = \int (\mathrm{d}x) \int (\mathrm{d}x') \frac{1}{4\pi\,|\mathbf{x}-\mathbf{x}'|} \Big[ J_2^i(x)\, J_1^i(x') + \partial^i J_2^i(x)\, \partial'^j J_1^j(x') \Big(\frac{1}{\partial/\partial t}\Big)^2 \Big] \\ \times\ \delta\Big((t-t') - |\mathbf{x}-\mathbf{x}'|\Big), \tag{13.4.3}$$

where the $+$ sign in the middle of the square brackets occurs because $t, t'$ are multiplied by different signs in the Dirac delta distribution.

Upon carrying out a Fourier transform

$$\mathbf{J}_1(x') = \int_{-\infty}^{\infty} \frac{\mathrm{d}\omega}{2\pi} \mathrm{e}^{-\mathrm{i}\omega t'}\, \mathbf{J}_1(\omega, \mathbf{x}'), \qquad x' = (t', \mathbf{x}'), \tag{13.4.4}$$

and using the fact that $t' < t$, the $t'$ integral in Eq. (13.4.3) is given simply by

$$\int_{-\infty}^{t} \mathrm{d}t'\, \mathrm{e}^{-\mathrm{i}\omega t'} \delta\Big(t' - \big(t - |\mathbf{x}-\mathbf{x}'|\big)\Big) = \mathrm{e}^{-\mathrm{i}\,\omega t} \mathrm{e}^{\mathrm{i}\omega|\mathbf{x}-\mathbf{x}'|}, \tag{13.4.5}$$

using the fact that the value restricted to $t'$, as specified by the Dirac delta distribution, is included in interval $(-\infty, t)$: $-\infty < t - |\mathbf{x}-\mathbf{x}'| < t$.

Accordingly,

$$\mathcal{E} = \frac{1}{4\pi} \int_{-\infty}^{\infty} \mathrm{d}t \int_{-\infty}^{\infty} \frac{\mathrm{d}\omega}{2\pi} \mathrm{e}^{-\mathrm{i}\,\omega t} \int \mathrm{d}^3\mathbf{x} \int \mathrm{d}^3\mathbf{x}' \\ \times\ \Big[ J_2^i(x)\, J_1^i(\omega, \mathbf{x}') - \frac{\partial^i J_2^j(x)\, \partial'^j J_1^j(\omega, \mathbf{x}')}{\omega^2} \Big] \frac{\mathrm{e}^{\mathrm{i}\omega|\mathbf{x}-\mathbf{x}'|}}{|\mathbf{x}-\mathbf{x}'|}. \tag{13.4.6}$$

We integrate by parts over $\mathbf{x}$ and $\mathbf{x}'$ and use the facts that

$$\frac{\partial}{\partial x'^i}|\mathbf{x}-\mathbf{x}'| = -\frac{(x^i - x'^i)}{|\mathbf{x}-\mathbf{x}'|} = -\frac{\partial}{\partial x^i}|\mathbf{x}-\mathbf{x}'|, \tag{13.4.7}$$

$$\frac{\partial}{\partial x'^i}\frac{1}{|\mathbf{x}-\mathbf{x}'|} = \frac{(x^i - x'^i)}{|\mathbf{x}-\mathbf{x}'|^3} = -\frac{\partial}{\partial x^i}\frac{1}{|\mathbf{x}-\mathbf{x}'|}, \tag{13.4.8}$$

to obtain

$$\begin{aligned}\mathcal{E} &= \frac{1}{4\pi}\int_{-\infty}^{\infty}\mathrm{d}t\int_{-\infty}^{\infty}\frac{\mathrm{d}\omega}{2\pi}\,\mathrm{e}^{-\mathrm{i}\,\omega t}\int \mathrm{d}^3\mathbf{x}\,J_2^i(t,\mathbf{x})\\ &\quad\times\left[\delta^{ij}+\frac{\partial^i\partial^j}{\omega^2}\right]\int \mathrm{d}\mathbf{x}'\,J_1^j(\omega,\mathbf{x}')\,\frac{\mathrm{e}^{\mathrm{i}\omega|\mathbf{x}-\mathbf{x}'|}}{|\mathbf{x}-\mathbf{x}'|}.\end{aligned} \tag{13.4.9}$$

We introduce the inverse Fourier transform

$$\int_{-\infty}^{\infty}\mathrm{d}t\,\mathrm{e}^{-\mathrm{i}\omega t}\mathbf{J}_2(t,\mathbf{x}) = \mathbf{J}_2(-\omega,\mathbf{x}), \qquad \mathbf{J}_2(-\omega,\mathbf{x}) = \mathbf{J}_2^*(\omega,\mathbf{x}), \tag{13.4.10}$$

where (*) denotes complex conjugation, to re-express $\mathcal{E}$ as

$$\mathcal{E} = \frac{1}{4\pi}\int_{-\infty}^{\infty}\frac{\mathrm{d}\omega}{2\pi}\int \mathrm{d}^3\mathbf{x}\,J_2^{*i}(\omega,\mathbf{x})\left[\delta^{ij}+\frac{\partial^i\partial^j}{\omega^2}\right]\int \mathrm{d}\mathbf{x}'\,J_1^j(\omega,\mathbf{x}')\,\frac{\mathrm{e}^{\mathrm{i}\omega|\mathbf{x}-\mathbf{x}'|}}{|\mathbf{x}-\mathbf{x}'|}. \tag{13.4.11}$$

Physically $|\mathbf{x}|$ is much larger than $|\mathbf{x}'|$, with the former corresponding to macroscopic observational distances, while the latter corresponds to molecular sizes. Accordingly, $\mathcal{E}$ takes the form

$$\mathcal{E} \simeq \frac{1}{4\pi}\int_{-\infty}^{\infty}\frac{\mathrm{d}\omega}{2\pi}\int \mathrm{d}^3\mathbf{x}\,J_2^{*i}(\omega,\mathbf{x})\left[\delta^{ij}+\frac{\partial^i\partial^j}{\omega^2}\right]\frac{\mathrm{e}^{\mathrm{i}\omega|\mathbf{x}|}}{|\mathbf{x}|}\int \mathrm{d}\mathbf{x}'\,J_1^j(\omega,\mathbf{x}'). \tag{13.4.12}$$

We have seen earlier, in Eqs. (13.1.5)/(13.1.4), that a dipole may be defined by

$$\mathbf{p}(t) = \int \mathrm{d}^3\mathbf{x}\,\mathbf{x}\,\rho(x), \qquad \int \mathrm{d}^3\mathbf{x}\,\mathbf{J}(x) = \frac{\partial}{\partial t}\int \mathrm{d}^3\mathbf{x}\,\mathbf{x}\,\rho(x). \tag{13.4.13}$$

Upon carrying, quite generally, a Fourier transform

$$\rho(x) = \int_{-\infty}^{\infty}\mathrm{d}\omega'\,\mathrm{e}^{-\mathrm{i}\omega' t}\,\rho(\omega',\mathbf{x}), \tag{13.4.14}$$

we may also, conveniently, write

$$\int \mathrm{d}^3\mathbf{x}' \int_{-\infty}^{\infty} \mathrm{d}\omega'\, \mathrm{e}^{-\mathrm{i}\,\omega' t}\,\mathbf{J}_1(\omega', \mathbf{x}') = -\int_{-\infty}^{\infty} \mathrm{i}\,\omega'\, \mathrm{d}\omega'\, \mathrm{e}^{-\mathrm{i}\omega' t} \int \mathrm{d}^3\mathbf{x}\,\mathbf{x}\,\rho_1(\omega', \mathbf{x})$$
$$= -\int_{-\infty}^{\infty} \mathrm{i}\,\omega'\, \mathrm{d}\omega'\, \mathrm{e}^{-\mathrm{i}\omega' t}\,\mathbf{p}_1(\omega'). \qquad (13.4.15)$$

That is,

$$\int \mathrm{d}^3\mathbf{x}'\mathbf{J}_1(\omega', \mathbf{x}') = -\,\mathrm{i}\,\omega\,\mathbf{p}_1(\omega'). \qquad (13.4.16)$$

By inserting in the process the speed of light c, we note that

$$\frac{\omega}{\mathrm{c}} = \frac{2\pi}{\lambda} \gg \frac{1}{|\mathbf{x}|}, \qquad (13.4.17)$$

for the wavelength $\lambda$ in the visible region, and $|\mathbf{x}|$ corresponding to macroscopic distances. Hence,

$$\Big[\delta^{ij} + \frac{\partial^i \partial^j}{\omega^2}\Big] \frac{\mathrm{e}^{\mathrm{i}\omega|\mathbf{x}|}}{|\mathbf{x}|} \simeq \frac{\mathrm{e}^{\mathrm{i}\omega|\mathbf{x}|}}{|\mathbf{x}|} \big[\delta^{ij} - \mathbf{n}^i\mathbf{n}^j\big], \qquad \mathbf{n}^i = \frac{x^i}{|\mathbf{x}|}. \qquad (13.4.18)$$

From Eqs. (13.4.10), (13.4.16), (13.4.17), we get

$$\mathcal{E} \simeq \frac{1}{4\pi} \int_{-\infty}^{\infty} \frac{\mathrm{d}\omega}{2\pi} (-\mathrm{i})\,\omega \int \mathrm{d}^3\mathbf{x}\,\mathbf{J}_2^*(\omega, \mathbf{x}) \cdot \frac{\mathrm{e}^{\mathrm{i}\omega|\mathbf{x}|}}{|\mathbf{x}|} \big[\mathbf{p}_1(\omega) - \mathbf{n}\,(\mathbf{n}\cdot\mathbf{p}_1(\omega))\big]. \qquad (13.4.19)$$

Or, more interestingly, as

$$\mathcal{E} \simeq \int_{-\infty}^{\infty} \frac{\mathrm{d}\omega}{2\pi} \Big[\int \mathrm{d}^3\mathbf{x}\,(-\mathrm{i})\,\frac{\omega}{4\pi} \frac{\mathrm{e}^{\mathrm{i}\omega|\mathbf{x}|}}{|\mathbf{x}|} \Big(\mathbf{n} \times \mathbf{J}_2^*(\omega, \mathbf{x})\Big)\Big] \cdot \mathbf{n} \times \mathbf{p}_1(\omega), \qquad (13.4.20)$$

where for a unit vector $\mathbf{A}$, $\mathbf{A}\cdot\mathbf{A} = 1$, we have used the identity

$$(\mathbf{A} \times \mathbf{B}) \cdot (\mathbf{A} \times \mathbf{C}) = \mathbf{B}\cdot\mathbf{C} - (\mathbf{A}\cdot\mathbf{B})\,(\mathbf{A}\cdot\mathbf{C}). \qquad (13.4.21)$$

The physical interpretation of Eq. (13.4.20) is clear. We recall that the unit vector $\mathbf{n}$ specifies the direction of the scattered light, i.e., the direction of light taken to reach the observer. That is, at a given point in space, one may direct one's observation axis in such a way that the component of polarization of the sun light, specified by the direction of the vector $\mathbf{p}_1$, turns up to be aligned with the direction of observation and, in turn, annihilates it from view as will give zero contribution to the energy as a consequence the consequence of the fact that $\mathbf{n} \times \mathbf{p}_1(\omega) = 0$. Since the polarisation direction is perpendicular to the direction of incidence, the direction of scattered light would necessarily, in this case, make a 90° with the direction of incidence. (See

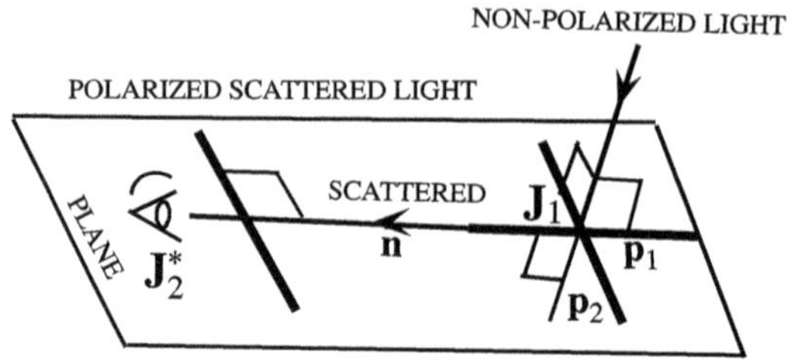


**Fig. 13.3** Figure shows that for a specific case of observation, the component of polarization of the sun light, specified by the direction of the vector $\mathbf{p}_1$, which turns up to be along the direction of observation, specified by the unit vector **n**, only the other polarization is transmitted to the observer. Also note that in *this* case, **n** is orthogonal to the direction of incident light from the sun

Fig. 13.3)

The expression within the square brackets in Eq. (13.4.20), as an operator, creates the role of a polarizer to block a given polarization. It is remarkable, that it is possible to observe a *polarized* light from the *unpolarised* light from the sun, as one may readily try with a polarizer, and be able to describe it here mathematically on how to do so.

In Chap. 16, we will consider a generalization of the process depicted in Fig. 13.3 where, in the latter process, light is incident from two directions perpendicular to each other.

## 13.5 Mie Radiation

From Eq. (7.5.1), and as done in deriving Eq. (7.5.3), we may infer, that in the temporal gage, the vector potential satisfies the equation

$$[-\boldsymbol{\nabla}^2 + \varepsilon\mu\,\partial_0^2]A^i(x) \;=\; \mu\Big[\delta^{ij} - \frac{\partial^i\partial^j}{\varepsilon\mu\partial_0^2}\Big]J^j(x). \tag{13.5.1}$$

In particular, for harmonic fields involving the factor $\exp[-\mathrm{i}\omega t]$, this gives

$$[-\boldsymbol{\nabla}^2 - \varepsilon\mu\,\omega^2]A^i(\omega,\mathbf{r}) \;=\; \mu\Big[\delta^{ij} + \frac{\partial^i\partial^j}{\varepsilon\mu\omega^2}\Big]J^j(\omega,\mathbf{r}). \tag{13.5.2}$$

For the electric field **E**, with components $E^i = \partial^0 A^i$, and the magnetic field **B**, with component $B^\ell = \epsilon^{\ell m i}\partial^m A^i$, satisfy, respectively, the equations

$$[-\nabla^2 - \varepsilon\mu\,\omega^2]E^i(\omega,\mathbf{r}) \;=\; \mu\Big[\delta^{ij} + \frac{\partial^i\partial^j}{\varepsilon\mu\omega^2}\Big|\,\partial^0\,J^j(\omega,\mathbf{r}), \tag{13.5.3}$$

$$[-\nabla^2 - \varepsilon\mu\,\omega^2]B^\ell(\omega,\mathbf{r}) \;=\; \mu\Big[\delta^{ij} + \frac{\partial^i\partial^j}{\varepsilon\mu\omega^2}\Big|\,\epsilon^{\ell mi}\partial^m J^j(\omega,\mathbf{r}). \tag{13.5.4}$$

In the absence of a current $J^\mu$, they both satisfy the equations

$$[\nabla^2 + \varepsilon\mu\,\omega^2]\mathbf{E} \;=\; 0, \qquad \nabla\cdot\mathbf{E} = 0, \tag{13.5.5}$$

$$[\nabla^2 + \varepsilon\mu\,\omega^2]\mathbf{B} \;=\; 0, \qquad \nabla\cdot\mathbf{B} = 0, \tag{13.5.6}$$

with the very last relation, of course, being always true.

In the problem at hand, considering dielectric spheres in free space, we put $\mu = 1$, and set

$$k_{\text{sph}} \;=\; \sqrt{\varepsilon}\,\omega, \qquad k \;=\; \omega. \tag{13.5.7}$$

In spherical coordinates $(r, \theta, \phi)$,

$$\nabla^2 \;=\; \frac{1}{r^2}\frac{\partial}{\partial r}\Big(r^2\frac{\partial}{\partial r}\Big) + \frac{1}{r^2\sin\theta}\frac{\partial}{\partial\theta}\Big(\sin\theta\frac{\partial}{\partial\theta}\Big) + \frac{1}{r^2\sin^2\theta}\frac{\partial^2}{\partial\phi^2}, \tag{13.5.8}$$

and the two equations on the left-hand sides of Eqs. (13.5.5), (13.5.6) are referred to as Helmholtz equations in spherical coordinates.

With the vector $\mathbf{r} = r\,\hat{\mathbf{r}}$, we introduce the relevant unit vectors involved in the theory:

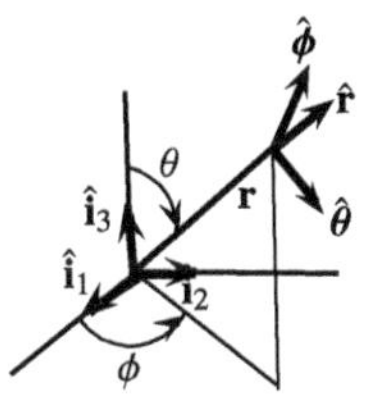


$$\hat{\mathbf{r}} \;=\; \sin\theta\cos\phi\,\hat{\mathbf{i}}_1 \;+\; \sin\theta\sin\phi\,\hat{\mathbf{i}}_2 \;+\; \cos\phi\,\hat{\mathbf{i}}_3, \tag{13.5.9}$$

$$\hat{\boldsymbol{\theta}} \;=\; \cos\theta\cos\phi\,\hat{\mathbf{i}}_1 \;+\; \cos\theta\sin\phi\,\hat{\mathbf{i}}_2 \;-\; \sin\phi\,\hat{\mathbf{i}}_3, \tag{13.5.10}$$

$$\hat{\boldsymbol{\phi}} \;=\; -\sin\phi\,\hat{\mathbf{i}}_1 \;+\; \cos\phi\,\hat{\mathbf{i}}_2. \tag{13.5.11}$$

and

$$\begin{aligned}\hat{\mathbf{i}}_1 &= (\hat{\mathbf{i}}_1\cdot\hat{\mathbf{r}})\,\hat{\mathbf{r}} + (\hat{\mathbf{i}}_1\cdot\hat{\boldsymbol{\theta}})\,\hat{\boldsymbol{\theta}} + (\hat{\mathbf{i}}_1\cdot\hat{\boldsymbol{\phi}})\,\hat{\boldsymbol{\phi}}\\ &= \sin\theta\cos\phi\,\hat{\mathbf{r}} + \cos\theta\cos\phi\,\hat{\boldsymbol{\theta}} \;-\; \sin\phi\,\hat{\boldsymbol{\phi}},\end{aligned} \tag{13.5.12}$$

$$\begin{aligned}\hat{\mathbf{i}}_2 &= (\hat{\mathbf{i}}_2\cdot\hat{\mathbf{r}})\,\hat{\mathbf{r}} + (\hat{\mathbf{i}}_2\cdot\hat{\boldsymbol{\theta}})\,\hat{\boldsymbol{\theta}} + (\hat{\mathbf{i}}_2\cdot\hat{\boldsymbol{\phi}})\,\hat{\boldsymbol{\phi}}\\ &= \sin\theta\sin\phi\,\hat{\mathbf{r}} + \cos\theta\sin\phi\,\hat{\boldsymbol{\theta}} + \cos\phi\,\hat{\boldsymbol{\phi}},\end{aligned} \tag{13.5.13}$$

$$\begin{aligned}\hat{\mathbf{i}}_3 &= (\hat{\mathbf{i}}_3\cdot\hat{\mathbf{r}})\,\hat{\mathbf{r}} + (\hat{\mathbf{i}}_3\cdot\hat{\boldsymbol{\theta}})\,\hat{\boldsymbol{\theta}} + (\hat{\mathbf{i}}_3\cdot\hat{\boldsymbol{\phi}})\,\hat{\boldsymbol{\phi}}\\ &= \cos\theta\,\hat{\mathbf{r}} \;-\; \sin\theta\,\hat{\boldsymbol{\theta}}.\end{aligned} \tag{13.5.14}$$

We will see that solutions of the vectors in Eqs. (13.5.5), (13.5.6), may be generated from scalar fields satisfying, the equation

$$[\boldsymbol{\nabla}^2 + k^2]\,\psi(\mathbf{r}) = 0, \tag{13.5.15}$$

with $\boldsymbol{\nabla}^2$ given in Eq. (13.5.8).

The reader has undoubtedly studied such equations, involving the operator in Eq. (13.5.8), in his/her undergraduate courses in quantum mechanics to find their solutions by the method of separation of variables, e.g., in the Hydrogen atom problem.

To the above end, by setting, $\psi(\mathbf{r}) = R(r)\Theta(\theta)\Phi(\phi)$, two independent solutions may be given in terms of well know functions spelled out below:

$$\psi^{(\mathrm{e})}_{\ell m} = \cos(m\phi)\,P^m_\ell(\cos\theta)\,z_\ell(kr),\quad \psi^{(\mathrm{o})}_{\ell m} = \sin(m\phi)\,P^m_\ell(\cos\theta)\,z_\ell(kr), \tag{13.5.16}$$

$$m = -\ell, -\ell+1, \ldots, 0, 1, \ldots, \ell-1, \ell;\quad \ell = 0, 1, \ldots, \tag{13.5.17}$$

where $P^m_\ell(x)$ are the associated Legendre functions which may be expressed in terms of the Legendre polynomials $P_\ell(x)$ as follows:

$$P^m_\ell(x) = (1-x^2)^{m/2}\Big(\frac{\mathrm{d}}{\mathrm{d}x}\Big)^m P_\ell(x), \quad m \geq 0, \tag{13.5.18}$$

$$P^{-m}_\ell(x) = (-1)^m \frac{(\ell-m)!}{(\ell+m)!}\,P^m_\ell(x), \tag{13.5.19}$$

$$P_\ell(x) = \frac{1}{2^\ell \ell!}\Big(\frac{\mathrm{d}}{\mathrm{d}x}\Big)^\ell (x^2-1)^\ell. \tag{13.5.20}$$

The $z_\ell(x)$ are so-called spherical Bessel functions. Within a sphere, we choose the spherical Bessel function of first kind, well behaved at the center (origin) of a sphere, explicitly given by[7]

$$j_\ell(x) = (-x)^\ell\Big(\frac{\mathrm{d}}{x\,\mathrm{d}x}\Big)^\ell \frac{\sin x}{x}, \tag{13.5.21}$$

while for outgoing radiation, outside the dielectric spheres in free space, we choose a so-called Hankel function $h_\ell(x)$, given explicitly by

$$h_\ell(x) = (-\mathrm{i})^{(\ell+1)}\,\frac{\mathrm{e}^{\mathrm{i}\,x}}{x}\sum_{n=0}^{\ell}\frac{\mathrm{i}^n(\ell+n)!}{n!\,(2x)^n(\ell-n)!}. \tag{13.5.22}$$

If we multiply the exponential factor $\exp[\mathrm{i}x]$ above by the time development factor $\exp[-\mathrm{i}\omega t]$, thus obtaining $\exp[\mathrm{i}(x-\omega t)]$, note that as we move forward in time, $x$ also increases, for a constant phase, thus defining an outgoing wave.

Given the functions in Eq. (13.5.16), we may generate vector solutions of equa-

[7] Abramowitz and Stegun (1964) [1], p. 439.

tions given in Eqs. (13.5.5), (13.5.6) as follows, by introducing two vector fields $\mathbf{V}(\mathbf{r}), \mathbf{U}(\mathbf{r})$

$$\mathbf{V}(\mathbf{r}) = \nabla \times (\mathbf{r}\psi(\mathbf{r})), \quad \nabla \cdot \mathbf{V}(\mathbf{r}) = 0, \quad \nabla \times (\nabla \times \mathbf{V}) = -\nabla^2 \mathbf{V} \tag{13.5.23}$$

$$\mathbf{U}(\mathbf{r}) = \frac{1}{k} \nabla \times \mathbf{V}(\mathbf{r}), \quad \nabla \times \mathbf{U}(\mathbf{r}) = k\mathbf{V}(\mathbf{r}), \quad \nabla \cdot \mathbf{U}(\mathbf{r}) = 0. \tag{13.5.24}$$

That these vector fields are, so-called, solenoidal, i.e., have zero divergence, is immediate. The fact they satisfy the equations Eqs. (13.5.5), (13.5.6)

$$[\nabla^2 + k^2]\mathbf{V}(\mathbf{r}) = 0, \qquad [\nabla^2 + k^2]\mathbf{U}(\mathbf{r}) = 0, \tag{13.5.25}$$

follow by using the facts

$$V^i = \epsilon^{ij\ell}\partial^j\big(\mathbf{r}^\ell \psi\big), \qquad \nabla^2 = \partial^s\partial^s, \qquad \epsilon^{ij\ell}\partial^j\partial^\ell = 0, \qquad \nabla^2\psi = -k^2\psi,$$

and noting that

$$\begin{aligned} \nabla^2 V^i &= \epsilon^{ij\ell}\partial^j \nabla^2(\mathbf{r}^\ell\,\psi) = \epsilon^{ij\ell}\partial^j\partial^s\partial^s\big(\mathbf{r}^\ell\psi\big) = \epsilon^{ij\ell}\partial^j\partial^s\big(\delta^{s\ell}\psi + \mathbf{r}^\ell\partial^s\psi\big) \\ &= \epsilon^{ij\ell}\partial^j(2\,\partial^\ell\psi + \mathbf{r}^\ell\nabla^2\psi\big) = -k^2\epsilon^{ij\ell}\partial^j(\mathbf{r}^\ell\psi\big) = -k^2 V^i. \end{aligned} \tag{13.5.26}$$

While for $\mathbf{U}$, we have $\nabla^2\mathbf{U} = (1/k)\nabla \times \big(\nabla^2\mathbf{V}\big) = -k^2\mathbf{U}$.

The vector fields $\mathbf{V}$, $\mathbf{U}$, corresponding to the scalar fields in Eq. (13.5.16), together with their orthogonality properties are spelled out in the Appendix, in the next Sect. 13.6 of the chapter.

The electric and magnetic fields in Eqs. (13.5.5), (13.5.6), may be expanded in terms of these generated vector fields. Before doing that, we consider the initial incident electric field which will be taken with polarization along the unit vector, say, $\hat{\mathbf{i}}_1$ propagating along the unit vector $\hat{\mathbf{i}}_3$, given by

$$\mathbf{E}_{\text{inc}}(\mathbf{r}) = E_0\,\hat{\mathbf{i}}_1\,\mathrm{e}^{\mathrm{i}\mathbf{k}\cdot\mathbf{r}}, \qquad \mathbf{r} = r\hat{\mathbf{r}}, \qquad \mathbf{k} = k\,\hat{\mathbf{i}}_3 = k\,\big(\cos\theta\,\hat{\mathbf{r}} - \sin\theta\hat{\boldsymbol{\theta}}\big), \tag{13.5.27}$$

$$\hat{\mathbf{i}}_1 = \sin\theta\cos\phi\,\hat{\mathbf{r}} + \cos\theta\cos\phi\,\hat{\boldsymbol{\theta}} - \sin\phi\,\hat{\boldsymbol{\phi}}, \tag{13.5.28}$$

with $\hat{\mathbf{i}}_1$, $\hat{\mathbf{i}}_3$, given in Eqs. (13.5.12), (13.5.14), respectively.

Such a description of an incident electric field, here, is quite appropriate as a shower of a uniform field emerging everywhere in space, rather than being localized.

Using the fact that $\mathbf{k}\cdot\mathbf{r} = \omega\, r\cos\theta$, and applying the Bauer Formula for the expansion of exp[i$\rho\cos\theta$], given in the Appendix Sect. 13.6, Eq. (13.6.6), we may express the incident field as

$$
\begin{aligned}
\mathbf{E}_{\text{inc}}(\mathbf{r}) &= E_0 \sum_{\ell=0}^{\infty} (\mathrm{i})^{\ell}\,(2\ell+1)\, j_\ell(\omega r)\, P_\ell(\cos\theta) \\
&\times\ [\sin\theta\cos\phi\,\hat{\mathbf{r}} + \cos\theta\cos\phi\,\hat{\boldsymbol{\theta}} - \sin\phi\,\hat{\boldsymbol{\phi}}]. \qquad (13.5.29)
\end{aligned}
$$

Now we have to re-express the latter in terms of the vector solutions $\mathbf{V}(\mathrm{r})$, $\mathbf{U}(\mathrm{r})$, satisfying the same equations Eq. (13.5.25) as the electric and magnetic fields in Eqs. (13.5.5), (13.5.6), with appropriate values for $k$. In particular, we note that the above expression involves $\sin m\phi$, $\cos m\phi$, *only* with $m = 1$.

We recall the definition of the components of the curl of a vector $\mathbf{W}(\mathbf{r})$ in spherical coordinates:

$$
\mathbf{W}(\mathbf{r}) = W_r\,\hat{\mathbf{r}} + W_\theta\,\hat{\boldsymbol{\theta}} + W_\phi\,\hat{\boldsymbol{\phi}}, \qquad (13.5.30)
$$

$$
\nabla\times\mathbf{W} = \big(\nabla\times\mathbf{W}\big)_r\,\hat{\mathbf{r}} + \big(\nabla\times\mathbf{W}\big)_\theta\,\hat{\boldsymbol{\theta}} + \big(\nabla\times\mathbf{W}\big)_\phi\,\hat{\boldsymbol{\phi}} \qquad (13.5.31)
$$

and

$$
\big(\nabla\times\mathbf{W}\big)_r = \frac{1}{r\sin\theta}\Big[\frac{\partial}{\partial\theta}\big(\sin\theta\,W_\phi\big) - \frac{\partial W_\theta}{\partial\phi}\Big], \qquad (13.5.32)
$$

$$
\big(\nabla\times\mathbf{W}\big)_\theta = \frac{1}{r}\Big[\frac{1}{\sin\theta}\frac{\partial W_r}{\partial\phi} - \frac{\partial}{\partial r}(r\,W_\phi)\Big], \qquad (13.5.33)
$$

$$
\big(\nabla\times\mathbf{W}\big)_\phi = \frac{1}{r}\Big[\frac{\partial}{\partial r}(r\,W_\theta) - \frac{\partial W_r}{\partial\theta}\Big]. \qquad (13.5.34)
$$

Regarding $\mathbf{V}(\mathbf{r}) = \nabla\times(\mathbf{r}\psi)$, with, $\mathbf{r}\psi \equiv \hat{\mathbf{r}}(r\psi)$, we note from Eqs. (13.5.32)–(13.5.34) that

$$
\big(\nabla\times\mathbf{r}\psi\big)_r = 0, \qquad (13.5.35)
$$

$$
\big(\nabla\times\mathbf{r}\psi\big)_\theta = \frac{1}{\sin\theta}\frac{\partial\psi}{\partial\phi}, \qquad (13.5.36)
$$

$$
\big(\nabla\times\mathbf{r}\psi\big)_\phi = -\frac{\partial\psi}{\partial\theta}. \qquad (13.5.37)
$$

Therefore, Eq. (13.5.35) implies that with one differentiation with respect to $\phi$ of a $\cos m\phi$ or a $\sin m\phi$ corresponding to $\psi^{(\mathrm{e})}$ or $\psi^{(\mathrm{o})}$, as given in Eq. (13.5.16), and one differentiation with respect to $\theta$, we may generate two distinct vector fields given by

$$
\mathbf{V}^{(\mathrm{e})}_{\ell m} = -m\,(\sin m\phi)\,\frac{P^m_\ell}{\sin\theta}\,z_\ell(kr)\,\hat{\boldsymbol{\theta}} - (\cos m\phi)\,\frac{\mathrm{d}P^m_\ell}{\mathrm{d}\theta}\,z_\ell(kr)\,\hat{\boldsymbol{\phi}}, \qquad (13.5.38)
$$

$$
\mathbf{V}^{(\mathrm{o})}_{\ell m} = +m(\cos m\phi)\,\frac{P^m_\ell}{\sin\theta}\,z_\ell(kr)\,\hat{\boldsymbol{\theta}} - (\sin m\phi)\,\frac{\mathrm{d}P^m_\ell}{\mathrm{d}\theta}\,z_\ell(kr)\,\hat{\boldsymbol{\phi}}, \qquad (13.5.39)
$$

where $P_\ell^m = P_\ell^m(\cos\theta)$.

In particular for $m = 1$, which will turn up to be the only relevant case in the subsequence analysis, we have, in a convenient notation,

$$\mathbf{v}_\ell^{(\mathrm{e})}(\mathbf{r}) = -(\sin\phi)\,\frac{P_\ell^1}{\sin\theta}\,z_\ell(kr)\,\hat{\boldsymbol{\theta}} - (\cos\phi)\,\frac{\mathrm{d}P_\ell^1}{\mathrm{d}\theta}\,z_\ell(kr)\,\hat{\boldsymbol{\phi}}, \tag{13.5.40}$$

$$\mathbf{v}_\ell^{(\mathrm{o})}(\mathbf{r}) = +(\cos\phi)\,\frac{P_\ell^1}{\sin\theta}\,z_\ell(kr)\,\hat{\boldsymbol{\theta}} - (\sin\phi)\,\frac{\mathrm{d}P_\ell^1}{\mathrm{d}\theta}\,z_\ell(kr)\,\hat{\boldsymbol{\phi}}. \tag{13.5.41}$$

As a matter of fact, from the summand in Eq. (13.5.29), corresponding to the expression of the incident electric field $\mathbf{E}_{\mathrm{inc}}$, only the terms with $m = 1$, (i.e., also $\ell \geq 1$), are relevant here, and the coefficient of $\hat{\boldsymbol{\theta}}$ has, in particular, $\cos\phi$, while the coefficient of $\hat{\boldsymbol{\phi}}$ has $\sin\phi$. Accordingly, we may infer that the vector fields $\mathbf{v}_\ell^{(\mathrm{o})}$ rather than $\mathbf{v}_\ell^{(\mathrm{e})}$, are relevant in Eq. (13.5.29).

In the same manner, it is straightforward, though a bit longer task, to infer that the vector fields generated by $\mathbf{U}$ in Eq. (13.5.24) for $m = 1$ are given by

$$\mathbf{u}_\ell^{(\mathrm{e})}(\mathbf{r}) = (\cos\phi)\,\ell(\ell+1)\,\frac{z_\ell(kr)}{kr}\,P_\ell^1\,\hat{\mathbf{r}}$$
$$+ (\cos\phi)\,\frac{[kr\,z_\ell(kr)]'}{kr}\,\frac{\mathrm{d}P_\ell^1}{\mathrm{d}\theta}\,\hat{\boldsymbol{\theta}} - (\sin\phi)\,\frac{[kr\,z_\ell(kr)]'}{kr\sin\theta}\,P_\ell^1\,\hat{\boldsymbol{\phi}}, \tag{13.5.42}$$

where $[krz_\ell(kr)]' = \partial[\rho z_\ell(\rho)]/\partial\rho$, with $\rho = kr$, and

$$\mathbf{u}_\ell^{(0)}(\mathbf{r}) = (\sin\phi)\,\ell(\ell+1)\,\frac{z_\ell(kr)}{kr}\,P_\ell^1\,\hat{\mathbf{r}}$$
$$+ (\sin\phi)\,\frac{[kr\,z_\ell(kr)]'}{kr}\,\frac{\mathrm{d}P_\ell^1}{\mathrm{d}\theta}\,\hat{\boldsymbol{\theta}} + (\cos\phi)\,\frac{[kr\,z_\ell(kr)]'}{kr\sin\theta}\,P_\ell^1\,\hat{\boldsymbol{\phi}}, \tag{13.5.43}$$

Here, on the other hand, the $\mathbf{u}_\ell^{(\mathrm{e})}$, fields, rather than the $\mathbf{u}_\ell^{(0)}$ fields, have $\cos\phi$ in the coefficient of $\hat{\boldsymbol{\theta}}$ and $\sin\phi$ in the coefficient of $\hat{\boldsymbol{\phi}}$. Accordingly, the latter are the relevant vector fields in Eq. (13.5.29).

We may thus re-express $\mathbf{E}_{\mathrm{inc}}$ in Eq. (13.5.29) as

$$\mathbf{E}_{\mathrm{inc}} = E_0 \sum_{\ell=1}^{\infty} \big[a_\ell\,\mathbf{v}_\ell^{(\mathrm{o})}(\mathbf{r}) + b_\ell\,\mathbf{u}_\ell^{(\mathrm{e})}(\mathbf{r})\big]. \tag{13.5.44}$$

In the Appendix, Eq. (13.6.12), in the next Sect. 13.6, we show that

$$\int_0^\pi \sin\theta\,\mathrm{d}\theta \int_0^{2\pi} \mathrm{d}\phi\,\mathbf{v}_\ell^{(\mathrm{o})}(\mathbf{r})\cdot\mathbf{u}_{\ell'}^{(\mathrm{e})}(\mathbf{r}) = 0, \quad \text{for all } \ell, \ell'. \tag{13.5.45}$$

Upon multiplying Eq. (13.5.44) by $\mathbf{v}^{(\mathrm{o})}_{\ell'}(\mathbf{r})\cdot$, using the above orthogonality property, and integrate over $\theta$ and $\phi$, we obtain from Eq. (13.6.24), and the normalization property in Eq. (13.6.17) in the Appendix, next Sect. 13.6, give

$$a_\ell E_0 = \Big[\int \mathrm{d}\Omega\, \mathbf{E}_{\mathrm{inc}} \cdot \mathbf{v}^{(\mathrm{o})}_\ell\Big]\Big[\int \mathrm{d}\Omega\, \mathbf{v}^{(\mathrm{o})}_\ell \cdot \mathbf{v}^{(\mathrm{o})}_\ell\Big]^{-1} = (\mathrm{i})^\ell \frac{2\ell+1}{\ell(\ell+1)} E_0, \tag{13.5.46}$$

where $\mathrm{d}\Omega = \sin\theta\, \mathrm{d}\theta\, \mathrm{d}\phi$, and, similarly, upon multiplying Eq. (13.5.44) again, now by $\mathbf{u}^{(\mathrm{e})}_{\ell'}(\mathbf{r})$, using Eq. (13.6.25), and the normalization property in Eq. (13.6.18), give

$$b_\ell E_0 = \Big[\int \mathrm{d}\Omega\, \mathbf{E}_{\mathrm{inc}} \cdot \mathbf{u}^{(\mathrm{e})}_\ell\Big]\Big[\int \mathrm{d}\Omega\, \mathbf{u}^{(\mathrm{e})}_\ell \cdot \mathbf{u}^{(\mathrm{e})}_\ell\Big]^{-1} = -\,(\mathrm{i})^{\ell+1} \frac{2\ell+1}{\ell(\ell+1)} E_0. \tag{13.5.47}$$

Accordingly, we have the final expression for the incident electric field given by

$$\mathbf{E}_{\mathrm{inc}} = E_0 \sum_{\ell=1}^{\infty} (\mathrm{i})^\ell \frac{2\ell+1}{\ell(\ell+1)} \big[\mathbf{v}^{(\mathrm{o})}_\ell(\mathbf{r}) \;-\; \mathrm{i}\, \mathbf{u}^{(\mathrm{e})}_\ell(\mathbf{r})\big], \quad z_\ell(kr) = j_\ell(kr) \tag{13.5.48}$$

We have to consider the radiated electric field $\mathbf{E}_{\mathrm{rad}}$ as well as the transmitted, here, within the sphere, one $\mathbf{E}_{\mathrm{tr}}$. As boundary conditions at the surface of a dielectric sphere $r = a$ may be taken, corresponding to tangential components, as

$$[\mathbf{E}_{\mathrm{inc}} + \mathbf{E}_{\mathrm{rad}}] \times \hat{\mathbf{r}} = \mathbf{E}_{\mathrm{tr}} \times \hat{\mathbf{r}}, \qquad r = a, \tag{13.5.49}$$

$$[\mathbf{B}_{\mathrm{inc}} + \mathbf{B}_{\mathrm{rad}}] \times \hat{\mathbf{r}} = \mathbf{B}_{\mathrm{tr}} \times \hat{\mathbf{r}}, \qquad r = a, \tag{13.5.50}$$

with $\mathbf{B}_{\mathrm{inc}}$, $\mathbf{B}_{\mathrm{rad}}$, $\mathbf{B}_{\mathrm{tr}}$ denoting the corresponding magnetic fields ($\mu = 1$).

To satisfy, the boundary conditions, we may carry an expansion of the latter fields in terms of the vector fields $\mathbf{v}^{(\mathrm{o})}_\ell$ $\mathbf{u}^{(\mathrm{e})}_\ell$ in the expansion of $\mathbf{E}_{\mathrm{inc}}$ having the same angular dependence. That is, we may write, with coefficients to be determined below,

$$\mathbf{E}_{\mathrm{rad}} = E_0 \sum_{\ell=1}^{\infty} (\mathrm{i})^\ell \frac{2\ell+1}{\ell(\ell+1)} \big[\alpha_\ell \mathbf{v}^{(\mathrm{o})}_\ell - \mathrm{i}\, \beta_\ell\, \mathbf{u}^{(\mathrm{e})}_\ell\big],$$

$$z_\ell(kr) = h_\ell(kr), \quad k = \omega, \tag{13.5.51}$$

where, we have taken $z_\ell(kr) = h_\ell(kr)$ corresponding to outgoing radiation, as discussed above. On the other hand,

$$\mathbf{E}_{\mathrm{tr}} = E_0 \sum_{\ell=1}^{\infty} (\mathrm{i})^\ell \frac{2\ell+1}{\ell(\ell+1)} \big[\gamma_\ell \mathbf{v}^{(\mathrm{o})}_\ell - \mathrm{i}\, \zeta_\ell\, \mathbf{u}^{(\mathrm{e})}_\ell\big],$$

$$z_\ell(k_{\mathrm{sph}} r) = j_\ell(k_{\mathrm{sph}} r), \quad k_{\mathrm{sph}} = \sqrt{\varepsilon}\,\omega, \tag{13.5.52}$$

*within* the sphere.

To generate the summation formula for the magnetic fields, we recall that $\nabla \times \mathbf{E} = \mathrm{i}\omega\mathbf{B}$. Also from Eq. (13.5.24), note the explicit $k$ dependence on the right-hand sides of the two equations, $\nabla \times \mathbf{U} = k\mathbf{V}$, $\nabla \times \mathbf{V} = k\mathbf{U}$. Hence from Eqs. (13.5.48), (13.5.49), (13.5.52), we have from Eq. (13.5.7) and $\mathbf{B} = -\mathrm{i}\nabla \times \mathbf{E}/\omega$,

$$\mathbf{B}_{\text{inc}} = -E_0 \sum_{\ell=1}^{\infty} (\mathrm{i})^{\ell} \frac{2\ell+1}{\ell(\ell+1)} \big[\mathbf{v}_{\ell}^{(\mathrm{e})}(\mathbf{r}) + \mathrm{i}\,\mathbf{u}_{\ell}^{(\mathrm{o})}(\mathbf{r})\big], \quad z_{\ell}(kr) = j_{\ell}(kr), \tag{13.5.53}$$

and, on the other hand,

$$\begin{aligned} \mathbf{B}_{\text{rad}} &= -E_0 \sum_{\ell=1}^{\infty} (\mathrm{i})^{\ell} \frac{2\ell+1}{\ell(\ell+1)} \big[\beta_{\ell}\mathbf{v}_{\ell}^{(\mathrm{e})} + \mathrm{i}\,\alpha_{\ell}\,\mathbf{u}_{\ell}^{(0)}\big], \\ &\quad z_{\ell}(kr) = h_{\ell}(kr), \quad k = \omega, \end{aligned} \tag{13.5.54}$$

$$\begin{aligned} \mathbf{B}_{\text{tr}} &= -\sqrt{\varepsilon}\, E_0 \sum_{\ell=1}^{\infty} (\mathrm{i})^{\ell} \frac{2\ell+1}{\ell(\ell+1)} \big[\zeta_{\ell}\mathbf{v}_{\ell}^{(\mathrm{e})} + \mathrm{i}\,\gamma_{\ell}\,\mathbf{u}_{\ell}^{(\mathrm{o})}\big], \\ &\quad z_{\ell}(k_{\text{sph}}r) = j_{\ell}(k_{\text{sph}}r), \quad k_{\text{sph}} = \sqrt{\varepsilon}\omega, \end{aligned} \tag{13.5.55}$$

Because of the different angular dependence in the summations of each of the electric and magnetic fields, the two boundary conditions in Eqs. (13.5.49), (13.5.50) will give rise to four equations, from which all the expansion coefficients $\alpha_{\ell}$, $\beta_{\ell}$, $\gamma_{\ell}$,$\zeta_{\ell}$, may be determined.

Let us consider first the boundary conditions implied by Eq. (13.5.49) involving the electric fields in Eqs. (13.5.48), (13.5.51), (13.5.52):

Here from the expression of $\mathbf{v}_{\ell}^{(0)}$ in Eq. (13.5.41), the coefficient $\alpha_{\ell}$ is multiplied by $h_{\ell}(\omega a)$ in Eq. (13.5.51), while $\gamma_{\ell}$ is multiplied by $j_{\ell}(\sqrt{\varepsilon}\omega a)$ in Eq. (13.5.52).

On the other hand, from the expression of $\mathbf{u}_{\ell}^{(\mathrm{e})}$ in Eq. (13.5.42), $\beta_{\ell}$ is multiplied by $[krh_{\ell}(kr)]'/kr$ for $kr = \omega a$ in Eq. (13.5.51), while $\zeta_{\ell}$ is multiplied by $[k_{\text{sph}}r j_{\ell}(k_{\text{sph}}r)]'/k_{\text{sph}}r$ for $k_{\text{sph}}r = \sqrt{\varepsilon}\,\omega\,a$ in Eq. (13.5.52). Accordingly, from Eq. (13.5.49), we obtain

$$j_{\ell}(x) + \alpha_{\ell}\, h_{\ell}(x) = \gamma_{\ell}\, j_{\ell}(y), \tag{13.5.56}$$

$$[x j_{\ell}(x)]' + \beta_{\ell}\,[x h_{\ell}(x)]' = \zeta_{\ell}\,\frac{[y j_{\ell}(y)]'}{\sqrt{\varepsilon}}, \tag{13.5.57}$$

with $x = \omega\, a$, and $y = \sqrt{\varepsilon}\,\omega\, a$.

On the other hand, consider the boundary conditions set by Eq. (13.5.50), by the magnetic fields in Eqs. (13.5.53)–(13.5.55):

First note the overall $\sqrt{\varepsilon}$ factor multiplying $\mathbf{B}_{\text{tr}}$. Here, from the expression of $\mathbf{v}_{\ell}^{(\mathrm{e})}$ in Eq. (13.5.40), the coefficient $\beta_{\ell}$ is multiplied by $h_{\ell}(\omega a)$ in Eq. (13.5.54), while $\zeta_{\ell}$ is multiplied by $\sqrt{\varepsilon}\, j_{\ell}(\sqrt{\varepsilon}\,\omega\, a)$ in Eq. (13.5.55). From the expression of $\mathbf{u}_{\ell}^{(0)}$ in Eq. (13.5.43), $\alpha_{\ell}$ is multiplied by $[krh_{\ell}(kr)]'/kr$ for $kr = \omega\, a$ in Eq. (13.5.54), while $\gamma_{\ell}$ is multiplied by $\sqrt{\varepsilon}\,[k_{\text{sph}}r j_{\ell}(k_{\text{sph}}r)]'/k_{\text{sph}}r$, where $k_{\text{sph}}r = \sqrt{\varepsilon}\,\omega\, a$ in Eq.

(13.5.55). Accordingly, Eq. (13.5.50) leads to

$$j_\ell(x) + \beta_\ell\, h_\ell(x) = \sqrt{\varepsilon}\,\zeta_\ell\, j_\ell(y), \tag{13.5.58}$$

$$[xj_\ell(x)]' + \alpha_\ell\,[xh_\ell(x)]' = \gamma_\ell\,[yj_\ell(y)]', \tag{13.5.59}$$

with $x = \omega\, a$, and $y = \sqrt{\varepsilon}\,\omega\, a$.

The coefficients $\gamma_\ell$, $\zeta_\ell$ pertinent to the fields within a dielectric sphere are directly eliminated from Eqs. (13.5.56)–(13.5.59), leading to two separate linear equations for $\alpha_\ell$ and $\beta_\ell$ given by

$$\alpha_\ell = \frac{j_\ell(x)\,[yj_\ell(y)]' - j_\ell(y)\,[xj_\ell(x)]'}{j_\ell(y)\,[xh_\ell(x)]' - h_\ell(x)\,[yj_\ell(y)]'}, \tag{13.5.60}$$

$$\beta_\ell = \frac{\varepsilon\, j_\ell(y)\,[x\, j_\ell(x)]' - j_\ell(x)\,[y\, j_\ell(y)]'}{h_\ell(x)\,[y\, j_\ell(y)]' - \varepsilon\, j_\ell(y)\,[x\, h_\ell(x)]'}, \tag{13.5.61}$$

with $x = \omega\, a$, and $y = \sqrt{\varepsilon}\,\omega\, a$.

The following expressions which occur in these coefficients

$$S_\ell(x) = xj_\ell(x), \qquad \xi_\ell(x) = xh_\ell(x), \tag{13.5.62}$$

are referred to as Riccati-Bessel functions. This suggests to multiply the numerators and denominators of $\alpha_\ell$, and $\beta_\ell$, respectively, by $y = \sqrt{\varepsilon}\, x$, and $x$. This leads to the more manageable expressions[8]

$$\alpha_\ell = \frac{\sqrt{\varepsilon}\, S_\ell(x)\, S'_\ell(\sqrt{\varepsilon}\, x) - S_\ell(\sqrt{\varepsilon}\, x)\, S'_\ell(x)}{S_\ell(\sqrt{\varepsilon}\, x)\, \xi'_\ell(x) - \sqrt{\varepsilon}\, \xi_\ell(x)\, S'_\ell(\sqrt{\varepsilon}\, x)}, \tag{13.5.63}$$

$$\beta_\ell = \frac{\sqrt{\varepsilon}\, S_\ell(\sqrt{\varepsilon}\, x)\, S'_\ell(x) - S_\ell(x)\, S'_\ell(\sqrt{\varepsilon}\, x)}{\xi_\ell(x)\, S'_\ell(\sqrt{\varepsilon}\, x) - \sqrt{\varepsilon}\, S_\ell(\sqrt{\varepsilon}\, x)\, \xi'_\ell(x)}, \tag{13.5.64}$$

where $x = \omega a = 2\pi a/\lambda$, referred to as the size parameter of the sphere.

It is important to note that when $\varepsilon \to 1$ the coefficients $\alpha_\ell$, $\beta_\ell \to 0$, as expected, and we are merely dealing with free space.

In order to evaluate the explicit averaged power of radiation, per unit frequency , $\mathrm{P}(\omega)$ given in Eq. (7.5.13) in Chap. 7, we have first to determine,

$$\lim_{r\to\infty} \Big[ r^2 \int \mathrm{d}\Omega\, |\mathbf{E}_{\mathrm{rad}}|^2(r,\theta,\phi) \Big]. \tag{13.5.65}$$

From Eq. (13.5.51), and the orthogonality property in Eq. (13.5.45), and the normalization properties in Eqs. (13.5.46), (13.5.47),

---

[8] Some authors call our $\alpha_\ell$ $b_\ell$, and our $\beta_\ell$ $a_\ell$. Using Greek letters, this should not confuse you.

$$\int d\Omega\, |\mathbf{E}_{\mathrm{rad}}|^2(r,\theta,\phi)] = |E_0|^2 \sum_{\ell=1}^{\infty} \Big(\frac{2\ell+1}{\ell(\ell+1)}\Big)^2 \Big[|\alpha_\ell|^2\, I_\ell^{(1)}(\omega r) + |\beta_\ell|^2\, I_\ell^{(2)}(\omega r)\Big], \tag{13.5.66}$$

where

$$I_\ell^{(1)}(\omega r) = 2\pi\, [h_\ell(kr)]^2\, \frac{[\ell(\ell+1)]^2}{(2\ell+1)}, \tag{13.5.67}$$

$$I_\ell^{(2)}(\omega r) = 2\pi \Big[\frac{\ell(\ell+1)}{(2\ell+1)}\Big]^2 \Big[\ell\, [h_{\ell+1}(kr)]^2 + (\ell+1)\, [h_{\ell-1}(kr)]^2\Big]. \tag{13.5.68}$$

From the explicit expression of $h_\ell(x)$ in Eq. (13.5.22), we have

$$\lim_{x\to\infty} [x^2\, |h_\ell(x)|^2] = 1, \tag{13.5.69}$$

$$\Rightarrow \lim_{r\to\infty} [r^2\, |h_\ell(\omega r)|^2] = \frac{1}{\omega^2} \lim_{\rho\to\infty} [\rho^2\, |h_\ell(\rho)|^2] = \frac{1}{\omega^2}. \tag{13.5.70}$$

That is,

$$\lim_{r\to\infty} [r^2\, I_\ell^{(1)}(\omega r)] = \frac{2\pi}{\omega^2} \frac{[\ell(\ell+1)]^2}{(2\ell+1)}, \tag{13.5.71}$$

$$\lim_{r\to\infty} [r^2\, I_\ell^{(2)}(\omega r)] = \frac{2\pi}{\omega^2} \Big[\frac{\ell(\ell+1)}{(2\ell+1)}\Big]^2 (2\ell+1) = \frac{2\pi}{\omega^2} \frac{[\ell(\ell+1)]^2}{(2\ell+1)}, \tag{13.5.72}$$

Now we use our elegant, derived, result in Eq. (7.5.13) which states

$$\mathrm{P}(\omega) = \frac{1}{2} \lim_{r\to\infty} \Big[r^2 \int d\Omega\, |\mathbf{E}_{\mathrm{rad}}|^2(r,\theta,\phi)\Big], \tag{13.5.73}$$

for radiation outside a dielectric sphere (free space), and from Eqs. (13.5.66)–(13.5.73), that

$$\mathrm{P}(\omega) = \frac{\pi\, |E_0|^2}{\omega^2} \sum_{\ell=1}^{\infty} (2\ell+1)\big[|\alpha_\ell|^2 + |\beta_\ell|^2\big]. \tag{13.5.74}$$

Interesting numerical work[9] of Mie radiation was applied, in particular, to water with index of refraction taken to be 4/3, in the visible light spectrum, for a wide range of large value of the size parameter $x = 2\pi a/\lambda$, which supports the negligible variation of scattering with the wavelength for $a$ comparable in size to the wavelength $\lambda$, or larger, of the order of micrometers. This is, of course, confirmed by observing white light scattered from water drops in clouds, for which the size parameter is too

[9] Houghton and Chalker (1949) [9]. See also Kerker (1969) [13], Hergert and Wriedt (2012) [10], Grandy (2000) [8].

large for the applicability of Rayleigh theory. As a rule thumb, for the applicability of the Rayleigh theory, the particle radius is expected not to be larger than 1/20 of the wavelength of light.[10]

The generality of Mie's theory of radiation, is expected to go over to the Rayleigh theory of radiation in the limit of a sufficiently small size parameter. This is indeed the case as we now show.

To this end, we have to consider the limiting cases of the Ricatti-Bessel functions $S_\ell(x)$, $\xi_\ell(x)$ functions occurring in the expansion coefficients $\alpha_\ell$, $\beta_\ell$ in Eqs. (13.5.63), (13.5.64).

Using the elementary identity,

$$\frac{1}{(2\ell+1)\ldots 5.3.1} = \frac{2^\ell\,\ell!}{(2\ell+1)!}, \tag{13.5.75}$$

we may write directly from[11] Eq. (13.5.21), that

$$S_\ell(x) = x\, j_\ell(x) = A_\ell\, x^{\ell+1}\big[1 + \mathcal{O}(x^2)\big], \qquad A_\ell = \frac{2^\ell\,\ell!}{(2\ell+1)!}. \tag{13.5.76}$$

Moreover, directly from Eq. (13.5.22) we also have

$$\xi(x) = x\, h_\ell(x) = B_\ell\, \frac{1}{x^\ell}\big[1 + \mathcal{O}(x)\big], \qquad B_\ell = -\,\mathrm{i}\,\frac{(2\ell)!}{2^\ell\,\ell!}. \tag{13.5.77}$$

The denominator of $\beta_\ell$ in Eq. (13.5.64) for $0 < x \ll 1$ is given by

$$\begin{aligned} &\xi_\ell(x)\, S'_\ell(\sqrt{\varepsilon}\, x) - \sqrt{\varepsilon}\, S_\ell(\sqrt{\varepsilon}\, x)\, \xi'_\ell(x) = \\ &A_\ell\, B_\ell\left(\frac{1}{x^\ell}\,(\ell+1)\,(\sqrt{\varepsilon}x)^\ell + \sqrt{\varepsilon}(\sqrt{\varepsilon}x)^{\ell+1}\frac{\ell}{x^{\ell+1}}\right)\big[1 + \mathcal{O}(x)\big]. \end{aligned} \tag{13.5.78}$$

Or,

$$\begin{aligned} &\frac{1}{\xi_\ell(x)\, S'_\ell(\sqrt{\varepsilon}\, x) - \sqrt{\varepsilon}\, S_\ell(\sqrt{\varepsilon}\, x)\, \xi'_\ell(x)} = \\ &\frac{1}{\big(A_\ell\, B_\ell\,(\sqrt{\varepsilon})^\ell\big)}\,\frac{1}{\big(\ell+1+\ell\,\varepsilon\big)}\big[1 + \mathcal{O}(x)\big]. \end{aligned} \tag{13.5.79}$$

Similarly, for the denominator of $\alpha_\ell$ in Eq. (13.5.63), we have

---

[10] See, e.g., Kerker (1969) [13], Bohren and Huffman (1983) [2], for more precise estimates.

[11] Or directly from Abramowitz and Stegun (1964) [1], p. 437.

$$\frac{1}{S_\ell(\sqrt{\varepsilon}\,x)\,\xi_\ell'(x) - \sqrt{\varepsilon}\,\xi_\ell(x)\,S_\ell'(\sqrt{\varepsilon}\,x)} = \\ -\frac{1}{\big(A_\ell\,B_\ell(\sqrt{\varepsilon})^{\ell+1}\big)}\frac{1}{(2\ell+1)}\big[1 + \mathcal{O}(x)\big]. \tag{13.5.80}$$

For the numerators of $\beta_\ell$, $\alpha_\ell$ in Eqs. (13.5.64), (13.5.63), we have, respectively,

$$\sqrt{\varepsilon}\,S_\ell(\sqrt{\varepsilon}\,x)\,S_\ell'(x) - S_\ell(x)\,S_\ell'(\sqrt{\varepsilon}\,x) = \\ A^2(\sqrt{\varepsilon})^\ell(\varepsilon-1)x^{2\ell+1}(\ell+1)\big[1 + \mathcal{O}(x^2)\big]. \tag{13.5.81}$$

and

$$\sqrt{\varepsilon}\,S_\ell(x)\,S_\ell'(\sqrt{\varepsilon}\,x) - S_\ell(\sqrt{\varepsilon}\,x)\,S_\ell'(x)] = \\ A^2\Big[(\ell+1)\Big((\sqrt{\varepsilon})^{\ell+1}x^{\ell+1}x^\ell - (\sqrt{\varepsilon})^{\ell+1}x^\ell x^{\ell+1}\Big) + \mathcal{O}\big(x^{2\ell+3}\big)\Big] = \mathcal{O}\big(x^{2\ell+3}\big). \tag{13.5.82}$$

Therefore, from Eqs. (13.5.81)/(13.5.79) and Eqs. (13.5.82)/(13.5.80), we have, respectively

$$\beta_\ell \simeq \frac{A_\ell}{B_\ell}(\ell+1)x^{2\ell+1}\frac{(\varepsilon-1)}{\ell\varepsilon+\ell+1} = \mathrm{i}\frac{(2^\ell\,\ell!)^2\,(\ell+1)}{(2\ell)!\,(2\ell+1)!}\frac{x^{2\ell+1}(\varepsilon-1)}{(\ell\varepsilon+\ell+1)}, \tag{13.5.83}$$

$$\alpha_\ell = \mathcal{O}\big(x^{2\ell+3}\big). \tag{13.5.84}$$

Clearly, for $0 < x \ll 1$, the term $\ell = 1$ dominates, allowing one to consider only the coefficient

$$\beta_1 = \mathrm{i}\frac{2}{3}\frac{(\varepsilon-1)}{(\varepsilon+2)}(\omega a)^3, \tag{13.5.85}$$

where $\alpha_1 = \mathcal{O}\big((\omega a)^5\big)$.

From Eq. (13.5.74), the following explicit expression emerges for $\mathrm{P}(\omega)$:

$$\mathrm{P}(\omega) = \frac{4\pi}{3}|E_0|^2a^6(\omega)^4\Big(\frac{\varepsilon-1}{\varepsilon+2}\Big)^2 = \frac{\omega^4}{12\pi}|\mathbf{p}|^2, \tag{13.5.86}$$

where $\mathbf{p}$ is the electric dipole moment in Eq. (13.1.16), thus verifying the formula in Eq. (13.1.10) of Rayleigh radiation in flying colours.

As discussed earlier, water droplets that make up clouds are of a comparable size to the wavelengths in visible light, and may be even larger, and the scattering is described by Mie's theory for which the size parameter $x = 2\pi a/\lambda$ is not necessarily small, rather than by that of Rayleigh's. The interesting aspect in this case, is that the power of radiation is not strongly dependent on wavelength,[12] as it is in the Rayleigh theory, and all wavelengths of visible light are scattered almost equally. Accordingly,

[12] Houghton and Chalker (1949) [9].

clouds appear white. The Mie Scattering is responsible for the white appearance of clouds.

For further reading on Mie radiation, general recommended references are given below.[13]

## 13.6 Appendix: Mathematical Aspects of Special Functions in Mie Radiation

After separation of variables, the differential equations satisfied by the functions $R$, $\Theta$, $\Phi$, introduced above Eq. (13.5.16), are given by

$$\frac{\mathrm{d}}{\mathrm{d}r}\left(r^2\frac{\mathrm{d}R}{\mathrm{d}r}\right) + \left[k^2r^2 - \ell(\ell+1)\right]R = 0, \tag{13.6.1}$$

$$\frac{1}{\sin\theta}\frac{\mathrm{d}}{\mathrm{d}\theta}\left(\sin\theta\frac{\mathrm{d}\Theta}{\mathrm{d}\theta}\right) + \left[\ell(\ell+1) - \frac{m^2}{\sin^2\theta}\right]\Theta = 0, \tag{13.6.2}$$

$$\frac{d^2\Phi}{\mathrm{d}\phi^2} + m^2\Phi = 0. \tag{13.6.3}$$

The so-called spherical Bessel function of the first kind $j_\ell(x)$, as a solution of Eq. (13.6.1), with $x = kr$, is explicitly given in Eq. (13.5.21), and may be defined in terms of the Bessel function $J_\ell(x)$ as follows

$$j_\ell(x) = \sqrt{\frac{\pi}{2x}}\,J_{\ell+1/2}(x). \tag{13.6.4}$$

For physical considerations, the solution $z_\ell$ is given by the spherical Bessel function $j_\ell(x)$ within a dielectric sphere, which is well behaved at its center (at the origin), given in Eq. (13.5.21), where $x = \sqrt{\varepsilon}\omega r$, while for outgoing radiation, outside a dielectric sphere in free space, the solution taken is the Hankel function $h_\ell(x)$, given in Eq. (13.5.22), where $x = \omega r$. $\Theta(\theta)$ is given by the associated Legendre functions in Eqs. (13.5.18)/(13.5.19), and $\cos m\phi$, $\sin m\phi$ are the solutions of Eq. (13.6.3).

In particular, the associated Legendre functions satisfy the orthogonality condition

$$\int_{-1}^{1}(\mathrm{d}\cos\theta)\,P_\ell^m(\cos\theta)\,P_{\ell'}^m(\cos\theta) = \delta_{\ell\ell'}\frac{2}{2\ell+1}\frac{(\ell+m)!}{(\ell-m)!}. \tag{13.6.5}$$

---

[13] Stratton (2007) [18], Bohren and Huffman (1983) [2], van Hulst (1981) [12], Kerker (1969) [13], Born and Wolf, E. (2004) [4], Houghton Chalker (1949) [9], Hergert and Wriedt (2012) [10], Grandy (2000) [8], Mie (1908) [16].

Actually, we will be dealing with $m = 1$.

An important summation formula of an exponential, referred to as the "Bauer Formula", is given by

$$e^{i\rho\cos\theta} = \sum_{\ell=0}^{\infty} (i)^{\ell}\,(2\ell+1)\,j_\ell(\rho)\,P_\ell(\cos\theta). \tag{13.6.6}$$

To see this, first note that

$$\frac{1}{\rho^2}\frac{\partial}{\partial\rho}\Big(\rho^2\frac{\partial}{\partial\rho}\Big)\,e^{i\rho\cos\theta} = \Big(+\frac{2i}{\rho}\cos\theta - \cos^2\theta\Big)\,e^{i\rho\cos\theta}, \tag{13.6.7}$$

$$\frac{1}{\rho^2\sin\theta}\frac{\partial}{\partial\theta}\Big(\sin\theta\frac{\partial}{\partial\theta}\Big)\,e^{i\rho\cos\theta} = \Big(-\frac{2i}{\rho}\cos\theta - \sin^2\theta\Big)\,e^{i\rho\cos\theta}. \tag{13.6.8}$$

By adding the two equations, using $\sin^2\theta + \cos^2\theta = 1$, and referring to the expression of part of $\nabla^2$, e.g., in Eq. (13.1.11), in spherical coordinates, in the variables $\rho, \theta$, we may infer that

$$[\nabla^2 + 1]\,e^{i\rho\cos\theta} = 0, \;\Rightarrow\; e^{i\rho\cos\theta} = \sum_{\ell=0}^{\infty} a_\ell\, j_\ell(\rho)\,P_\ell(\cos\theta). \tag{13.6.9}$$

providing an expansion formula for $\exp[i\rho\cos\theta]$, with the coefficients $a_\ell$ are to be determined.

To the above end, we may use the integral[14]

$$\int_{-1}^{1} dq\, e^{i\rho q}\,P_\ell(q) = 2\,(i)^{\ell}\,j_\ell(\rho), \qquad q \equiv \cos\theta, \tag{13.6.10}$$

and note that upon multiplying the expression on the right-hand side of Eq. (13.6.9) by $P_{\ell'}(\cos\theta)$, integrating over $\cos\theta$ from $-1$ to 1, and using, in the process, Eq. (13.6.5) for $m = 0$, we obtain for a given $\ell$,

$$j_\ell(\rho)\frac{2}{2\ell+1}\,a_\ell = \int_{-1}^{1} dq\, e^{i\rho q}\,P_\ell(q), \qquad q \equiv \cos\theta, \tag{13.6.11}$$

from this and Eqs. (13.6.10), (13.6.6) follows, with $a_\ell = (i)^{\ell}(2\ell+1)$. This may be also be derived, independently of using the equality in Eq. (13.6.10), in a standard manner, by considering the limit of the variable $\rho \to \infty$ of the integral in Eq. (13.6.10), and showing that it approaches the expression on the right-hand side of the equation in this limit. From this, the expression of $a_\ell$ again follows.

We establish the orthogonality of $\mathbf{v}_\ell^{(o)}(\mathbf{r})$ in Eq. (13.5.41), and $\mathbf{u}_{\ell'}^{(e)}(\mathbf{r})$ in Eq. (13.5.42), as recorded in Eq. (13.5.45)

---

[14] Gradshteyn and Ryzhik (2000) [7], p.786, 5./(7.243), where $j_\ell(x)$ is defined in Eq. (13.6.4).

$$\int_0^{\pi} \sin\theta \, \mathrm{d}\theta \int_0^{2\pi} \mathrm{d}\phi \, \mathbf{v}_\ell^{(\mathrm{o})}(\mathbf{r}) \cdot \mathbf{u}_{\ell'}^{(\mathrm{e})}(\mathbf{r}) \; = \; 0, \quad \text{for all } \; \ell, \ell'. \tag{13.6.12}$$

To show this, the above integral is explicitly given by

$$z_\ell(kr) \frac{[kr \, z_{\ell'}(kr)]'}{kr} \int_0^{2\pi} \mathrm{d}\phi \int_0^{\pi} \sin\theta \, \mathrm{d}\theta \left[ P_\ell^1 \frac{\mathrm{d}P_{\ell'}^1}{\mathrm{d}\theta} \frac{\sin^2\phi}{\sin\theta} + P_{\ell'}^1 \frac{\mathrm{d}P_\ell^1}{\mathrm{d}\theta} \frac{\cos^2\phi}{\sin\theta} \right]. \tag{13.6.13}$$

The $\phi$ integrals, over $\sin^2\phi$ and $\cos^2\phi$, give $\pi$ for both, also $\sin\theta$ cancels out from the numerator and the denominator. Thus we are led to consider the simple integral

$$\begin{aligned} \int_0^{\pi} \mathrm{d}\theta \left[ P_\ell^1 \frac{\mathrm{d}P_{\ell'}^1}{\mathrm{d}\theta} + P_{\ell'}^1 \frac{\mathrm{d}P_\ell^1}{\mathrm{d}\theta} \right] &= \int_0^{\pi} \mathrm{d}\theta \, \frac{\mathrm{d}}{\mathrm{d}\theta} \Big[ P_\ell^1(\cos\theta) P_{\ell'}^1(\cos\theta) \Big] \\ &= \Big[ P_\ell^1(1) P_{\ell'}^1(1) - P_\ell^1(-1) P_{\ell'}^1(-1) \Big] = 0, \end{aligned} \tag{13.6.14}$$

as seen from the explicit express of $P_\ell^1(x)$ in Eq. (13.5.18), e.g., for $m = 1$.

Consider the integral

$$\begin{aligned} &\int_0^{2\pi} \mathrm{d}\phi \int_0^{\pi} \sin\theta \, \mathrm{d}\theta \, \mathbf{v}_{\ell'}^{(\mathrm{o})}(\mathbf{r}) \cdot \mathbf{v}_\ell^{(\mathrm{o})}(\mathbf{r}) \\ &= z_{\ell'}(kr) z_\ell(kr) \int_0^{2\pi} \mathrm{d}\phi \int_0^{\pi} \sin\theta \, \mathrm{d}\theta \left[ \frac{\cos^2\phi}{\sin^2\theta} P_{\ell'}^1 \, P_\ell^1 + \sin^2\phi \frac{\mathrm{d}P_{\ell'}^1}{\mathrm{d}\theta} \frac{\mathrm{d}P_\ell^1}{\mathrm{d}\theta} \right]. \end{aligned} \tag{13.6.15}$$

Upon integrating over $\phi$, giving a common factor of $\pi$, using Eq. (13.6.2) for $P_\ell^1$ and the identity

$$\sin\theta \left[ \frac{P_{\ell'}^1 P_\ell^1}{\sin^2\theta} + \frac{\mathrm{d}P_{\ell'}^1}{\mathrm{d}\theta} \frac{\mathrm{d}P_\ell^1}{\mathrm{d}\theta} \right] = \sin\theta \, \ell(\ell+1) P_{\ell'}^1 P_\ell^1 + \frac{\mathrm{d}}{\mathrm{d}\theta} \left[ \sin\theta \, P_{\ell'}^1 \frac{\mathrm{d}P_\ell^1}{\mathrm{d}\theta} \right], \tag{13.6.16}$$

give

$$\begin{aligned} \int_0^{2\pi} \mathrm{d}\phi \int_0^{\pi} \sin\theta \, \mathrm{d}\theta \, \mathbf{v}_{\ell'}^{(\mathrm{o})}(\mathbf{r}) \cdot \mathbf{v}_\ell^{(\mathrm{o})}(\mathbf{r}) \; &= \; \pi \, z_{\ell'}(kr) z_\ell(kr) \, \ell(\ell+1) \\ \times \int_0^{\pi} \sin\theta \, \mathrm{d}\theta \, P_{\ell'}^1 P_\ell^1 \; &= \; 2\pi \, [z_\ell(kr)]^2 \, \frac{[\ell(\ell+1)]^2}{(2\ell+1)} \, \delta_{\ell'\ell}. \end{aligned} \tag{13.6.17}$$

The equivalent relation for $\mathbf{u}_\ell^{(\mathrm{e})}$, in Eq. (13.5.42), with some additional effort may be handled similarly, giving

$$\int_0^{2\pi} \mathrm{d}\phi \int_0^{\pi} \sin\theta \,\mathrm{d}\theta\ \mathbf{u}_{\ell'}^{(\mathrm{e})}(\mathbf{r}) \cdot \mathbf{u}_{\ell}^{(\mathrm{e})}(\mathbf{r})$$
$$= 2\pi \Big[\frac{\ell(\ell+1)}{(2\ell+1)}\Big]^2 \big[\ell\,[z_{\ell+1}(kr)]^2 + (\ell+1)\,[z_{\ell-1}(kr)]^2\big]\,\delta_{\ell'\ell}. \qquad (13.6.18)$$

To find the expansion coefficient $a_\ell, b_\ell$ in the summation expression of the incident electric field, $\mathbf{E}_{\mathrm{inc}}$ in Eq. (13.5.44), with $\mathbf{E}_{\mathrm{inc}}$, itself, defined in Eqs. (13.5.27)/(13.5.28), we note that in detail,

$$\mathbf{E}_{\mathrm{inc}} = E_0\, \mathrm{e}^{\mathrm{i}kr\cos\theta}[\sin\theta\cos\phi\,\hat{\mathbf{r}} + \cos\theta\cos\phi\,\hat{\boldsymbol{\theta}} - \sin\phi\,\hat{\boldsymbol{\phi}}], \qquad (13.6.19)$$

and *with* $z_\ell(kr) = j_\ell(kr)$,

$$[\sin\theta\cos\phi\,\hat{\mathbf{r}} + \cos\theta\cos\phi\,\hat{\boldsymbol{\theta}} - \sin\phi\,\hat{\boldsymbol{\phi}}] \cdot \mathbf{v}_\ell^{(\mathrm{o})}$$
$$= j_\ell(kr)\Big[\cos^2\phi\,\cos\theta\,\frac{P_\ell^1(\cos\theta)}{\sin\theta} + \sin^2\phi\frac{\mathrm{d}P_\ell^1}{\mathrm{d}\theta}\Big]. \qquad (13.6.20)$$

Accordingly, upon integration over $\phi$,

$$\int_0^{\pi} \sin\theta\,\mathrm{d}\theta \int_0^{2\pi} \mathrm{d}\phi\ \mathbf{E}_{\mathrm{inc}} \cdot \mathbf{v}_\ell^{(\mathrm{o})}$$
$$= \pi\, j_\ell(kr)E_0 \int_0^{\pi} \mathrm{d}\theta\,\mathrm{e}^{\mathrm{i}kr\cos\theta}\Big[\cos\theta\,P_\ell^1 + \sin\theta\,\frac{\mathrm{d}P_\ell^1}{\mathrm{d}\theta}\Big]. \qquad (13.6.21)$$

On the other hand, using in the process Eq. (13.5.18) for $m = 1$, and Eq. (13.6.2) for $m = 0$, we have

$$P_\ell^1 = -\frac{\mathrm{d}P_\ell}{\mathrm{d}\theta}, \qquad \Big[\cos\theta\,P_\ell^1 + \sin\theta\,\frac{\mathrm{d}P_\ell^1}{\mathrm{d}\theta}\Big] = \frac{\mathrm{d}}{\mathrm{d}\theta}\big[\sin\theta\,P_\ell^1\big], \qquad (13.6.22)$$
$$\frac{\mathrm{d}}{\mathrm{d}\theta}\Big[\sin\theta\,\frac{\mathrm{d}P_\ell}{\mathrm{d}\theta}\Big] + \ell(\ell+1)\sin\theta\,P_\ell = 0. \qquad (13.6.23)$$

Therefore,

$$\int_0^{\pi} \sin\theta\,\mathrm{d}\theta \int_0^{2\pi} \mathrm{d}\phi\ \mathbf{E}_{\mathrm{inc}} \cdot \mathbf{v}_\ell^{(\mathrm{o})} = \pi\,\ell(\ell+1)\,j_\ell(kr)E_0 \int_0^{\pi} \sin\theta\,\mathrm{d}\theta\,\mathrm{e}^{\mathrm{i}kr\cos\theta}\,P_\ell(\cos\theta)$$
$$= 2\pi\,E_0(\mathrm{i})^\ell\,\ell(\ell+1)[j_\ell(kr)]^2, \qquad (13.6.24)$$

where we have finally used the integral in Eq. (13.6.10).

In a similar fashion,

$$
\begin{aligned}
&\int_0^{\pi} \sin\theta \, \mathrm{d}\theta \int_0^{2\pi} \mathrm{d}\phi \; \mathbf{E}_{\mathrm{inc}} \cdot \mathbf{u}_\ell^{(\mathrm{e})} \\
&= -2\pi (\mathrm{i})^{\ell+1} E_0 \, \frac{\ell(\ell+1)}{2\ell+1} \left[\ell \, [z_{\ell+1}(kr)]^2 + (\ell+1) \, [z_{\ell-1}(kr)]^2\right]. \qquad (13.6.25)
\end{aligned}
$$

## References


1. Abramowitz, M., & Stegun (Eds.) (1964). *Handbook of mathematical functions, National Bureau of Standards*, Washington, D.C.
2. Bohren, C. F., & Huffman, D. R. (1983). *Absorption and scattering of light by small particles.* New York: John Wiley & Sons.
3. Bohren, C. F. , & Alistair, F. (1985). Colors of the sky. *Physics Teacher 23*, 267-272.
4. Born, M., & Wolf, E. (2004). *Electromagnetic theory of propagation, interference and diffraction of light. 7th Edition*. Cambridge: Cambridge university Press.
5. Clausius, R. (1879). Die Mechanische Behandlung der Elektricität, *Teubner Verlag*, Braunschweig: Germany.
6. Eom, H. J. (2004). *Electromagnetic wave theory for boundary-value problems*. Berlin: Springer.
7. Gradshteyn, I. S., & Ryzhik, I. M. (2000). *Tables of integrals, series and products*, $6^{\mathrm{th}}$ Edition. New York: Academic Press.
8. Grandy, W. T. (2000). *Scattering of waves from large spheres.* Cambridge: Cambridge University Press.
9. Houghton, H. G., & Chalker, W. R. (1949). The scattering cross section of watter drops in air for visible light. *Journal of The Optical Society of America, 39 (11)*, 955-957.
10. Hergert, W., & Wriedt, T. (Editors) (2012). *The Mie theory: Basics and Applications.* Heidelberg: Springer.
11. Hoeppe, G. (2007). *Why the sky is blue: Discovering the Color of Life*. Princeton: Princeton University Press.
12. van Hulst, H. C. (1981). *Light scattering by small particles*. Now York: Dover Publications.
13. Kerker, M. (1969). *The Scattering of Light.* New York: Academic Press.
14. Kerker, M., Wang, D.-S., & Chew, H. (1980). Surface enhanced Raman scattering (SERS) by molecules adsorbed at spherical particles. *Applied Optics, 19 (19)*, 3373-3388
15. Mackowski, D. W. (1991). Analysis of radiative scattering for multiple sphere configurations. *Proceedings of Royal Society A, 433*, 599-614.
16. Mie, G., (1908). Beiträge zur Optik trüber Medien, speziell kolloidaler Metallösungen. *Annalen der Physik, 25 (3)*, 377–445.
17. Mossotti, O. F. (1850). Discussione analitica sull'influenza che l'azione di un mezzo dielettrico ha sulla distribuzione dell'elettricità alla superficie di più corpi electrici disseminati in esso. *Memorie di Matematica e di Fisica della Societa Italiana delle Scienze Residente, Modena 24 (Part 2)*, 49–74.
18. Stratton, J. A. (2007). *Electromagnetic theory*. New Jersey: IEEE Press.
19. Strutton J. W. (aka Lord Rayleigh) (1899). On the transmission of light through an atmosphere containing small particles in suspension. *Philosophical Magazine, 5th ser., 47*, 375-384.
20. Wallace, J. M. (2006). *Atmospheric Science*. New York: Academic Press.

# Chapter 14
# Radiation by Diffraction

*A treatment of radiation by diffraction is carried out by using*
*Green's Theorem involving vector fields generated from tensor ones -*
*a key ingredient in the problem that cannot be dismissed*

In this chapter, we consider radiation emission as a result of diffraction off a perfectly conducting plane screen through a circular hole. As a scattering problem, diffraction is considered as bending of light around obstacles. The analysis is based on using a so-called Green's Theorem involving vector fields generated from tensor ones. A highly recommended early source on this approach is a paper[1] by Harold Levine and Julian Schwinger in 1950. Unfortunately, some readers may find this paper too difficult to follow. I assure you, however, that this will not be the case here when you go through the detailed steps carried out in the chapter.

## 14.1 Radiation by Diffraction from a Circular Aperture

In the Appendix, in the next section, we establish the following, so-called, Green's Theorem, involving two vector fields $\mathbf{M}(\mathbf{r})$, $\mathbf{N}(\mathbf{r})$, embodying a general relation between surface and volume integrals (see Fig. 14.1),

$$\int_V \mathrm{d}^3\mathbf{r}\Big[\mathbf{M}\cdot\big(\nabla\times(\nabla\times\mathbf{N})\big) - \mathbf{N}\cdot\big(\nabla\times(\nabla\times\mathbf{M})\big)\Big]$$
$$= -\int_S \mathrm{d}S\,\mathbf{n}\cdot\Big[\mathbf{N}\times\big(\nabla\times\mathbf{M}\big) - \mathbf{M}\times\big(\nabla\times\mathbf{N}\big)\Big]. \tag{14.1.1}$$

[1] Levine and Schwinger (1950) [5]. See also Born and Wolf (2013) [1], Bouwkamp (1954) [2]. Some recent interesting investigations on the problem, in general, are also given, e.g., by Drezet, Woehl and Huant in (2006) [3], as well as by Jung and Keller in (2014) [4].

 
E. B. Manoukian, *Radiation*,
https://doi.org/10.1007/978-3-032-21506-2_14

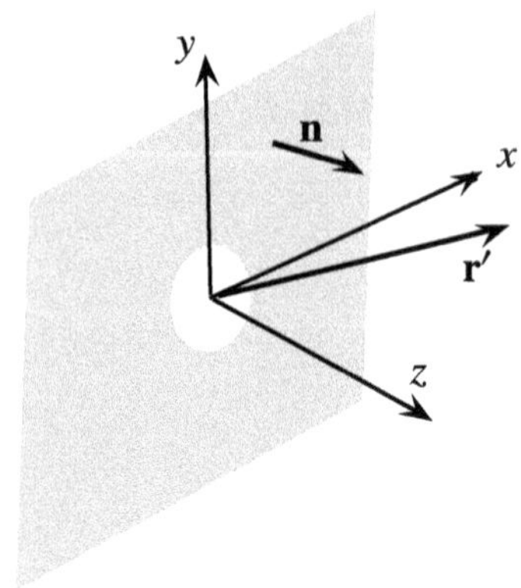


**Fig. 14.1** The figure refers to the problem considered of diffraction from a circular aperture in a perfectly conducting thin plate of infinite extension set in the $x - y$ plane located in free space. The volume in Eq. (14.1.1) consists the entire half-space to the right of the screen, and the surface consists of the surface of this entire half space

where the minus sign multiplying the surface integral arises because we have taken the unit vector $\mathbf{n}$ to point within the surface, i.e., within the region of observation (see Fig. 14.1), with observation made along the vector $\mathbf{r}'$.

To make use of this result, relevant to the diffraction process, we define the Green function,

$$G(\mathbf{r} - \mathbf{r}') = \frac{1}{4\pi} \frac{\mathrm{e}^{\mathrm{i}\,k|\mathbf{r}-\mathbf{r}'|}}{|\mathbf{r} - \mathbf{r}'|} \equiv G(z - z'), \tag{14.1.2}$$

$$|\mathbf{r} - \mathbf{r}'| = \big((x - x')^2 + (y - y')^2 + (z - z')^2\big)^{1/2}, \tag{14.1.3}$$

with the notation $G(z - z')$, introduced for convenience. It satisfies the equation

$$(\boldsymbol{\nabla}^2 + k^2)\Big[\frac{1}{4\pi} \frac{\mathrm{e}^{\mathrm{i}\,k|\mathbf{r}-\mathbf{r}'|}}{|\mathbf{r} - \mathbf{r}'|}\Big] = -\,\delta^{(3)}(\mathbf{r} - \mathbf{r}'), \tag{14.1.4}$$

already established in Eq. (7.5.20) of Chap. 7, with a Dirac delta distribution on the right-hand-side of the equation. In particular,

$$\big[\boldsymbol{\nabla}^2 + k^2\big]\, G(z - z') = -\,\delta^{(2)}(\mathbf{r}_\perp - \mathbf{r}'_\perp)\,\delta(z - z'), \tag{14.1.5}$$

$$\big[\boldsymbol{\nabla}^2 + k^2\big]\, G(z + z') = -\,\delta^{(2)}(\mathbf{r}_\perp - \mathbf{r}'_\perp)\,\delta(z + z'). \tag{14.1.6}$$

We introduce the tensor defined by

$$\Gamma_0^{ij}(z, \pm z') = \Big[\delta^{ij} + \frac{\partial^i \partial^j}{k^2}\Big] G(z \mp z'), \tag{14.1.7}$$

where, we use the notation $i, j = 1, 2, 3,\ a, b = 1, 2$, and the restriction that $z' > 0$. In detail, now with derivatives with respect to the components of $\mathbf{r}'$, we have

$$\Gamma_0^{ab}(z,z') = \left[\delta^{ab} + \frac{\partial'^a \partial'^b}{k^2}\right] G(z-z'), \tag{14.1.8}$$

$$\Gamma_0^{a3}(z,-z') = \left[0 \;-\; \frac{\partial'^a \partial'^3}{k^2}\right] G(z+z'), \tag{14.1.9}$$

$$\Gamma_0^{3a}(z,-z') = \left[0 \;-\; \frac{\partial'^3 \partial'^a}{k^2}\right] G(z+z'), \tag{14.1.10}$$

$$\Gamma_0^{33}(z,-z') = \left[1 \;+\; \frac{\partial'^3 \partial'^3}{k^2}\right] G(z+z'). \tag{14.1.11}$$

We may, in turn, combine these components, and introduce a tensor which will turn out to be relevant to the subsequent work of physical importance

$$\Gamma^{ij}(z,z') = \Gamma_0^{ij}(z,z') + \Gamma_0^{ik}(z,-z')\big[\delta^{kj} - 2\,\delta^{k3}\delta^{j3}\big]. \tag{14.1.12}$$

In detail, note that by replacing $(z', \partial/\partial z')$ by $(-z', -\partial/\partial z')$ whenever required in the above four equations,

$$\Gamma^{33}(z,z') = \Gamma_0^{33}(z,z') - \Gamma_0^{33}(z,-z') = \Big(1 + \frac{\partial'^3\partial'^3}{k^2}\Big)\big[G(z-z') - G(z+z')\big],$$

$$\Gamma^{3a}(z,z') = \Gamma_0^{3a}(z,z') + \Gamma_0^{3a}(z,-z') = \frac{\partial'^3\partial'^a}{k^2}\big[G(z-z') - G(z+z')\big],$$

$$\Gamma^{a3}(z,z') = \Gamma_0^{a3}(z,z') - \Gamma_0^{a3}(z,-z') = \frac{\partial'^a\partial'^3}{k^2}\big[G(z-z') + G(z+z')\big],$$

$$\Gamma^{ab}(z,z') = \Gamma_0^{ab}(z,z') + \Gamma_0^{ab}(z,-z') = \Big(\delta^{ab} + \frac{\partial'^a\partial'^b}{k^2}\Big)\big[G(z-z') + G(z+z')\big].$$

From the very definition of $\Gamma_0^{ij}(z,z')$ in Eq. (14.1.7), and Eqs. (14.1.2), (14.1.4), we may infer that

$$\big[\boldsymbol{\nabla}^2 + k^2\big]\boldsymbol{\Gamma}_0^{ij}(\mathbf{r},\mathbf{r}') = -\Big[\delta^{ij} + \frac{\partial^i\partial^j}{k^2}\Big]\delta^{(3)}(\mathbf{r}-\mathbf{r}'), \tag{14.1.13}$$

$$\begin{aligned} k^2\partial^i\,\boldsymbol{\Gamma}_0^{ij}(\mathbf{r},\mathbf{r}') &= \big[\boldsymbol{\nabla}^2 + k^2\big]\partial^j G(z-z') \\ &= -\partial^j\delta^{(3)}(\mathbf{r}-\mathbf{r}'). \end{aligned} \tag{14.1.14}$$

On the other hand,

$$\begin{aligned} \partial^\ell(\partial^i\Gamma_0^{ij}) - \big[\boldsymbol{\nabla}^2 + k^2\big]\boldsymbol{\Gamma}_0^{\ell j} &= -\frac{\partial^\ell\partial^j}{k^2}\delta^{(3)}(\mathbf{r}-\mathbf{r}') + \Big[\delta^{\ell j} + \frac{\partial^\ell\partial^j}{k^2}\Big]\delta^{(3)}(\mathbf{r}-\mathbf{r}') \\ &= \delta^{\ell j}\delta^{(3)}(\mathbf{r}-\mathbf{r}'). \end{aligned} \tag{14.1.15}$$

We introduce a *constant* vector $\mathbf{v}$, to be specified later, as well the vector

$$\boldsymbol{\Gamma}_0 \text{ with components } \boldsymbol{\Gamma}_0^i = \boldsymbol{\Gamma}_0^{ij}\mathbf{v}^j. \tag{14.1.16}$$

to re-write Eq. (14.1.15) as

$$\nabla\big(\nabla\cdot\boldsymbol{\Gamma}_0\big)-\big[\nabla^2+k^2\big]\boldsymbol{\Gamma}_0 = \mathbf{v}\,\delta^{(3)}(\mathbf{r}-\mathbf{r}'), \quad (14.1.17)$$

$$\Rightarrow\quad \nabla\times\big(\nabla\times\boldsymbol{\Gamma}_0\big)-k^2\boldsymbol{\Gamma}_0 = \mathbf{v}\,\delta^{(3)}(\mathbf{r}-\mathbf{r}'), \quad (14.1.18)$$

where we have used the vector identity $\nabla\times\big(\nabla\times\boldsymbol{\Gamma}_0\big)=\nabla\big(\nabla\cdot\boldsymbol{\Gamma}_0\big)-\nabla^2\boldsymbol{\Gamma}_0$.

Because of the linear relation between $\boldsymbol{\Gamma}^{ij}$ and $\boldsymbol{\Gamma}_0^{ij}$, the following fundamental equation emerges from Eqs. (14.1.12), (14.1.18),

$$\nabla\times\big(\nabla\times\boldsymbol{\Gamma}(\mathbf{r},\mathbf{r}')\big)-k^2\boldsymbol{\Gamma}(\mathbf{r},\mathbf{r}') =$$
$$\mathbf{v}\,\delta^{(2)}(\mathbf{r}_\perp-\mathbf{r}'_\perp)\delta(z-z')+\delta^{(2)}(\mathbf{r}_\perp-\mathbf{r}'_\perp)\big(\mathbf{v}-2\,\mathbf{v}^3\mathbf{u}\big)\delta(z+z'), \quad (14.1.19)$$

where

$$\boldsymbol{\Gamma}\quad\text{has components}\quad \boldsymbol{\Gamma}^i=\boldsymbol{\Gamma}^{ij}\mathbf{v}^j,\quad\text{and}\quad \mathbf{u}\ \text{ is a } \textit{unit} \text{ vector along }\ \boldsymbol{\Gamma}. \quad (14.1.20)$$

An important property of $\boldsymbol{\Gamma}$, for the subsequent work, is the following, evaluated at $z=0$,

$$[\nabla\times\boldsymbol{\Gamma}(\mathbf{r},\mathbf{r}')]^a\Big|_{z=0} = 0, \qquad a=1,2. \quad (14.1.21)$$

This is proved in the Appendix, in the next section below Eq. (14.2.10).

Now we are ready to use the Green's Theorem for vector fields stated in Eq. (14.1.1). To this end, we set

$$M(\mathbf{r}) = \mathbf{B}(\mathbf{r}), \qquad N(\mathbf{r},\mathbf{r}') = \boldsymbol{\Gamma}(\mathbf{r},\mathbf{r}'), \quad (14.1.22)$$

to obtain,

$$\int_V \mathrm{d}^3\mathbf{r}\Big[\mathbf{B}\cdot(\nabla\times(\nabla\times\boldsymbol{\Gamma}))\;-\;\boldsymbol{\Gamma}\cdot(\nabla\times(\nabla\times\mathbf{B}))\Big]$$
$$= -\int_S \mathrm{d}S\;\mathbf{n}\cdot\Big[\boldsymbol{\Gamma}\times\big(\nabla\times\mathbf{B}\big)\;-\;\mathbf{B}\times\big(\nabla\times\boldsymbol{\Gamma}\big)\Big], \quad (14.1.23)$$

where the minus sign multiplying the surface integral arises because we have taken the unit vector $\mathbf{n}$ to point within the surface, i.e., within the region of observation.

Within the volume which is the region of observations, there are no charges or currents, and hence from Eqs. (5.3.19)–(5.3.22), with harmonic fields, i.e., involving the factor $\exp[-\mathrm{i}\omega t]$, we have

$$\nabla\times\mathbf{E}=\mathrm{i}\,\omega\,\mathbf{B},\qquad \nabla\times\big(\nabla\times\mathbf{E}\big)-k^2\,\mathbf{E}=0,\qquad k^2=\omega^2, \quad (14.1.24)$$

$$\nabla\times\big(\nabla\times\mathbf{B}\big)-k^2\,\mathbf{B}=0,\qquad \nabla\times\mathbf{B}=-\,\mathrm{i}\,\omega\,\mathbf{E}, \quad (14.1.25)$$

and

$$\mathbf{B}\cdot\big(\nabla\times\big(\nabla\times\boldsymbol{\Gamma}\big)\big) - \boldsymbol{\Gamma}\cdot\big(\nabla\times\big(\nabla\times\mathbf{B}\big)$$
$$= \mathbf{B}\cdot\big[\nabla\times\big(\nabla\times\boldsymbol{\Gamma}\big) - k^2\,\boldsymbol{\Gamma}\big]. \tag{14.1.26}$$

Now in the region of observation $z$, as an integration variable in the volume integral in Eq. (14.1.23), is such that $z \geq 0$ and the variable $z' > 0$, associated with observation, i.e., $\delta(z+z') = 0$ in Eqs. (14.1.19), and (14.1.26) implies, from Eq. (14.1.19), that

$$\mathbf{B}\cdot\big(\nabla\times\big(\nabla\times\boldsymbol{\Gamma}\big)\big) - \boldsymbol{\Gamma}\cdot\big(\nabla\times\big(\nabla\times\mathbf{B}\big) = \mathbf{B}(\mathbf{r})\cdot\mathbf{v}\;\delta^{(3)}(\mathbf{r}-\mathbf{r}'), \tag{14.1.27}$$

where we recall the $\mathbf{v}$ is an arbitrary *constant* vector. Upon replacing the latter equation in the volume integral in Eq. (14.1.23) and integrating over $\mathbf{r}$, we obtain:

$$\int_S \mathrm{d}S\;\mathbf{n}\cdot\big[\boldsymbol{\Gamma}\times\big(\nabla\times\mathbf{B}\big) - \mathbf{B}\times\big(\nabla\times\boldsymbol{\Gamma}\big)\big] = -\,\mathbf{B}(\mathbf{r}')\cdot\mathbf{v}. \tag{14.1.28}$$

Considering the second term in the above surface integral, note that as a consequence of the vector identity,

$$\mathbf{n}\cdot\big[\mathbf{B}\times\big(\nabla\times\boldsymbol{\Gamma}\big)\big] = [\mathbf{n}\times\mathbf{B}]\cdot[\nabla\cdot\boldsymbol{\Gamma}] = 0, \qquad \text{at } z=0, \tag{14.1.29}$$

as a consequence of the fact that $\mathbf{n}\times\mathbf{B}$ is a vector *lying* on the plane, and we know from the basic result in Eq. (14.1.21) that $[\nabla\cdot\boldsymbol{\Gamma}]^a = 0,\ a = 1, 2,$ for $z = 0$.

On the other hand, concerning the first term in the surface integral in Eq. (14.1.28), note that,

$$\nabla\times\mathbf{B} = -\,\mathrm{i}\,\omega\mathbf{E}, \tag{14.1.30}$$
$$\Rightarrow\quad \mathbf{n}\cdot[\boldsymbol{\Gamma}\times(\nabla\times\mathbf{B})] = -\,\mathrm{i}\,\omega\,\mathbf{n}\cdot[\boldsymbol{\Gamma}\times\mathbf{E}] = \mathrm{i}\,\omega\,\boldsymbol{\Gamma}\cdot[\mathbf{n}\times\mathbf{E}], \tag{14.1.31}$$

where in the last step, we have used the vector identity $\mathbf{W}\cdot[\mathbf{U}\times\mathbf{V}] = -\mathbf{U}\cdot[\mathbf{W}\times\mathbf{V}]$. We also know from Eq. (5.4.32) that on the plate itself, for $z = 0, \mathbf{n}\times\mathbf{E} = 0$. That is, we learn that the entire surface integral on the plate is restricted to the *aperture*, and only the first term in the summand of the surface integral in Eq. (14.1.28) may be retained. From the rapid radial damping of the electric field in the wave zone, as well as of the Green's function, one may infer that the surface at infinity does not contribute.

All told, the following expression emerges from Eq. (14.1.28),

$$\mathbf{B}(\mathbf{r}')\cdot\mathbf{v} = -\,\mathrm{i}\,\omega\int_{\text{apert.}} \mathrm{d}S\;\boldsymbol{\Gamma}\cdot\big[\mathbf{n}\times\mathbf{E}\big]. \tag{14.1.32}$$

Let $\mathbf{N} = \mathbf{r}'/|\mathbf{r}'|$ denote direction of which radiation is being detected. From the definition of $\boldsymbol{\Gamma}$ in Eq. (14.1.20), together with Eqs. (14.1.2)/(14.1.7)/(14.1.12)/(7.5.9), we may infer that for large $|\mathbf{r}'| \equiv r'$, and $z = 0$, i.e., at the aperture in the $x$, $y$ plane corresponding *to* the integral,

$$\boldsymbol{\Gamma}^{ij}[\mathbf{n}\times\mathbf{E}]^i \simeq \frac{\mathrm{e}^{\mathrm{i}kr'}}{4\pi r'}\Big[\delta^{ij}\mathrm{e}^{-\mathrm{i}k\mathbf{N}\cdot\mathbf{r}} + \big(\delta^{ij} - 2\,\mathbf{N}^i\mathbf{N}^j\big)\,\mathrm{e}^{-\,\mathrm{i}k\mathbf{N}\cdot\mathbf{r}}\Big][\mathbf{n}\times\mathbf{E}]^i + \dots, \tag{14.1.33}$$

$$\simeq \frac{\mathrm{e}^{\mathrm{i}kr'}}{2\pi r'}\mathrm{e}^{-\mathrm{i}k\mathbf{N}\cdot\mathbf{r}}\Big[(\mathbf{n}\times\mathbf{E})^j - \big(\mathbf{N}\cdot(\mathbf{n}\times\mathbf{E})\big)\mathbf{N}^j\Big] + \dots, \tag{14.1.34}$$

where any term involving the following structure

$$\delta^{i3}\,\delta^{j3}\,\epsilon^{i\ell m}\,\mathbf{n}^\ell\,\mathbf{E}^m \tag{14.1.35}$$

gives zero, on account of the fact that $\mathbf{n} = \hat{\mathbf{z}}$, i.e., along the 3-axis, which leads to the factor $\epsilon^{33m} = 0$.

Accordingly, from the definition of the vector components $\Gamma^i = \Gamma^{ij}v^j$ in Eq. (14.1.20),

$$\mathbf{B}(\mathbf{r}')\cdot\mathbf{v} = \mathrm{i}\,\omega\,\frac{\mathrm{e}^{\mathrm{i}kr'}}{2\pi r'}\int_{\text{apert.}}\mathrm{d}S\;\mathrm{e}^{-\mathrm{i}k\mathbf{N}\cdot\mathbf{r}}\Big[\mathbf{N}\cdot\big(\mathbf{n}\times\mathbf{E}\big)\;\mathbf{N} - \big(\mathbf{n}\times\mathbf{E}\big)\Big]\cdot\mathbf{v}. \tag{14.1.36}$$

Since $\mathbf{v}$ is an arbitrary constant vector, we may choose it, in turn, to be given by

$$\mathbf{v} \;=\; (1,0,0), \quad \text{or} \quad = \;(0,1,0), \quad \text{or,} \quad = \;(0,0,1), \tag{14.1.37}$$

from which all the components of magnetic field are readily extracted and emerge as

$$\mathbf{B}(\mathbf{r}') = \mathrm{i}\,\omega\,\frac{\mathrm{e}^{\mathrm{i}kr'}}{2\pi r'}\int_{\text{apert.}}\mathrm{d}S\;\mathrm{e}^{-\mathrm{i}k\mathbf{N}\cdot\mathbf{r}}\Big[\mathbf{N}\cdot\big(\mathbf{n}\times\mathbf{E}\big)\;\mathbf{N} - \big(\mathbf{n}\times\mathbf{E}\big)\Big], \tag{14.1.38}$$

and for a right-handed system with $\mathbf{N} = \mathbf{r}'/|\mathbf{r}'|$ the unit vector specifying the direction of observation defined below Eq. (14.1.32), the electric field $\mathbf{E}$ is given by

$$\mathbf{E}(\mathbf{r}') = -\mathbf{N}\times\mathbf{B}(\mathbf{r}'), \tag{14.1.39}$$

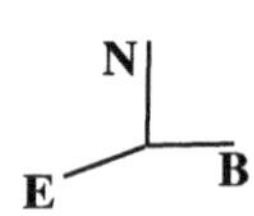


For a circular aperture of radius $a$, in particular, with normal incidence, i.e., along the unit vector $\mathbf{n} = \hat{\mathbf{z}}$, we consider, as an illustration, an input to the surface integral, at the aperture for $z = 0$, an incident electric field $\mathbf{E}_{\text{inc.}}$ along the $x$-axis simply given by

$$\mathbf{E}_{\text{inc.}}\Big|_{z=0} = \hat{\mathbf{x}}\, E_0 \tag{14.1.40}$$

That is, with

$$\mathbf{N}\times(\mathbf{n}\times\hat{\mathbf{x}}) \;=\; \mathbf{n}(\mathbf{N}\cdot\hat{\mathbf{x}}) \;-\; \hat{\mathbf{x}}(\mathbf{n}\cdot\mathbf{N}), \tag{14.1.41}$$

we get

$$\mathbf{E}(\mathbf{r}') \;=\; \mathrm{i}\,\omega\, E_0\, \frac{\mathrm{e}^{\mathrm{i}\,\omega\, r'}}{2\pi r'} \int_{\text{apert.}} \mathrm{d}S\;\mathrm{e}^{-ik\mathbf{N}\cdot\mathbf{r}}\,[\mathbf{n}(\mathbf{N}\cdot\hat{\mathbf{x}}) \;-\; \hat{\mathbf{x}}(\mathbf{n}\cdot\mathbf{N})], \tag{14.1.42}$$

Here the relevant vectors and their products, with $z = 0 \to \theta = \pi/2$, are:

$$\mathbf{N} = (\sin\theta'\cos\phi', \sin\theta'\sin\phi', \cos\theta'),\quad \mathbf{n} = (0,0,1),\quad \hat{\mathbf{x}} = (1,0,0), \tag{14.1.43}$$

$$[\mathbf{n}(\mathbf{N}\cdot\hat{\mathbf{x}}) \;-\; \hat{\mathbf{x}}(\mathbf{n}\cdot\mathbf{N})] \;=\; (-\cos\theta', 0, \sin\theta'\cos\theta'). \tag{14.1.44}$$

$$\begin{aligned}\mathbf{N}\cdot\mathbf{r}\Big|_{z=0} &= r(\sin\theta'\cos\phi', \sin\theta'\sin\phi', \cos\theta')\cdot(\cos\phi, \sin\phi, 0)\\ &= r\sin\theta'\cos(\phi-\phi').\end{aligned} \tag{14.1.45}$$

The following expression emerges for the asymptotic electric field,

$$\mathbf{E}(\mathbf{r}') = \mathrm{i}\,\omega\, E_0\,(-\cos\theta', 0, \sin\theta'\cos\theta')\,\frac{\mathrm{e}^{\mathrm{i}\,\omega\, r'}}{2\pi r'}\int_0^a r\,\mathrm{d}r\int_0^{2\pi}\mathrm{d}\phi\,\mathrm{e}^{-\mathrm{i}\omega\sin\theta'\cos(\phi-\phi')r}, \tag{14.1.46}$$

where $a$ is the radius of the aperture.

The $\phi-$ integral is given by

$$\int_0^{2\pi}\mathrm{d}\phi\,\mathrm{e}^{-\mathrm{i}\omega\sin\theta'\cos(\phi-\phi')\,r} \;=\; 2\pi\, J_0^{\mathrm{B}}(\omega r\sin\theta'), \tag{14.1.47}$$

defining the Bessel function of order zero. Upon setting $\omega r\sin\theta' = \rho$, and using the integral

$$\int_0^b \rho\,\mathrm{d}\rho\, J_0^{\mathrm{B}}(\rho) \;=\; b\, J_1^{\mathrm{B}}(b),$$

$$\Rightarrow\quad \frac{2\pi}{\omega^2\sin^2\theta'}\int_0^{\omega a\sin\theta'}\rho\,\mathrm{d}\rho\, J_0^{\mathrm{B}}(\rho) = \frac{2\pi\, a}{\omega\sin\theta'}\, J_1^{\mathrm{B}}(\omega a\sin\theta'), \tag{14.1.48}$$

defining the Bessel function of order one.

All told, the asymptotic electric field emerges as

$$\mathbf{E}(\mathbf{r}') = \mathrm{i}\,\omega\,a^2\,E_0\,(-\cos\theta', 0, \sin\theta'\cos\phi')\,\frac{J_1(\omega\,a\sin\theta')}{\omega\,a\sin\theta'}\,\frac{\mathrm{e}^{\mathrm{i}\,k\,r'}}{r'}. \tag{14.1.49}$$

Most importantly, we have the following limit

$$\lim_{r'\to\infty}\,[r'^2\mathbf{E}^*(\mathbf{r}')\cdot\mathbf{E}(\mathbf{r}')]$$
$$= \frac{(\omega\,a)^2\,(\pi a^2)}{\pi}\,(\cos^2\theta' + \sin^2\theta'\cos^2\phi')\left|\frac{J_1^{\mathrm{B}}(\omega\,a\sin\theta')}{\omega\,a\sin\theta'}\right|^2. \tag{14.1.50}$$

Now we refer to the elegant relation derived in Sect. 7.5, Eq. (7.5.12), expressing the time averaged power of radiation per unit (angular) frequency per unit solid angle, in terms of the asymptotic electric field, given by

$$\frac{\mathrm{dP}(\omega)}{\mathrm{d}\omega} = \frac{1}{2}\lim_{r'\to\infty}\,[r'^2\mathbf{E}^*(\mathbf{r}')\cdot\mathbf{E}(\mathbf{r}')], \tag{14.1.51}$$

where in this equation, $\varepsilon = 1$, $\mu = 1$. From this, one obtains the celebrated textbook expression

$$\frac{\mathrm{d}P(\omega)}{\mathrm{d}\omega} = \frac{(\omega\,a)^2\,(\pi a^2)}{2\pi}\,(\cos^2\theta' + \sin^2\theta'\cos^2\phi')\left|\frac{J_1^{\mathrm{B}}(\omega\,a\sin\theta')}{\omega\,a\sin\theta'}\right|^2. \tag{14.1.52}$$

## 14.2 Appendix: Mathematical Aspects of the Diffraction Process

**Green's Theorem for Vector Fields**:
We establish the following so-called Green's Theorem involving two vector fields $\mathbf{M}(\mathbf{r})$, $\mathbf{N}(\mathbf{r})$, giving a general relation between surface and volume integrals,

$$\int_V \mathrm{d}^3\mathbf{r}\Big[\mathbf{M}\cdot\big(\nabla\times(\nabla\times\mathbf{N})\big) - \mathbf{N}\cdot\big(\nabla\times(\nabla\times\mathbf{M})\big)\Big]$$
$$= -\int_S \mathrm{d}S\;\mathbf{n}\cdot\Big[\mathbf{N}\times\big(\nabla\times\mathbf{M}\big) - \mathbf{M}\times\big(\nabla\times\mathbf{N}\big)\Big]. \tag{14.2.1}$$

where the minus sign in the surface integral arises because we have taken the unit vector $\mathbf{n}$ to point within the surface (see Fig. 14.1 in the beginning of Sect. 14.1).

To establish this, consider the following vector identities:

$$\nabla\times\big(\nabla\times\mathbf{N}\big) = \nabla\big(\nabla\cdot\mathbf{N}\big) - \nabla^2\mathbf{N}, \tag{14.2.2}$$
$$\Rightarrow\; \mathbf{M}\cdot\big[\nabla\times\big(\nabla\times\mathbf{N}\big)\big] = \big(\mathbf{M}\cdot\nabla\big)\big(\nabla\cdot\mathbf{N}\big) - \mathbf{M}\cdot\nabla^2\mathbf{N}, \tag{14.2.3}$$
$$\Rightarrow\; \mathbf{N}\cdot\big[\nabla\times\big(\nabla\times\mathbf{M}\big)\big] = \big(\mathbf{N}\cdot\nabla\big)\big(\nabla\cdot\mathbf{M}\big) - \mathbf{N}\cdot\nabla^2\mathbf{M}.. \tag{14.2.4}$$

Accordingly, from Eqs. (14.2.3) and (14.2.4), we have

$$\mathbf{M}\cdot\big[\nabla\times\big(\nabla\times\mathbf{N}\big)\big]-\mathbf{N}\cdot\big[\nabla\times\big(\nabla\times\mathbf{M}\big) = \\ \mathbf{N}\cdot\nabla^2\mathbf{M}-\mathbf{M}\cdot\nabla^2\mathbf{N}-(\mathbf{N}\cdot\nabla)(\nabla\cdot\mathbf{M})+(\mathbf{M}\cdot\nabla)(\nabla\cdot\mathbf{N}). \tag{14.2.5}$$

On the other hand,

$$\big[\mathbf{M}\times\big(\nabla\times\mathbf{N}\big)\big]^i=\mathbf{M}\cdot(\partial^i\mathbf{N})-(\mathbf{M}\cdot\nabla)\mathbf{N}^i, \tag{14.2.6}$$

and

$$\begin{aligned}&\nabla\cdot\big[\mathbf{N}\times\big(\nabla\times\mathbf{M}\big)-\mathbf{M}\times\big(\nabla\times\mathbf{N}\big)\big]\\ &=\partial^i\Big[\mathbf{N}\cdot(\partial^i\mathbf{M})-\big(\mathbf{N}\cdot\nabla\big)\mathbf{M}^i-\mathbf{M}\cdot(\partial^i\mathbf{N})+\big(\mathbf{M}\cdot\nabla\big)\mathbf{N}^i\Big]\\ &=\mathbf{N}\cdot(\nabla^2\mathbf{M})-(\mathbf{N}\cdot\nabla)(\nabla\cdot\mathbf{M})-\mathbf{M}\cdot(\nabla^2\mathbf{N})+(\mathbf{M}\cdot\nabla)(\nabla\cdot\mathbf{N}),\end{aligned} \tag{14.2.7}$$

where the two terms $-(\partial^i N^k)(\partial^k M^i)$, $(\partial^i M^k)(\partial^k N^i)$ cancel out, as seen upon interchanging $i$ and $k$, e.g., in the latter term, as well as the two terms $(\partial^i N^k)(\partial^i M^k)$, $-(\partial^i M^k)(\partial^i N^k)$.

That is from Eqs. (14.2.5), (14.2.7), we have

$$\begin{aligned}&\mathbf{M}\cdot\big[\nabla\times\big(\nabla\times\mathbf{N}\big)-\mathbf{N}\cdot\big[\nabla\times\big(\nabla\times\mathbf{M}\big)\big]\\ &=\nabla\cdot\Big[\mathbf{N}\times\big(\nabla\times\mathbf{M}\big)-\mathbf{M}\times\big(\nabla\times\mathbf{N}\big)\Big],\end{aligned} \tag{14.2.8}$$

and the proof follows from Gauss' Theorem.

Another critical piece of information needed is the equation satisfied by the Green function $G(z-z')$, introduced in the previous section,

$$(\nabla^2+k^2)\Big[\frac{1}{4\pi}\frac{\mathrm{e}^{\mathrm{i}\,k|\mathbf{r}-\mathbf{r}'|}}{|\mathbf{r}-\mathbf{r}'|}\Big]=-\,\delta^{(3)}(\mathbf{r}-\mathbf{r}'). \tag{14.2.9}$$

and was proved in Sect. 7.5, Eq. (7.5.20).

An important property of $\boldsymbol{\Gamma}$, defined through Eqs. (14.1.20), (14.1.12), (14.1.7), is the following expression evaluated at $z=0$,

$$[\nabla\times\boldsymbol{\Gamma}(\mathbf{r},\mathbf{r}')]^a\Big|_{z=0}\equiv\epsilon^{a\ell m}\partial^\ell\boldsymbol{\Gamma}^m(z,z')\Big|_{z=0}=0,\qquad a=1,2. \tag{14.2.10}$$

suppressing, in the process, on the dependence of $\boldsymbol{\Gamma}(\mathbf{r},\mathbf{r}')$ on $\mathbf{r}_\perp$, $\mathbf{r}'_\perp$. For convenience, we use the notation $G(z\pm z')|_{z=0}=G(z')$, and suppress the overall constant vector $\mathbf{v}$ with components $\mathbf{v}^j$ multiplying $\boldsymbol{\Gamma}^{ij}$ in the proof below. Note also the properties of $\boldsymbol{\Gamma}^{ij}$ spelled out below Eq. (14.1.12) in following the proof.

The proof follows by noting that

$$\epsilon^{ab3}\,\partial^b\,\Gamma^{3c}(z,z')\Big|_{z=0} = \epsilon^{ab3}\,\partial^b\,\frac{\partial'^3\partial'^c}{k^2}\,\big[G(z-z') - G(z+z')\big]\Big|_{z=0}$$
$$= -\,\epsilon^{ab3}\,\partial'^b\,\frac{\partial'^3\partial'^c}{k^2}\,\big[G(z') - G(z')\big] = 0, \qquad (14.2.11)$$

as well

$$\epsilon^{ab3}\,\partial^b\,\Gamma^{33}(z,z')\Big|_{z=0} = -\,\epsilon^{ab3}\,\partial'^b\Big(1 + \frac{\partial'^3\partial'^3}{k^2}\Big)\big[G(z') - G(z')\big] = 0, \quad (14.2.12)$$

$$\epsilon^{abc}\,\partial^b\,\Gamma^{cd}(z,z')\Big|_{z=0} = -\,\epsilon^{abc}\partial'^b\Big[0 + \frac{\partial'^c\partial'^d}{k^2}\Big]\big[G(z') - G(z')\big] = 0. \quad (14.2.13)$$

Using the fact that $\epsilon^{abc}\partial'^b\partial'^c = 0$, for $a, b, c = 1, 2$, we also have

$$\epsilon^{abc}\,\partial^b\,\Gamma^{c3}(z,z')\Big|_{z=0} = -\,\epsilon^{abc}\,\partial'^b\frac{\partial'^c\partial'^3}{k^2}2\,G(z') = 0, \qquad (14.2.14)$$

$$\epsilon^{a3c}\,\partial^3\,\Gamma^{cj}(z,z')\Big|_{z=0} = \epsilon^{a3c}\,\partial^3\Big[\delta^{cj} + \frac{\partial'^c\partial'^j}{k^2}\Big]\big[G(z-z') + G(z+z')\big]\Big|_{z=0},$$
$$\rightarrow\quad -\,\epsilon^{a3j}\,\partial'^3\big[G(z') - G(z')\big] = 0, \quad j = 1,2,3, \qquad (14.2.15)$$

thus establishing the result stated in Eq. (14.2.10).

## References


1. Born, M. & E. Wolf, E. (2013), *Principles of optics: electromagnetic theory of propagation, interference and diffraction of light.* Amsterdam: Elsevier.
2. Bouwkamp, C. J. (1954). Diffraction theory. *Reports on Progress in Physics, 17 (1),* 35-100.
3. Drezet, A., Woehl, J. C., & S. Huant, S. (2006). Diffraction of light by a planar aperture in a metallic screen. *Journal of Mathematical Physics, 47,* 072901.
4. Jung, J. & Keller, O. (2014). Microscopic theory of diffraction. *Physical Review A, 90 (4),* 043830.
5. Levine, H. & Schwinger, J. (1950). On the theory of electromagnetic wave diffraction by an aperture in an infinite plane conducting screen. *Communications in Pure and Applied Mathematics, 3,* 355-391.

# Chapter 15
# Radiation by X-Ray Diffraction Off Crystals

*X-rays having very small wavelengths are able to unravel the structure of matter by interacting with the electrons of atoms in it, and no need arises to introduce phenomenological parameters, as in chemistry, in such a description. As Feynman puts it, it thus becomes a problem of pure physics*

As we consider light of smaller and smaller wavelengths, than those in the visible region, and reach wavelengths of the order of an Angstrom $\text{Å} = 10^{-10}$ m or so, light is referred to as X-ray. When a metal like sliver is exposed to these rays, one starts to unravel the structure of the metal with the atoms in it regularly placed, with their "centers" separated by distances of the order 4.08 Å, just as of the order of an X-ray wavelength. Such a structure, is what one calls a crystal. The need then arises to consider the X-rays as they are scattered off the electrons in atoms making up the crystal.

As a matter of fact, Richard Feynman, in his famous lectures on "The strange theory of light and matter",[1] when applying it to the subject matter of the present chapter, states that one gets the impression that the scattered rays are all from the front or back of a crystal, but in reality they arise from the scattering of light by the electrons in the atoms making up the crystal.

Wavelengths of visible light, about $4 \times 10^{-7}$ to $7 \times 10^{-7}$ m, are too large to unravel the structure of a metal like silver. It then becomes appropriate to treat it as a continuum, and, thanks to precise measurements of its so-called complex index of refraction,[2] one may, carefully, study the problem of scattering of light from such a metal, as we have already done in Chap. 11. For even smaller wavelengths than those of X-rays, ultimately the need, in particular, arises to develop a theory to describe the structure of nuclei—a subject matter which is beyond the scope of the present book.[3]

---

[1] Feynman (1983) [3], p. 107.

[2] Nash and Sanders (1996) [9].

[3] For detailed treatment of this, see Manoukian (2016) [7], Sects. 6.9–6.11.



E. B. Manoukian, *Radiation*,
https://doi.org/10.1007/978-3-032-21506-2_15

In Chap. 6, we have already generated X-rays from the synchrotron which may be considered as an initial source of X-rays to strike a crystal. In the present chapter, we derive, in particular, the power of radiation of the diffracted X-rays as a scattering off crystals, with the latter treated as sources of X-rays, taking into consideration experimentally important aspects, such as of thermal vibrations of atoms about their equilibrium positions, the practical aspect that incident X-rays impinging on the diffracting surface may not be perfectly parallel, as well as of treating unpolarized incident radiation.[4]

The “International Tables for Crystallography” publications[5] constitute a series of important references for all those interested in the intricacies of crystallography, in general.

The reader may be also interested in different applications of X-ray scattering and diffraction from materials, in general,[6] as well as of imaging of biological materials and cells,[7] and of structures of biological macromolecules, such as proteins and DNA crucial for understanding the functioning of life.[8]

## 15.1 Crystals and Miller Planes

Given three basis vectors $\mathbf{a}_1$, $\mathbf{a}_1$, $\mathbf{a}_1$, in 3D, of sizes $a_1, a_2, a_3$ and angles $\alpha = \mathbf{a}_2 \wedge \mathbf{a}_3$, $\beta = \mathbf{a}_3 \wedge \mathbf{a}_1$, $\gamma = \mathbf{a}_1 \wedge \mathbf{a}_2$, between them, one may construct different parallelepiped figures of six sides. The latter, as building blocks, referred to as unit cells, may be repeatedly connected together, applying, in the process, rules of translational symmetry, to generate a periodic array that one calls a crystal.

Such a process acquires a physical significance, when atoms, molecules, ions are set in these geometrical constructions, characterized by translational symmetry, an arrangement referred as crystal lattices, at points in the crystals, called lattice points, such as at the corners of the unit cells. The resulting configurations of such lattices are referred to as Bravais lattices, in honor of the French physicist Auguste Bravais.[9] The conventional Bravais lattices are described in the Appendix Sect. 15.2.

The simplest, and excellent to describe for pedagogical reasons, are crystals with cubic cells of equal sides and basis vectors at right angles to one and other, i,e., $a_1 = a_2 = a_3 \equiv a$, and $\alpha = \beta = \gamma = 90^{\circ}$. These involve three types, referred to as simple-cubic (SC), body-centered-cubic (BCC), and face-centered-cubic (FCC), where $a$ is referred to as a *lattice constant*. Their unit cells, as building blocks of a given element, may be represented as follows:

---

[4] Fine recommended references for this chapter, I am familiar with, are by James (1982) [6]; Warren (1990) [11]; Zolotoyabko (2014) [13].

[5] Hahn (2005) [4], and other volumes in the series.

[6] Schwartz and Cohen (2012) [10].

[7] Hémonnot and Köster [5].

[8] Ares (2017) [1].

[9] The publication of Bravais, on the so-called Bravais lattices, appeared in 1850: [2].

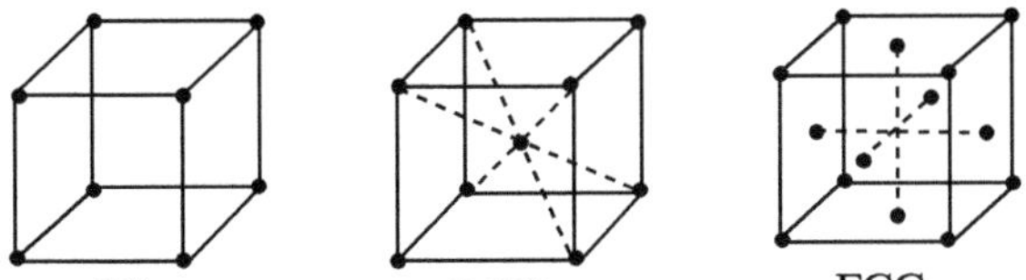


where a dark sphere, in particular, at the corner or on the surface, as the case may be, of a cube denotes the location of an atom, with the exception of the atom located at the center of a BCC crystal unit cell, that is shared between the unit cells in contact.

There are many BCC and FCC elements in nature, but only one SC, with the latter known as the polonium (Po) element, having a lattice constant (side of the unit cell) $a \cong 3.35$ Å, where Å, denotes the Angstrom, equal to $10^{-10}$ m, as mentioned above. For example, some FCC elements are copper (Cu) with $a \cong 3.60$ Å, silver (Ag) with $a \cong 4.08$ Å, and gold (Au) with $a \cong 4.07$ Å. On the other hand, for example, chromium (Cr) is a BCC element with $a \cong 2.90$ Å, lithium (Li) with $a \cong 3.51$ Å, and sodium (Na) with $a \cong 4.09$ Å.[10]

On the other hand, quite generally, if the angles $\alpha$, $\beta$, $\gamma$, are not necessarily equal, and $a_1$, $a_2$, $a_3$, as well, are not, then given such a crystal, one may conveniently introduce layers of scattering sites in a crystal by introducing parallel equidistant planes in it, referred to as Miller planes, with a standard notation, denoted by a triplet $(hk\ell)$, within round brackets and without commas, introduced as follows:

Referring to Fig. 15.1, pick any corner point of the crystal, setting it as the origin. Let $\mathbf{a}_1, \mathbf{a}_2, \mathbf{a}_3$ represent basis vectors of a unit cell. Introduce a plane which passes through this origin, and a next one parallel to it that intercepts, the three axes, along these basis vectors, respectively, at $a_1/h, a_2/k, a_3/\ell$, where $h, k, \ell$ are non-negative integers. If you choose, say, one of $h, k, \ell$ to be zero, say, $\ell = 0$, then the plane intercepts the axes along $\mathbf{a}_1, \mathbf{a}_3$ axes at $a_1/h$, $a_2/k$, and so on. Cubic cells are of $90^\circ$ angles and equal sides, $\mathbf{a}_1 = a\hat{\mathbf{i}}_1$, $\mathbf{a}_2 = a\hat{\mathbf{i}}_2$, $\mathbf{a}_3 = a\hat{\mathbf{i}}_3$.

Given, the three basis vectors $\mathbf{a}_1$, $\mathbf{a}_2$, $\mathbf{a}_3$ of a unit cell, we may introduce, quite generally for a given crystal, three vectors $\mathbf{b}_1$, $\mathbf{b}_2$ $\mathbf{b}_3$, defined by

$$\mathbf{b}_1 = \frac{\mathbf{a}_2 \times \mathbf{a}_3}{\mathbf{a}_1 \cdot (\mathbf{a}_2 \times \mathbf{a}_3)}, \quad \mathbf{b}_2 = \frac{\mathbf{a}_3 \times \mathbf{a}_1}{\mathbf{a}_2 \cdot (\mathbf{a}_3 \times \mathbf{a}_1)}, \quad \mathbf{b}_3 = \frac{\mathbf{a}_1 \times \mathbf{a}_2}{\mathbf{a}_3 \cdot (\mathbf{a}_1 \times \mathbf{a}_2)}, \tag{15.1.1}$$

referred to as reciprocal lattice vectors, satisfying the following orthogonality relations

$$\mathbf{b}_i \cdot \mathbf{a}_j = \delta_{ij}. \tag{15.1.2}$$

We introduce, in turn, the vector

$$\mathbf{H} = h\,\mathbf{b}_1 + k\,\mathbf{b}_2 + \ell\,\mathbf{b}_3, \tag{15.1.3}$$

[10] See, e.g., Weast (1986) [12]. The lattice constants may vary slightly with temperature.

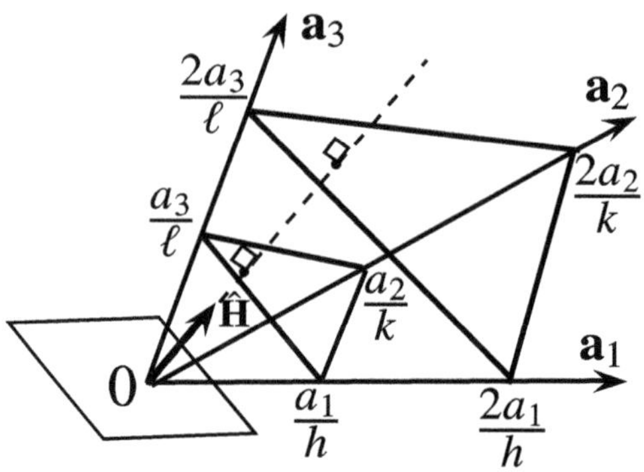


**Fig. 15.1** The first plane, parallel to the plane set at the origin, intercepts the three axes, along the basis vectors of a unit cell $\mathbf{a}_1, \mathbf{a}_2, \mathbf{a}_3$, at $a_1/h,\ a_2/k, a_3/\ell$ separated by a distance $d_{hk\ell} = 1/|\mathbf{H}|$, where $\mathbf{H} = h\,\mathbf{b}_1 + k\,\mathbf{b}_2 + \ell\,\mathbf{b}_3$, and $\mathbf{b}_i \cdot \mathbf{a}_j = \delta_{ij}$. A second parallel plane to the first one is shown at a distant $d_{hk\ell}$ away from it, and so on

and note that the vectors $(\mathbf{a}_1/h - \mathbf{a}_2/k)$, $(\mathbf{a}_2/k - \mathbf{a}_3/\ell)$ are parallel to the $(hkl)$ planes, two of which are shown in Fig. 15.1, and due to the orthogonality conditions in Eq. 15.1.2, they are orthogonal to the vector **H**, i.e., $\mathbf{H} \cdot (\mathbf{a}_1/h - \mathbf{a}_2/k) = 1 - 1 = 0$, $\mathbf{H} \cdot (\mathbf{a}_2/k - \mathbf{a}_3/\ell) = 1 - 1 = 0$. That is, **H** is orthogonal to the $(hkl)$ planes.

$\mathbf{H}/|\mathbf{H}| = \hat{\mathbf{H}}$ denotes the unit vector along the vector **H**. From the figure, the distance between consecutive $(hk\ell)$ planes, denoted by $d_{hkl}$, is, from Eq. (15.1.3), simply given by

$$\begin{aligned} d_{hkl} &= \mathbf{a}_1 \cdot \frac{\hat{\mathbf{H}}}{h} = \mathbf{a}_2 \cdot \frac{\hat{\mathbf{H}}}{k} = \mathbf{a}_3 \cdot \frac{\hat{\mathbf{H}}}{\ell} \\ &= \frac{1}{|\mathbf{H}|}, \end{aligned} \tag{15.1.4}$$

as a consequence of the orthogonality relation in Eq. (15.1.2).

For additional clarity, we apply the above construction to the cubic lattices discussed earlier each of sides $a$. To this end,

$$\mathbf{a}_1 = a\,\hat{\mathbf{i}}_1, \quad \mathbf{a}_2 = a\,\hat{\mathbf{i}}_2, \quad \mathbf{a}_3 = a\,\hat{\mathbf{i}}_3, \tag{15.1.5}$$

$$\mathbf{b}_1 = \frac{1}{a}\,\hat{\mathbf{i}}_1, \quad \mathbf{b}_2 = \frac{1}{a}\,\hat{\mathbf{i}}_2, \quad \mathbf{b}_3 = \frac{1}{a}\,\hat{\mathbf{i}}_3, \tag{15.1.6}$$

and since $\mathbf{b}_i \cdot \mathbf{b}_j = \delta_{ij}/a^2$, for the cubic cases, and $\mathbf{H} = \frac{h}{a}\mathbf{i}_1 + \frac{k}{a}\mathbf{i}_2 + \frac{\ell}{a}\mathbf{i}_3$, the distance between the $(hk\ell)$ planes is simply given by

$$d_{hkl} = \frac{a}{\sqrt{h^2 + k^2 + \ell^2}}. \tag{15.1.7}$$

Why do we bother constructing such planes in a crystal?. The reason is that atoms are not point-like and are of finite extensions, essentially fill a crystal and thus these planes are relevant in scattering off a crystal. Most importantly, we will learn in Sect. 15.5 conditions for which parallel planes $(hk\ell)$ are appropriate for so-called Bragg reflection, discussed below, to occur from a crystal when considering the so-called *crystal structure* factor of a crystal. That is, we will look for *planes in a crystal that diffract* the X-rays.

These appropriate planes tell us what kind of a crystal we have and which element or compound we are dealing with. In the subsequent sections, we also spell out how atoms are located with respect to a given unit cell.

Now consider two $(hk\ell)$ planes, separated by a distance $d$, suppressing $(hk\ell)$ in it for simplicity of the notation. Most probably, you have, in your introductory physics courses, studied the following scattering problem of light off a crystal depicted here:

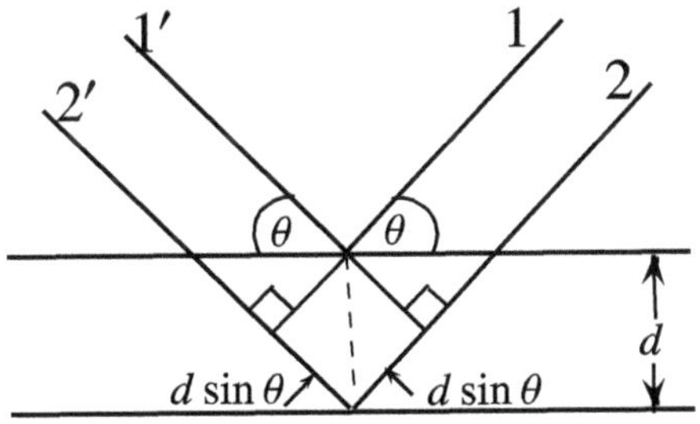


From the figure, we note that incident X-ray $2'$ travels an additional distance equal to $2d\sin\theta$ than the X-ray $1'$. Accordingly, in order that the amplitudes of the rays add up constructively, the extra distance $2d\sin\theta$, traveled by X-ray $2'$, must be equal to a full wavelength $\lambda$, or more generally, equal to $n\lambda$, $n = 1, 2, \ldots$. That is, the condition of constructive interference and generation of X-ray emission is given by

$$2d\sin\theta \;=\; n\lambda, \tag{15.1.8}$$

referred to as Bragg's law, with a phase shift developed upon scattering given by

$$\varphi \;=\; \frac{2\pi}{\lambda} \times \text{path difference} \;=\; \frac{4\pi d\sin\theta}{\lambda} \;=\; 2\pi, \tag{15.1.9}$$

for $n = 1$.

In particular, from the condition that $\lambda \leq 2d$, we note that visible light does not satisfy Bragg's law, since $d \sim$ Å.

Using the fact that for the wave vectors in question $\mathbf{k}, \mathbf{k}'$, with $|\mathbf{k}| = |\mathbf{k}'| = 2\pi/\lambda$, we have, from the figure, that $\mathbf{k}\cdot\mathbf{k}' = |\mathbf{k}|^2\cos(2\theta)$, $\cos(2\theta) = 1 - 2\sin^2\theta$, which give

$$\left|\mathbf{k}-\mathbf{k}'\right|^2 = 2[|\mathbf{k}|^2 - |\mathbf{k}|^2\cos(2\theta)], \quad \Rightarrow \quad 2\sin\theta = \frac{\left|\mathbf{k}-\mathbf{k}'\right|}{|\mathbf{k}|}. \tag{15.1.10}$$

That is, Bragg's law may be rewritten as

$$\left|\mathbf{k}-\mathbf{k}'\right| = \frac{2\pi n}{d}, \qquad \text{where } \left|\mathbf{k}\right| = \frac{2\pi}{\lambda}. \tag{15.1.11}$$

A vector form version of the Bragg's law may be also derived. Since the vectors $\mathbf{k}$ and $\mathbf{k}'$ are of equal magnitudes, in Bragg's law, the vector $\mathbf{k}-\mathbf{k}'$ is orthogonal to the crystallographic planes, but so is the vector $\mathbf{H}$, by construction. Accordingly,

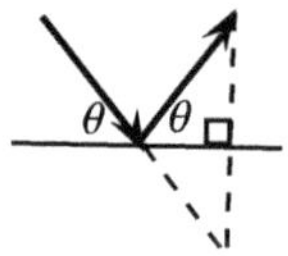


$$[\mathbf{k}-\mathbf{k}'] = \text{Const.}\,\mathbf{H},$$

from which we may infer, upon taking the magnitudes of the above equation, that according to Bragg's law in Eqs. (15.1.11) and (15.1.4) that Const. $= 2\pi n$, and in a vectorial form,

$$\mathbf{k}-\mathbf{k}' = 2\pi n\,\mathbf{H}. \tag{15.1.12}$$

It is worth repeating, that atoms are not point-like and are of finite extensions and essentially fill a crystal and thus the planes, discussed earlier, are relevant in scattering off a crystal, and we will learn in Sect. 15.5 conditions for which parallel planes $(hk\ell)$ are appropriate for so-called Bragg reflection to occur from a crystal when considering the so-called *crystal structure* factor of a crystal. That is, we will look for *planes in a crystal that diffract* the X-rays.

In the following section, we spell out how atoms are located with respect to a given unit cell.

## 15.2 Appendix: Conventional Cells of Bravais Lattices

We consider parallelepipeds, i.e., of faces being parallelograms, determined by basis vectors $\mathbf{a}_1, \mathbf{a}_2, \mathbf{a}_3$ such that,

$$|\mathbf{a}_1| = a_1, \qquad |\mathbf{a}_2| = a_2, \qquad |\mathbf{a}_3| = a_3, \tag{15.2.1}$$

$$\text{angle } \mathbf{a}_2 \wedge \mathbf{a}_3 = \alpha, \quad \text{angle } \mathbf{a}_3 \wedge \mathbf{a}_1 = \beta, \quad \text{angle } \mathbf{a}_1 \wedge \mathbf{a}_2 = \gamma. \tag{15.2.2}$$

A unit cell is the building block of a crystal as discussed earlier. Such arrangements of crystals, in terms of unit cells, involving atoms, molecules, or ions generate structures of lattices referred to as Bravais lattices. The effective number of atoms in a parallelepiped is counted as follows: Each corner is common to 8 units, thus each contributing 1/8th of an atom. An atom in its center is not shared by other units, thus contributing 1 atom. Face-centered, and base-centered ones, are common to 2 units, thus each containing 1/2 of an atom. The hexagonal contains 3 parallelepipeds in contact, and is easily worked out below. Conventional Bravais lattices are given below, and are defined in terms of unit cells.

---

**Cubic**: For this system, $a_1 = a_2 = a_3 \equiv a$, with $\alpha = \beta = \gamma = 90^o$, and there are three types, simple cubic (SC), body-centered cubic (BCC), face-centered cubic (FCC), with the atoms denoted by the black dots. With the exception of the atom in the center of a BCC unit, they denote fractions of atoms, which becomes evident as they are shared with other unit cells when one combines them together to form a crystal. These cubic cells are reproduced here from the previous section for completeness. The SC contains effectively 1 atom, the BCC 2 atoms, and FCC 4.

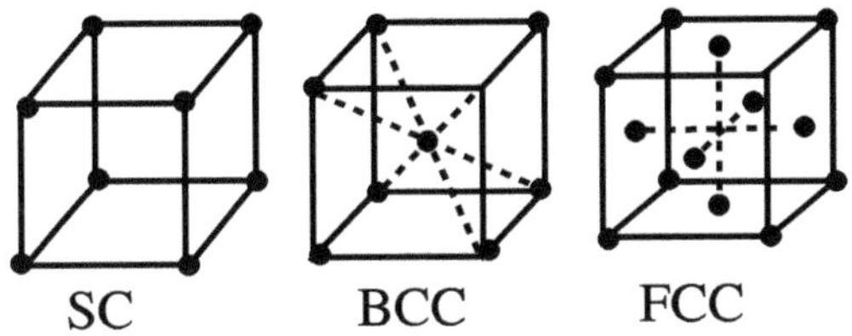


---

**Tetragonal**: Here $a_1 = a_2 \equiv a \neq a_3$, $\alpha = \beta = \gamma = 90^o$. The second is a body-centered one. The first contains effectively 1 atom, while the second 2.

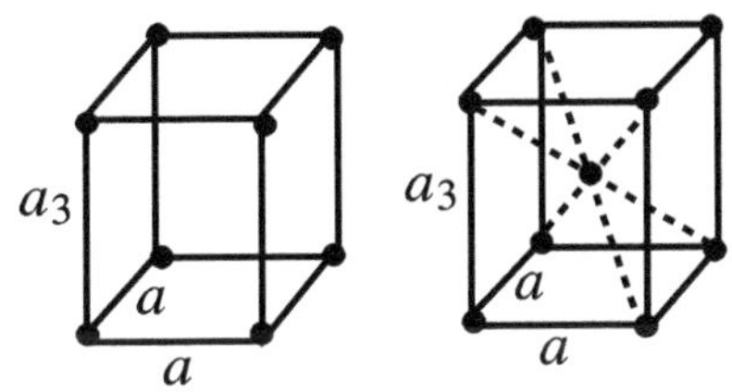


---

**Orthorhombic**: For this system $a_1, a_2, a_3$ are all unequal, but all the angles are equal to $90^o$, and involves the following four types. The third unit is referred to as base-centered in order not to confuse it with the four face-centered (or all face-centered). One may also consider the other two different opposite sides base-centered cases. By doing so, one defines these three possibilities as A, B, C base-centered. The first effectively contains 1 atom the second and the third 2, while the last one 4.

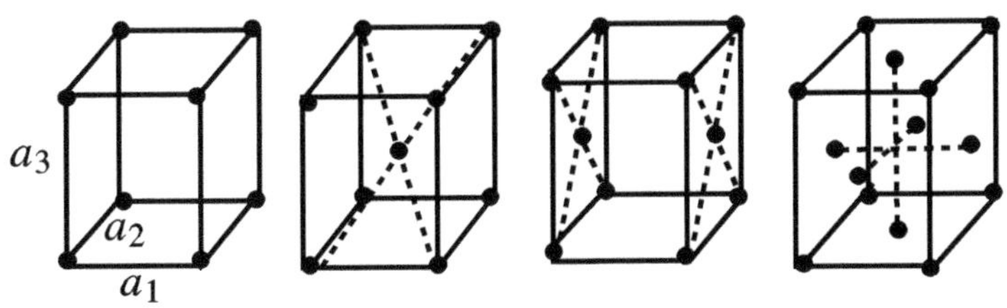


---

**Monoclinic**: Here $a_1, a_2, a_3$ are all unequal, $\alpha = \gamma = 90^o \neq \beta$. The second unit is referred to as base-centered. in order not to confuse it with the four face-centered (or all face-centered). The first contains effectively 1 atom, and the second 2.

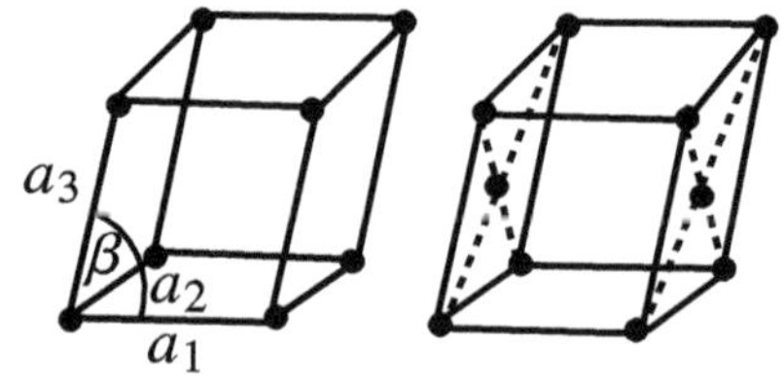


**Triclinic**: This is quite general as $a_1$, $a_2$, $a_3$ are all unequal, and all the angles are different and none equal to 90°. It effectively contains 1 atom.

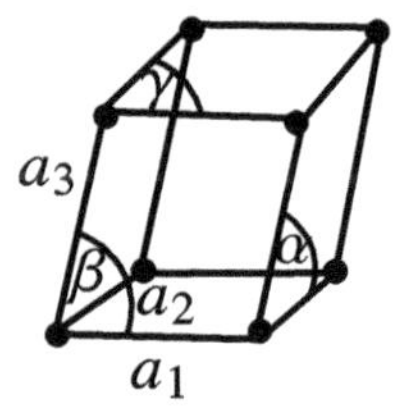


**Hexagonal**: Here the parallelepiped is to be considered with other two units as indicated with $\alpha = \beta = 90^{\circ}, \gamma = 120^{\circ}, a_1 = a_2 \equiv a \neq a_3$. When considered together, this simple hexagonal unit involves 12 corners, each common to 6 hexagons not 8, contributing each 1/6th of an atom, and two halves, thus containing all in all effectively 3 atoms. The hexagonal so-called close-packed contains additional 3 atoms within the body, see Sect. 15.5.

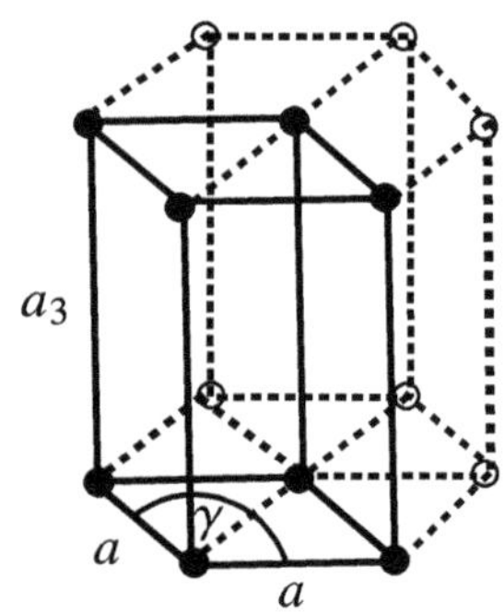


**Rhombohedral**: In this case $a_1 = a_2 = a_3 \equiv a$, and all the angles are equal but are different from 90°. It effectively contains 1 atom.

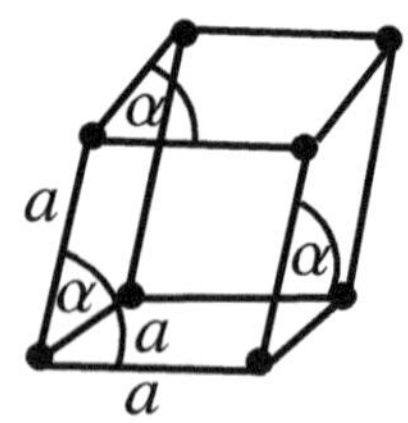

## 15.3 Irradiated Crystal as a Source of Radiation

In order to derive the expression of the power of radiation of an X-ray irradiated crystal, we need to introduce the atomic structure factor in the unit cell, also called the atomic form factor. If $\rho_s(\mathbf{x})$ denotes the electron density of the $s$th atom in a given unit cell, then, by definition, in units of the electron charge,

$$\int \mathrm{d}^3\mathbf{x}\, \rho_s(\mathbf{x}) \;=\; Z_s, \tag{15.3.1}$$

$$\int \mathrm{d}^3\mathbf{x}\, |\mathbf{x}|^2 \rho_s(\mathbf{x}) \;=< |\mathbf{x}|_s^2 >, \tag{15.3.2}$$

where $Z_s$ denotes the atomic number, i.e., the number of electrons in the $s$th atom, and $< |\mathbf{x}|_s^2 >$ defines the root mean squared charge radius of the atom.

The atomic structure factor is defined by

$$f_s(\mathbf{Q}) \;=\; \int \mathrm{d}^3\mathbf{x}\, \mathrm{e}^{-\mathrm{i}\,\mathbf{Q}\cdot\mathbf{x}}\, \rho_s(\mathbf{x}), \qquad \mathbf{Q} = \mathbf{k} - \mathbf{k}', \tag{15.3.3}$$

where $\mathbf{Q}$ is the momentum transfer of a scattering process, with $\mathbf{k}'$ denoting the incident momentum, and $\mathbf{k}$ denoting the scattered one. Because of this, the atomic structure factor is also called the atomic scattering factor.[11]

Assuming spherical symmetry of $\rho_s(\mathbf{x}) = \rho_s(r), |\mathbf{x}| \equiv r$, we explicitly have

$$f_s(\mathbf{Q}) \;=\; \int \mathrm{d}^3\mathbf{x}\, \rho_s(r)\, \mathrm{e}^{-\mathrm{i}\,\mathbf{Q}\cdot\mathbf{x}}$$

$$= 2\pi \int_0^\infty r^2 \mathrm{d}r \int_0^\pi \sin\vartheta \mathrm{d}\vartheta\, \rho(r)\, \mathrm{e}^{-\mathrm{i}Qr\cos\vartheta} = 4\pi \int_0^\infty r^2 \mathrm{d}r\, \rho(r) \frac{\sin(Qr)}{Qr}, \tag{15.3.4}$$

where, from Eqs. (15.1.10), (15.1.11), $Q = |\mathbf{k} - \mathbf{k}'| = 4\pi \sin\theta/\lambda$.

We recall from Eq. (7.4.10), that for a given current distribution $\mathbf{J}$, the power of radiation emitted in the direction of the unit vector $\mathbf{n}$ is given by

$$\mathrm{P}_{\mathbf{n}}(\omega) \;=\; \frac{\omega^2}{32\pi^2} \int \mathrm{d}^3\mathbf{x}\, \mathbf{J}^{*j}(\omega, \mathbf{x})[\delta^{ij} - \mathbf{n}^i\mathbf{n}^j] \int \mathrm{d}^3\mathbf{x}'\, \mathbf{J}^i(\omega, \mathbf{x}')\, \mathrm{e}^{\mathrm{i}\omega\mathbf{n}\cdot[\mathbf{x}-\mathbf{x}']} \tag{15.3.5}$$

invoking reality of $\mathrm{P}_{\mathbf{n}}(\omega)$, where no phenomenological parameters are introduced.

---

[11] In high energy physics, such form factors, e.g., of the proton have established, through so-called deep inelastic scattering, conformity with the existence of quarks and gluons considered as a subset of fundamental building blocks of matter in the so-called standard model. For a detailed pedagogical treatment of such form factors, see Manoukian (2016) [7], Chap. 6.

An electron[12] acquires an acceleration $\mathrm{e}\,\mathbf{E}_0/m$ in an electric field $\mathbf{E}_0$ and velocity $\mathbf{v} = \mathrm{i}\,\mathrm{e}\mathbf{E}_0/m\omega$, assuming harmonic fields, where $\omega = 2\pi/\lambda$. Then the current $\mathbf{J}(x^0, \mathbf{x})$ is not given by $\mathbf{v}\delta^{(3)}(\mathbf{x} - \mathbf{R})$ like in earlier cases treated, because we are considering electrons in atoms with given charge distributions $\rho$. Accordingly, let $\mathbf{x}_s$ denote the position of an atom, taken relative to a convenient origin chosen in a unit cell. The contribution of a unit cell to the current $\mathbf{J}(\omega, \mathbf{x}')$ is then given by

$$\mathrm{e}\left(\frac{\mathrm{i}\,\mathrm{e}\,\mathbf{E}_0}{m\omega}\right) \sum_s f_s(\mathbf{Q})\, \delta^3\big(\mathbf{x}' - \mathbf{x}_s\big)\, \mathrm{e}^{\mathrm{i}\mathbf{k}'\cdot\mathbf{x}'}, \tag{15.3.6}$$

as an emission source of radiation of momentum $\mathbf{k}'$, having an overall one Fourier component as a function of $\mathbf{k}'$. Here the sum is over all the atoms within a unit cell, with $f_s(\mathbf{Q})$ denoting the atomic scattering factors introduced above.

When the latter is substituted in Eq. (15.3.5) for the current, gives the following contribution to it

$$\int \mathrm{d}^3\mathbf{x}'\, \mathbf{J}(\omega, \mathbf{x}')\, \mathrm{e}^{-\mathrm{i}\mathbf{k}\cdot\mathbf{x}'} = \frac{\mathrm{i}\,\mathrm{e}^2\,\mathbf{E}_0}{m\omega} \sum_s f_s(\mathbf{Q})\, \mathrm{e}^{-\mathrm{i}\left(\mathbf{k}-\mathbf{k}'\right)\cdot\mathbf{x}_s}, \tag{15.3.7}$$

where $\mathbf{k} = \omega\mathbf{n}$ is fixed. The factor

$$F = \sum_s f_s(\mathbf{Q})\, \mathrm{e}^{-\mathrm{i}\left(\mathbf{k}-\mathbf{k}'\right)\cdot\mathbf{x}_s}, \quad \mathbf{k} = \omega\,\mathbf{n},\ \ \mathbf{k}' = \omega\,\mathbf{n}',\ \ \mathbf{Q} = \mathbf{k} - \mathbf{k}', \tag{15.3.8}$$

is called the crystal structure factor. In Sect. 15.5, we spell out the details on how the coordinate labels of an atom may be chosen in a unit cell, and thus generate the vectors $\mathbf{x}_s$ of interest.

To determine the remaining contributions to the current, consider parallel Miller planes $(hk\ell)$, distant $d_{hk\ell} \equiv d$ apart, in a crystal, consisting of $N$ planes:

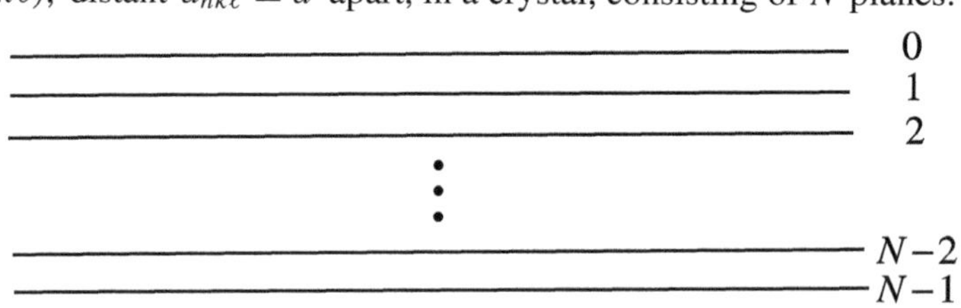


If $v_\mathrm{c}$ denotes the volume of a unit cell, then given an element of area $\mathrm{d}x\mathrm{d}y$ on the surface of the diffracting plane, the number of unit cells in an infinitesimal volume $(\mathrm{d}x\mathrm{d}y)d$ of a layer is given by $(\mathrm{d}x\mathrm{d}y)\,d/v_\mathrm{c}$.

The contribution, in turn, of incident radiation hitting all points on the diffracting surface is given by considering Fig. 15.2, and deriving the path difference of an

[12] Electron photon scattering is most appropriately referred to as Klein-Nishina scattering and in a limiting case as Thomson scattering.

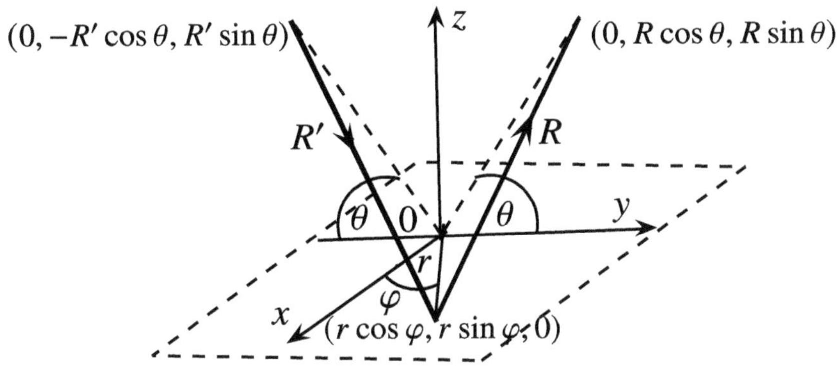


**Fig. 15.2** A figure to determine the accumulated phase due to a path taken by a ray hitting the plane at an infinitesimal distance from the origin at a point $(r, \phi)$, where the Bragg path, hitting the plane *at* the (origin) 0, denotes the shortest path

incident ray in the neighborhood of the path taken by the Bragg incident ray, thus obtaining the corresponding expression of phase shift. Referring to the figure, the path difference for a path in the neighboring of the Bragg path is given by

$$\sqrt{r^2\cos^2\varphi + (R'\cos\theta + r\sin\varphi)^2 + R'^2\sin^2\theta} + \sqrt{r^2\cos^2\varphi + (R\cos\theta - r\sin\varphi)^2 + R^2\sin^2\theta} - (R + R'). \tag{15.3.9}$$

We may, in turn, write, $x = r\cos\varphi,\ y = r\sin\phi$, for $x \approx 0,\ y \approx 0$, and define the function,

$$g(x, y) = \sqrt{x^2 + y^2 + R'^2 + 2yR'\cos\theta}. \tag{15.3.10}$$

We carry out a Taylor expansion in the two variables $x \approx 0,\ y \approx 0$, using the well known expression

$$g(x, y) \approx g(0, 0) + x\,g_x(0, 0) + y\,g_y(0, 0) + xy\,g_{xy}(0, 0) + \frac{x^2}{2}\,g_{xx}(0, 0) + \frac{y^2}{2}\,g_{yy}(0, 0), \tag{15.3.11}$$

where, e.g., $g_{xx}(0, 0) = \partial^2 f(x, y)/\partial x^2$ evaluated at $x = 0,\ y = 0$, and so on.

This gives, for the path difference the expression,

$$\frac{x^2 + y^2\sin^2\theta}{2}\Big(\frac{1}{R'} + \frac{1}{R}\Big), \tag{15.3.12}$$

and a phase shift, defined by the first equality in Eq. (15.1.9),

$$\Phi(x, y) = \frac{2\pi}{\lambda} \times \frac{x^2 + y^2 \sin^2\theta}{2}\Big(\frac{R + R'}{RR'}\Big) + 2\pi, \tag{15.3.13}$$

where the extra $2\pi$ is the phase shift from the initial Bragg reflection law corresponding to the limit $r \to 0$, as given by the third equality in Eq. (15.1.9).

In order to derive the total phase shift due to the full crystal, we have to integrate over $\mathrm{d}x\mathrm{d}y$. We also consider, in addition to the $2\pi$ phase shift occurring in Bragg's law, any deviation that may occur to it due to the full crystal by introducing, in the process, a small parameter $\kappa$ to take into account of a corresponding small deviation that may arise in the angle of incidence $\theta$, whose nature will be investigated in Sect. 15.4,

$$\begin{aligned} 2\pi = \frac{2\pi}{\lambda} d \sin\theta \ \to \ \frac{2\pi}{\lambda} d \sin(\theta + \kappa) &\simeq \frac{2\pi}{\lambda} d \sin\theta + \frac{2\pi}{\lambda} d \cos\theta\,\kappa \\ &= 2\pi + 2\pi \cot\theta\,\kappa, \end{aligned} \tag{15.3.14}$$

due to all of the scattering planes in the crystal.

According to Eqs. (7.5.12), (7.5.13), we need to take the limit $R \to \infty$ to determine the power of radiation. Therefore, the total phase accumulation factor of the traveling waves through the $N-1$ layers, and contributing to the current, including and up to a multiplicative contributing factor to be determined for generalizing the expression of the Fourier transform of the current in Eq. (15.3.7), with the phase factors obtained directly from Eqs. (15.3.13), (15.3.14), is given by

$$\frac{d}{v_{\mathrm{c}}} \sum_{\ell=0}^{N-1} \int \mathrm{d}x\, \mathrm{d}y \exp\Big[\mathrm{i}\frac{2\pi}{\lambda} \frac{(x^2 + y^2 \sin^2\theta)}{2R'}\Big] \exp\big[2\pi \mathrm{i}\,\ell \cot\theta\,\kappa\big]. \tag{15.3.15}$$

To sum up the series in Eq. (15.3.15), we make use of the following elementary, easily established, summation result

$$S_N = 1 + r + r^2 + \ldots + r^{N-1}, \Rightarrow r S_N = (r + r^2 + \ldots r^{N-1}) + r^N, \tag{15.3.16}$$

$$\begin{aligned} (1 + r + r^2 + \ldots r^{N-1}) + (r^N - 1) &= S_N + (r^N - 1) \Rightarrow \\ (r - 1) S_N = r^{N-1} - 1, \quad \Rightarrow \quad S_N &= \frac{r^N - 1}{r - 1}. \end{aligned} \tag{15.3.17}$$

That is,

$$\sum_{\ell=0}^{N-1} \mathrm{e}^{2\pi \mathrm{i}\ell \cot\theta\kappa} = \frac{\mathrm{e}^{2\pi \mathrm{i}\, N \cot\theta\kappa} - 1}{\mathrm{e}^{2\pi \mathrm{i} \cot\theta} - 1} = \frac{\sin\big(N\pi\kappa \cot\theta\big)}{\sin\big(\pi\kappa\cot\theta\big)}\, \mathrm{e}^{\pi \mathrm{i}\,(N-1)\kappa\cot\theta}. \tag{15.3.18}$$

We make a change of variables of integration

$$x = X\sqrt{\frac{\lambda}{2\pi R'}}, \quad y = \frac{Y}{\sin\theta}\sqrt{\frac{\lambda}{2\pi R'}}, \tag{15.3.19}$$

to rewrite the expression in Eq. (15.3.15) as

$$\left[\frac{\mathrm{i}R'\lambda d}{v_c \sin\theta}\,\frac{\sin\left(N\pi\kappa\cot\theta\right)}{\sin\left(\pi\kappa\cot\theta\right)}\,I\right]\mathrm{e}^{\pi\mathrm{i}\,(N-1)\kappa\cot\theta}, \tag{15.3.20}$$

where the dimensionless integral $I$ is given by

$$I = \frac{1}{2\pi\,\mathrm{i}R'^2}\int \mathrm{d}X\mathrm{d}Y\,\exp\left[-\frac{X^2+Y^2}{2\,\mathrm{i}R'^2}\right]. \tag{15.3.21}$$

In the radiation problem, one has to consider the entire irradiated surface of the crystal to determine the total energy of emitted radiation from the irradiated crystal. That is, one, eventually, has to take into account, in turn, of the total energy imparted to the crystal by the incident one, and, as a result, multiply the above expression by the square root of the cross sectional area of the incident radiation hitting the diffracting surface as part of the source of emitted radiation, i.e., of the current itself. So that when we take the square of the current, in computing power of radiation, the total energy imparted to the crystal is taken care of. If $A$ is the area of the diffracting surface, the cross section area of the incident radiation is $A\cos(\pi/2-\theta) = A\sin\theta$, using the fact that the angle between the cross section area and $A$ is $(\pi/2-\theta)$, where $\theta$ is the angle of incidence. That is, we must multiply the above expression by $\sqrt{A\sin\theta}$, as a factor of the source of emitted radiation:

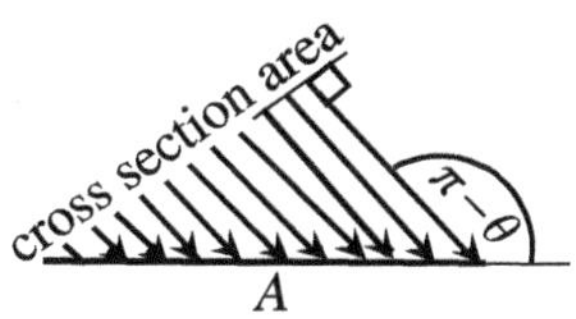


That is, Eq. (15.3.20) becomes

$$\left[\frac{\mathrm{i}R'\sqrt{A}\lambda d}{v_c\sqrt{\sin\theta}}\,\frac{\sin\left(N\pi\kappa\cot\theta\right)}{\sin\left(\pi\kappa\cot\theta\right)}\,I\right]\mathrm{e}^{\pi\mathrm{i}\,(N-1)\kappa\cot\theta}. \tag{15.3.22}$$

This expression is to be multiplied by the Fourier transform of the current $\mathbf{J}(\omega, \mathbf{x}')$ in Eq. (15.3.7), and, in our system of units, should be dimensionless.

With a given electric field $\mathbf{E}_0$, with incident radiation right at the diffracting surface making an angle $\theta$ with it, we have as an input, right *at* the surface of the crystal, that the above expression should be independent of $R'$ - the distance from the source of emission and point impact on the crystal. That is, we should extract the $R'$ independent part of Eq. (15.3.22).

On the other hand, since the intensity of incident radiation right *at* the diffracting surface is already known with an electric field $\mathbf{E}_0$, we may consider the limit $R' \sim 0$, for which, we may use the formal definition of a two dimensional delta distribution (see Eq. (3.6.44), in Chap. 3 for the one dimensional case)

$$\frac{1}{2\pi\,(\mathrm{i}\,R'^2)}\exp\Big[-\frac{X^2+Y^2}{2\,(\mathrm{i}R'^2)}\Big] \simeq \delta(X)\,\delta(Y), \tag{15.3.23}$$

as a product of two one dimensional distributions. We may thus replace the dimensionless integral $I$ by 1.

Accordingly, by replacing the integral $I$ by one, and extracting the dimensionless coefficient of $R'$ from the dimensionful expression in Eqs. (15.3.22), (15.3.7), for the Fourier transform of the current, becomes replaced by the

$$(\mathrm{i})^2\frac{\mathrm{e}^2\mathbf{E}_0}{m\omega}\,F\,\frac{\sqrt{A}\lambda d}{v_c\sqrt{\sin\theta}}\,\frac{\sin\big(N\pi\kappa\cot\theta\big)}{\sin\big(\pi\kappa\cot\theta\big)}\,\mathrm{e}^{\pi\mathrm{i}\,(N-1)\kappa\cot\theta}, \tag{15.3.24}$$

where the crystal structure factor $F$ is defined in Eq. (15.3.8).

We may also consider the thermal vibrations causing displacements of the atoms from their equilibrium positions. That is, in reference to the $s$th atom in a unit cell, we introduce a random variable $\mathbf{U}_s$, and consider displacements introduced as follows

$$[\mathbf{k}-\mathbf{k}']\cdot\mathbf{x}_s \;\rightarrow\; [\mathbf{k}-\mathbf{k}']\cdot\mathbf{x}_s + [\mathbf{k}-\mathbf{k}']\cdot\mathbf{U}_s. \tag{15.3.25}$$

Note that $[\mathbf{k}-\mathbf{k}']$ is perpendicular to the diffracting surface of the crystal, and from Eq. (15.1.10), $|\mathbf{k}-\mathbf{k}'| = 2|\mathbf{k}|\sin\theta$. Hence by denoting the component of $\mathbf{U}_s$ in a direction perpendicular to the diffracting surface by $u_s$, we have

$$[\mathbf{k}-\mathbf{k}']\cdot\mathbf{U}_s \;=\; 2|\mathbf{k}|\,\sin\theta\,u_s. \tag{15.3.26}$$

Accordingly, we may generalize the expression for the structure factor $F$ in Eq. (15.3.8) to

$$F_T \;=\; \sum_s \mathrm{e}^{-\mathrm{i}(\mathbf{k}-\mathbf{k}')\cdot\mathbf{x}_s}\,\mathrm{e}^{-\mathrm{i}Q\,u_s}, \qquad \text{where}\quad Q = 2|\mathbf{k}|\sin\theta = \frac{4\pi\sin\theta}{\lambda}, \tag{15.3.27}$$

where $Q$ was already given in Eq. (15.3.3), with $T$ in $F_T$, denoting temperature dependence through the random variables $u_s$.

## 15.4 Power of Radiation of X-Ray Irradiated Crystals

To determine the power of radiation, we integrate over $\kappa$ within a small interval $(-\epsilon/2, \epsilon/2)$, to take into account of deviations that may arise from the angle of incidence $\theta$ in considering all the Miller planes in the crystal. From the expression of the power radiation through Eq. (15.3.5) and Eqs. (15.3.7), (15.3.8) and the basic Eqs. (15.3.24), (15.3.27), we also average the displacements of the atoms from their equilibrium positions, due to thermal vibrations, over a bivariate normal distribution. We also consider, quite generally, initially unpolarized radiation. These lead us to examine the following two equations

$$\mathrm{P}_{\mathbf{n},\mathbf{n}'} = \frac{1}{2m^2}\Big(\frac{\mathrm{e}^2}{4\pi}\Big)^2 \mathbf{E}_0^{*i}[\delta^{ij} - \mathbf{n}^i\mathbf{n}^j]\mathbf{E}_0^j |F_T|^2 \frac{A\lambda^2 d^2}{\upsilon_{\mathrm{c}}^2 \sin\theta} \int_{-\epsilon/2}^{\epsilon/2} \mathrm{d}\kappa \, \frac{\sin^2\big(N\pi\kappa\cot\theta\big)}{\sin^2\big(\pi\kappa\cot\theta\big)}, \tag{15.4.1}$$

and

$$|F_T|^2 = \sum_s \sum_{s'} f_s\, f_{s'} [\mathrm{e}^{\mathrm{i}(\mathbf{k}-\mathbf{k}')\cdot\mathbf{x}_s}][\mathrm{e}^{-\mathrm{i}(\mathbf{k}-\mathbf{k}')\cdot\mathbf{x}_{s'}}]\Big\langle \mathrm{e}^{\mathrm{i}Q(u_s-u_{s'})}\Big\rangle. \tag{15.4.2}$$

The probability density of a pair of atoms of average displacements $< u_s >= 0$, $< u_{s'} >= 0$, from their equilibrium will be taken to be given by the bivariate normal distribution, defined by[13]

$$f\big(u_s, u_{s'}\big) = \frac{1}{2\pi\,\sigma_s\sigma_{s'}\sqrt{1-\rho_{ss'}^2}} \exp\Big[-\frac{q}{2}\Big], \tag{15.4.3}$$

$$q = \frac{1}{1-\rho_{ss'}^2}\Big[\Big(\frac{u_s}{\sigma_s}\Big)^2 - 2\rho_{ss'}\Big(\frac{u_s\,u_{s'}}{\sigma_s\,\sigma_{s'}}\Big) + \Big(\frac{u_{s'}}{\sigma_{s'}}\Big)^2\Big], \tag{15.4.4}$$

where,

$$< u_s^2 >= \sigma_s^2, \quad < u_{s'}^2 >= \sigma_{s'}^2, \quad < u_s\,u_{s'} >= \rho_{ss'}\,\sigma_s\,\sigma_{s'}, \quad \rho_{ss'}^2 < 1. \tag{15.4.5}$$

The parameters, assumed to be labeled by temperature, are the variances of the corresponding random variables $u_s$,and $u_{s'}$ are denoted, respectively, by $\sigma_s^2$, $\sigma_{s'}^2$, and the parameter $\rho_{ss'}$ is referred as the correlation coefficient . If $\rho_{ss'} = 0$, the random variables $u_s$, $u_{s'}$ are said to be uncorrelated. A basic property of this distribution is the following statistical average,[14]

$$\Big\langle \exp\big[\mathrm{i}Q(u_s-u_{s'})\big]\Big\rangle = \exp\Big[-\frac{Q^2}{2}\sigma_s^2 + Q^2\rho_{ss'}\,\sigma_s\,\sigma_{s'} - \frac{Q^2}{2}\sigma_{s'}^2\Big]. \tag{15.4.6}$$

[13] See, e.g., Manoukian (1986) [8], p. 127.
[14] See, e.g., Manoukian (1986) [8], p. 127.

On the other hand,

$$\int_{-\epsilon/2}^{\epsilon/2} \mathrm{d}\kappa \, \frac{\sin^2\left(N\pi \cot\theta\, \kappa\right)}{\sin^2\left(\pi \cot\theta\, \kappa\right)} = \frac{1}{N\pi\cot\theta} \int_{-N\epsilon\sin\theta/2\pi\cos\theta}^{N\epsilon\sin\theta/2\pi\cos\theta} \mathrm{d}y \, \frac{\sin^2 y}{\sin^2(y/N)}$$
$$\cong \frac{N}{\pi\cot\theta}\int_{-\infty}^{\infty} \mathrm{d}y \, \frac{\sin^2 y}{y^2} = \frac{N}{\cot\theta}, \qquad (15.4.7)$$

for $N$ large enough, which is the case even for a very small crystal.

Finally, recall that the unit vector $\mathbf{n}$ specifies the direction of scattered radiation, while $\mathbf{n}'$ specifies the direction of incidence. Thus upon choosing our coordinate system conveniently, with

$$\mathbf{n} = (\cos\theta, 0, \sin\theta), \quad \mathbf{n}' = (\cos\theta, 0, -\sin\theta), \qquad (15.4.8)$$
$$\mathbf{E}_0 = E_0\,(\sin\phi_0\sin\theta, \cos\phi_0, \sin\phi_0\cos\theta), \qquad (15.4.9)$$

give

$$|\mathbf{E}_0| = |E_0|, \qquad \mathbf{n}\cdot\mathbf{n}' = \cos(2\theta), \qquad \mathbf{E}_0\cdot\mathbf{n}' = 0, \qquad (15.4.10)$$
$$\mathbf{n}\cdot\mathbf{E}_0 = E_0\sin\phi_0\sin 2\theta \equiv E_0\,\cos\beta, \qquad (15.4.11)$$

thus defining $\beta$. Accordingly,

$$\mathbf{E}_0^{*i}\,[\delta^{ij} - \mathbf{n}^i\mathbf{n}^j]\,\mathbf{E}_0^j = |\mathbf{E}_0|^2\sin^2\beta = |\mathbf{E}_0|^2[1 - \sin^2\phi_0\sin^2 2\theta]. \qquad (15.4.12)$$

Hence by averaging over $\phi_0$ from 0 to $2\pi$, we have

$$< \sin^2\beta > = \frac{1}{2\pi}\int_0^{2\pi} \mathrm{d}\phi_0[1 - \sin^2\phi_0\,\sin^2(2\theta)] = \frac{1}{2}[1+\cos^2(2\theta)]. \qquad (15.4.13)$$

All told, the power of radiation for initially unpolarized radiation, is given by

$$\mathrm{P}_{\mathbf{n},\mathbf{n}'} = \frac{|\mathbf{E}_0|^2}{4m^2}\left(\frac{\mathrm{e}^2}{4\pi}\right)^2[1+\cos^2(2\theta)]\,|F_T|^2\,\frac{A\lambda^2 d^2}{v_c^2\sin\theta}\frac{N}{\cot\theta}. \qquad (15.4.14)$$

We may use the relation $d = \lambda/2\sin\theta$, to replace one of the $d$'s in the numerator by $\lambda/2\sin\theta$. Also for large $N$, as is the case even for a very small crystal, $ANd$ is equal to the volume of the crystal, which we denote by $V$. Finally, by using the identity $2\sin\theta\cos\theta = \sin(2\theta)$, we obtain a celebrated expression for the power of radiation of the X-ray diffracted crystal[15]

[15] Expressed in rationalized cgs units.

$$\mathrm{P}_{\mathbf{n},\mathbf{n}'} = \frac{|\mathbf{E}_0|^2}{4m^2}\Big(\frac{\mathrm{e}^2}{4\pi}\Big)^2 [1+\cos^2(2\theta)]\,|F_T|^2\,\frac{\lambda^3\,V}{\upsilon_{\mathrm{c}}^2\sin(2\theta)}, \tag{15.4.15}$$

where note that upon writing $1/\cot = \sin\theta/\cos\theta$, the sine factor cancels out with the sine factor in one of the $d = \lambda/2\sin\theta$ used, as mentioned above. Also, recall, that $\upsilon_{\mathrm{c}}$ denotes the volume of a unit cell and $|F_T|^2$, is given in Eqs. (15.4.2), (15.4.6). In particular, *if* one assumes that the different atoms in thermal vibrations are uncorrelated, i.e., $\rho_{ss'} = 0$, then the expression for $|F_T|^2$ simplifies to

$$|F_T|^2 = \Big|\sum_s \big(f_s(\mathbf{Q})\,\mathrm{e}^{-Q^2\sigma_s^2/2}\big)[\mathrm{e}^{\mathrm{i}\,\mathbf{Q}\cdot\mathbf{x}_s}]\Big|^2, \qquad \mathbf{Q} = \mathbf{k} - \mathbf{k}'. \tag{15.4.16}$$

One may develop various dynamical models to express the variances $\sigma_s^2$ as functions of temperature. Also the above expression, considered for a very small crystal, may be used to study more complicated materials of mosaic character, by putting such very small (infinitesimal) crystals together as their building blocks. Admittedly, the subject of X-ray diffraction is a fascinating one and hopefully the reader finds it interesting. It brings scientists, of various disciplines, and engineers together.

In the next section, we study further the structure factor of a crystal by assigning coordinate labels to atoms in a unit cell, a subject often neglected, and, in particular, investigate of particular Miller planes, set up in a crystal, for which Bragg radiation is observed. These appropriate Miller planes tell us something about a crystal structure and about the element or compound we are dealing with.

## 15.5 Structure Factor: Coordinate Labels of Atoms and Selection Rules

Consider first the simple cubic cells as building blocks. By duplicating the unit cell, systematically, by proper translations, we can construct the relevant crystals, *either* by having 1/8th of an atom at the eight corners of each cell, with one full atom in the BCC crystal unit cell, and with six half atoms in FCC case indicated by black dots, *OR* by having effectively one full atom in the SC crystal, effectively two full atoms in the BCC case, and four full atoms in the FCC as indicated by the circles below.[16]

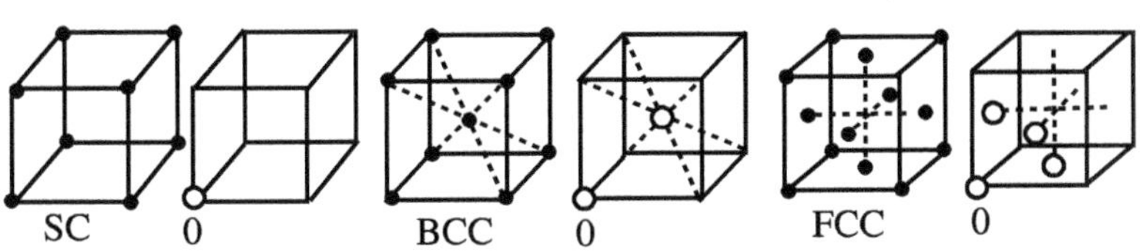


[16] Needless to say one has to add some fractions of atoms to the overall surface of a finite crystal, in the first case and some full atoms in the second one. But one is dealing with a very large number of unit cells and of atoms even for a small crystal.

Accordingly, once we set the origin of a cell, indicated by 0, the coordinate labels $x_1, x_2, x_3$ in $\mathbf{x} = x_1\mathbf{a}_1 + x_2\mathbf{a}_3 + x_3\mathbf{a}_3$, of the atoms in question, suppressing the label $s$ in $\mathbf{x}_s$, may be taken conveniently as follows

$$\text{SCC}: \quad 0,0,0 \qquad \text{BCC}: \quad 0,0,0 \quad \frac{1}{2},\frac{1}{2},\frac{1}{2}\ , \tag{15.5.1}$$

$$\text{FCC}: \quad 0,0,0 \quad \frac{1}{2},\frac{1}{2},0 \quad \frac{1}{2},0,\frac{1}{2} \quad 0,\frac{1}{2},\frac{1}{2}\ , \tag{15.5.2}$$

From Eq. (15.1.12), we know that $\mathbf{k} - \mathbf{k}' = 2\pi\mathbf{H}$, for $n = 1$. That is,

$$\begin{aligned} [\mathbf{k} - \mathbf{k}']\cdot\mathbf{x} &= 2\pi[h\,\mathbf{b}_1 + k\,\mathbf{b}_2 + \ell\,\mathbf{b}_3]\cdot[x_1\mathbf{a}_1 + x_2\mathbf{a}_3 + x_3\mathbf{a}_3] \\ &= 2\pi\,[hx_1 + kx_2 + \ell x_3], \\ \Rightarrow \quad F &= \sum f\mathrm{e}^{-\mathrm{i}2\pi[hx_1+kx_2+\ell x_3]}, \end{aligned} \tag{15.5.3}$$

for the cristal structure factor $F$ in Eq. (15.3.8), suppressing the label $s$th for simplicity of the notation, using, in the process, Eqs. (15.1.2), (15.1.3), where $x_1, x_2, x_3$ denote coordinate labels of full atoms of unit cells in units of $a_1, a_2, a_3$. In the sequel we eliminate the thermal radiation factors.

Consider first cells with identical atoms, to investigate the condition for Bragg reflection.

To this end, note that for the BCC case, with coordinate labels of the two atoms given in Eq. (15.5.1),

$$F = f\left[1 + \exp[-\mathrm{i}\pi(h + k + \ell)]\right], \tag{15.5.4}$$

which will be non-zero, only if $h + k + \ell$ is even, otherwise no Bragg reflection occurs.

For the FCC case

$$F = f\left[1 + \exp[-\mathrm{i}\pi(h + k)] + \exp[-\mathrm{i}\pi(h + \ell)] + \exp[-\mathrm{i}\pi(k + \ell)]\right]. \tag{15.5.5}$$

If two of $hk\ell$ are odd and one even, or two of $hk\ell$ are even and one odd, then a complete cancellation occurs and no Bragg reflection occurs. On the other if all of $hk\ell$ are even, or all are odd, i.e., unmixed, Bragg reflection occurs.

For a SC, we have only one atom, set at the origin, and no such restrictions are set.

The above rules of the structure factors apply equally to the body centered and faced centered orthorhombic crystal discussed in the Sect. 15.2.[17]

On the other hand, for the *base*-centered cubic unit cells in orthorhombic and monoclinic crystals, the figures indicated in Sect. 15.2, involve two atoms with coordinate labels

[17] Note that the base vectors $\mathbf{a}_i$ are not normalized.

$$0,0,0 \qquad 0,\frac{1}{2},\frac{1}{2} \quad . \tag{15.5.6}$$

One may also consider the two other *opposite* ends, in defining these unit cells, with coordinate labels,

$$0,0,0, \quad \frac{1}{2},0,\frac{1}{2} \quad \text{and} \quad 0,0,0, \quad \frac{1}{2},\frac{1}{2},0 \quad . \tag{15.5.7}$$

The structure functions for these base-centered cases[18] are, respectively,

$$f\Big[1+\exp[-\mathrm{i}\pi(k+\ell)]\Big],\ f\Big[1+\exp[-\mathrm{i}\pi(h+\ell)]\Big],\ f\Big[1+\exp[-\mathrm{i}\pi(h+k)]\Big]. \tag{15.5.8}$$

for which no Bragg reflections occur for $(k+\ell)$, $(h+\ell)$, $(h+k)$ are odd, and occur when they are even.

Now let us consider a different application of a crystal involving two ions $Na^+$, $Cl^-$ of rock salt NaCl. Here a unit cell in its entirety, i.e., the whole block, contains only 8 full ions, and may be represented as:

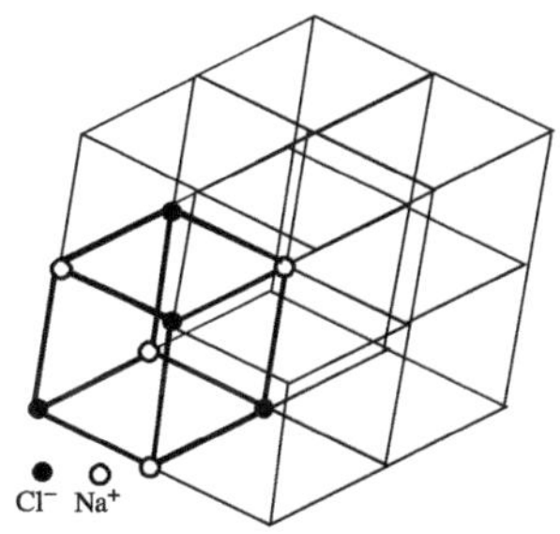


Again by duplicating the unit cell in its entirety, by proper translations, we can construct the crystal of rock salt. The coordinate labels of the full 8 ions may be read directly from the figure as

$$Cl^- : \ 0,0,0 \quad \frac{1}{2},0,\frac{1}{2}, \quad \frac{1}{2},\frac{1}{2},0 \quad 0,\frac{1}{2},\frac{1}{2} \quad , \tag{15.5.9}$$

$$Na^+ : \ \frac{1}{2},0,0 \quad 0,0,\frac{1}{2} \quad \frac{1}{2},\frac{1}{2},\frac{1}{2} \quad 0,\frac{1}{2},0 \quad . \tag{15.5.10}$$

Hence, we have

$$\begin{aligned} F &= f_{Cl^-}\Big[1+\exp[-\mathrm{i}\pi(h+\ell)]+\exp[-\mathrm{i}\pi(h+k)]+\exp[-\mathrm{i}\pi(k+\ell)]\Big] \\ &+ f_{Na^+}\Big[\exp[-\mathrm{i}\pi h]+\exp[-\mathrm{i}\pi\ell]+\exp[-\mathrm{i}\pi(h+k+\ell)]+\exp[-\mathrm{i}\pi k]\Big]. \end{aligned} \tag{15.5.11}$$

[18] These three cases are also referred to as A, B, C face-centered.

Clearly, if two of $hk\ell$ are odd and one even, or two are odd and one even, i.e., mixed, the coefficients of both, $f_{\mathrm{Cl}^-}$ $f_{\mathrm{Na}^+}$ are zero and no Bragg radiation occurs. On the other hand if all of $hk\ell$ are odd or are all even, we get $4(f_{\mathrm{Cl}^-} - f_{\mathrm{Na}^+})$, $4(f_{\mathrm{Cl}^-} + f_{\mathrm{Na}^+})$, respectively.

Another illuminating example, is the zinc blende ZnS with a unit cubic cell consisting of four full $Zn^{2+}$ ions and four of $S^{2-}$, with a unit cell:

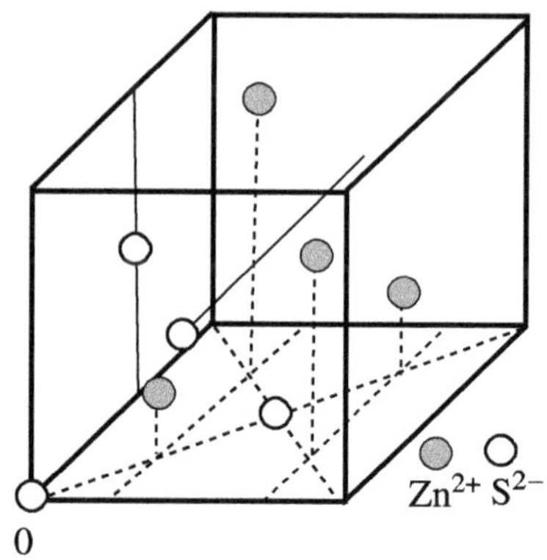


The coordinate labels are as follows:

$$\mathrm{Sn}^{2-}: \quad 0,0,0 \qquad \frac{1}{2},\frac{1}{2},0 \qquad \frac{1}{2},0,\frac{1}{2} \qquad 0,\frac{1}{2},\frac{1}{2}\ , \tag{15.5.12}$$

$$\mathrm{Zn}^{2+}: \quad \frac{1}{4},\frac{1}{4},\frac{1}{4} \qquad \frac{1}{4},\frac{3}{4},\frac{3}{4} \qquad \frac{3}{4},\frac{1}{4},\frac{3}{4} \qquad \frac{3}{4},\frac{3}{4},\frac{1}{4}\ . \tag{15.5.13}$$

Accordingly, the coefficient of $f_{\mathrm{Zn}^{2+}}$ is given by

$$\begin{aligned}&\exp\Big[-\mathrm{i}\pi\Big(\frac{h+k+\ell}{2}\Big)\Big] + \exp\Big[-\mathrm{i}\pi\Big(\frac{h+3k+3\ell}{2}\Big)\Big]\\&+\exp\Big[-\mathrm{i}\pi\Big(\frac{3h+k+3\ell}{2}\Big)\Big] + \exp\Big[-\mathrm{i}\pi\Big(\frac{3h+3k+\ell}{2}\Big)\Big],\end{aligned} \tag{15.5.14}$$

or given by

$$\begin{aligned}&\exp\Big[-\mathrm{i}\pi\Big(\frac{h+k+\ell}{2}\Big)\Big]\\&\times\Big[1+\exp[-\mathrm{i}(k+\ell)]+\exp[-\mathrm{i}(h+\ell)]+\exp[-\mathrm{i}(h+k)]\Big].\end{aligned} \tag{15.5.15}$$

While the coefficient of $f_{\mathrm{S}^{2-}}$ is

$$1+\exp[-\mathrm{i}(h+k)]+\exp[-\mathrm{i}(h+\ell)]+\exp[-\mathrm{i}(k+\ell)]. \tag{15.5.16}$$

That is,

$$F = \Big\{1 + \exp\big[-\mathrm{i}(k+\ell)\big] + \exp\big[-\mathrm{i}(h+\ell)\big] + \exp\big[-\mathrm{i}(h+k)\big]\Big\} \times \Big\{f_{S^{2-}} + f_{Zn^{2+}}\exp\Big[-\mathrm{i}\pi\Big(\frac{h+k+\ell}{2}\Big)\Big]\Big\}. \tag{15.5.17}$$

The first overall factor is zero if the $hk\ell$ are *mixed*, i.e., two are even, one odd, or two are odd and one even, and *no* Bragg radiation occurs. On the other hand, if they are all even or are all odd, i.e., unmixed, it is equal to 4.

On the other hand if $h\,k\,\ell$ are all even, i.e., $h+k+\ell = 2n$, the factor on the second line in Eq. (15.5.17) is simply $[f_{S^{2-}} + f_{Zn^{2+}}(-1)^n]$, by using the fact that $\exp[-\mathrm{i}\pi n] = (-1)^n$. Moreover, if $h\,k\,\ell$ are all odd, then $h+k+\ell = 2n+1$, and $\exp[-\mathrm{i}\pi(2n+1)/2] = -\mathrm{i}(-1)^n$. Accordingly,

$$F = 0, \quad \text{if } h\,k\,\ell \text{ are mixed, and there is no Bragg reflection,} \tag{15.5.18}$$

$$F = 4\,[f_{S^{2-}} + (-1)^n\, f_{Zn^{2+}}] \quad \text{if } h\,k\,\ell \text{ are all even,} \tag{15.5.19}$$

$$F = 4\,[f_{S^{2-}} - \mathrm{i}\,(-1)^n\, f_{Zn^{2+}}] \quad \text{if } h\,k\,\ell \text{ are all odd.} \tag{15.5.20}$$

Thus the physically relevant expression concerning the crystal structure factor $F$ is given by the following:

$$|F|^2 = 0, \quad h\,k\,\ell \text{ mixed, and no Bragg reflection,} \tag{15.5.21}$$

$$|F|^2 = 16\,[f_{S^{2-}} + f_{Zn^{2+}}]^2, \quad h\,k\,\ell \text{ all even, } h+k+\ell = 4n, \tag{15.5.22}$$

$$|F|^2 = 16\,[f_{S^{2-}} - f_{Zn^{2+}}]^2, \quad h\,k\,\ell \text{ all aven, i.e., } h+k+\ell\, = 2n,\ n \text{ odd,} \tag{15.5.23}$$

$$|F|^2 = 16\,[f^2_{S^{2-}} + f^2_{Zn^{2+}}], \quad h\,k\,\ell \text{ are all odd.} \tag{15.5.24}$$

As another application, we consider a so-called hexagonal close-packed unit cell, made up of 3 parallelepipeds, involving effectively 2 identical full atoms *per* parallelepiped, the so-called primitive cells, as one is shown here in the hexagon with dark lines:

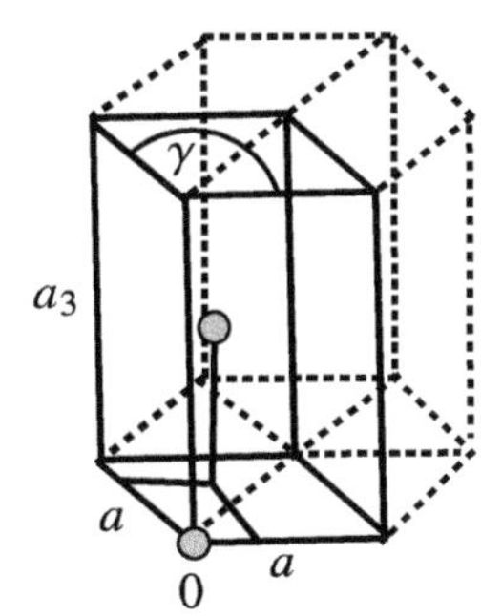

with $a_1 = a_2 \equiv a \neq a_3$, $\alpha = \beta = 90^\circ$, $\gamma = 120^\circ$, and coordinate labels of the two atoms given by

$$0, 0, 0 \qquad \frac{1}{3}, \frac{2}{3}, \frac{1}{2} \quad . \tag{15.5.25}$$

When the primitive cell is repeated forming the hexagon, as already derived in Sect. 15.2 in the hexagonal cell, the atom at (0, 0, 0), in the primitive unit, accounts for 1/6 of an atom at each of the twelve corners and base-centered 1/2 atoms, one 1/2 of an atom at the top and another 1/2 atom at the bottom of the hexagon, thus leading to $(12/6 + 2\times 1/2)$ making 3 atoms. What is new, now, is that the atom with coordinate label (1/3,2/3/1/2) accounts for 3 full atoms within the body of the hexagon, all at a hight equal to $a_3/2$. The 3 atoms would form vertices of a triangle. Thus all in all, the hexagon close-packed unit cell contains 6 atoms, as for pure zinc.

From Eq. (15.5.25), the structure factor takes the form

$$F = f\Big[1 + \exp\Big(-2\mathrm{i}\pi\Big[\frac{(h+2k)}{3} + \frac{\ell}{2}\Big]\Big)\Big]. \tag{15.5.26}$$

If $\ell$ is odd or even, we have, respectively

$$F = f\Big[1 - \exp\Big(-2\mathrm{i}\pi\Big[\frac{(h+2k)}{3}\Big)\Big], \quad \ell \text{ odd}, \tag{15.5.27}$$

$$F = f\Big[1 + \exp\Big(-2\mathrm{i}\pi\Big[\frac{(h+2k)}{3}\Big)\Big], \quad \ell \text{ even}. \tag{15.5.28}$$

Accordingly, for $h + 2k = 3n$, and $\ell$ odd, we may immediately infer that no Bragg reflection occurs. On the other hand, using the fact that the magnitude of a phase factor cannot exceed 1, the structure factor will contribute its maximum value $|F|^2 = 4f^2$, as obtained for $h + 2k = 3n$ and $\ell$ even.

These strict rules derived above to be satisfied by Miller planes, which are, in turn, conditions to be set on the parameters $h,\ k,\ \ell$ are pertinent to the structure of a given crystal. For such details and further studies, a reader may consult books devoted to crystallography and X-rays, a few of which are given in the references below.

## References

1. Ares, A. E. (Editor) (2017). *X-ray scattering*. Berlin: InTechOpen.
2. Bravais, A. (1850). Mémoire sur les systemes formés par des points distribués régulièrement sur un plan ou dans l'espace. *Journal de l'Ecole Polytechnique, 19*, 1-128.
3. Feynman, R. P. (1983). *The strange theory of light and matter*. New Jersey: Princeton University Press.
4. Hahn, T. (2005). *International Tables for Crystallography. Vol. A*. Dordrecht: Springer.
5. Hémonnot, C. Y. J, & Köster, S. (2017). Imaging of biological materials and cells by X-ray scattering and diffraction. *ACS Nano 2017, 11, American Chemical Society*, 8542-8559.

6. James, R. W. (1982). *Optical principles of the diffraction of X-rays*. Woodbridge: Ox-Bow Press.
7. Manoukian, E. B. (2016). *Quantum field theory I: Foundations and abelian and non-abelian gauge theories*. Dordrecht: Springer.
8. Manoukian, E. B. (1986). *Modern concepts and theorems of mathematical Statistics*. New York: Springer.
9. Nash, D. J., & Sanders, J. R. (1996). Surface plasmon-polariton study of the optical dielectric funcion of silver. *Journal of Modern Optics, 43*, 81-91.
10. Schwartz, L. H., & Cohen, J. B. (2012). *Diffraction from materials*. Berlin: Springer.
11. Warren, B. E. (1990). *X-ray diffraction*. New York: Dover Publications, Inc.
12. Weast, R. C. (Editor) (1986). *CRC Handbook of Chemistry and Physics*. Florida. CRC Press, Inc.
13. Zolotoyabko, E. (2014). *Basic concepts of X-ray diffraction*. Weinheim: Wiley-VCH.

# Chapter 16
# Cosmic Microwave Background (CMB) Radiation

*The ultimate radiation problem, with light coming at us from our right and left from the remote past of the Universe when it was still an infant.*
*[No previous knowledge of general relativity and cosmology is required]*

This chapter deals with the crucial radiation problem relevant to our Universe, our home. Every person who studied physics should be aware of the cosmic background radiation which originated when the Universe was about 380,000 years old, and cooled down enough so that neutral atoms were formed and photons became mere spectators free to move, and, now, they are right here around us as we observe them today. They come at us from our right and left and from all directions in a spherically symmetric way. We can determine the temperature of the Universe at that time, the wavelength of the liberated photons, their numbers, now, per unit volume, as well as many other parameters such as the age of our visible Universe today. We can study the temperature power spectrum of the cosmic radiation, and even compute the tiny angular variation of temperature fluctuations across the Universe consistent with experiments. The profile of radiation, that is, the energy of the cosmic background radiation per unit volume per unit frequency, is also quite precisely known as will be discussed.

It was an epoch in which the Universe was getting ready for matter to eventually come together and create galaxies and put us on one of them. As a matter fact from studying polarization aspects of the photons, we know these photons got scattered by matter on their way to us and this may provide us further insight about this transitional period. Polarization aspects may even provide us further information about the very early stage of our Universe.

You don't need to know cosmology or the theory of general relativity to follow the chapter. Everything is here—yes everything.



E. B. Manoukian, *Radiation*,
https://doi.org/10.1007/978-3-032-21506-2_16

## 16.1 Where Does the Cosmic Background Radiation Come From: Cosmology for All

The standard modern theory of our Universe is based on the hypothesis that it is homogeneous and isotropic, i.e., it looks the same in all directions, at large scales, with a mathematical description provided by Einstein's general theory of relativity.[1] Moreover, the latter takes into account not only of ordinary matter/energy but also called dark matter, as well as a vacuum-energy, also referred to as dark energy, expressed in terms of a cosmological constant $\Lambda$. This defines the general framework of the so-called standard model of cosmology which describes an expanding and accelerating Universe, originating from an initial state of high density and high temperature, referred to as the Big Bang (BB).

Moreover, at a very early stage, the Universe, and for a very short time, it is assumed that it has expanded enormously, not predicted by the classic BB theory, referred to as inflation, in order to have consistency with observations made of the present stage of the Universe.

Although almost nothing is known about its original state itself as such, one has learnt much about the evolution of our Universe a tiny fraction of a second after its beginning as an *initial value problem* provided by the inflationary state. Thus, more precisely, the standard model of cosmology is referred to as the *inflationary Big Bang theory*.

If one assumes that we do not occupy a privileged place in the Universe, that is, our place is in no way special, and the Universe looks the same to every observer anywhere in the Universe, and if local matter fluctuations are averaged out, an assumption referred to as the Copernican principle, then the Hubble Law, based on observations, asserts that the separation distance between any pair of galaxies divided by their relative velocity gives the time it took for the pair to reach their present separation as the Universe expands. According to Hubble's law, as inferred from observations, this time is the same for any pair of galaxies. One may then infer that they must have all been together some time in the remote past. This is the so-called Big Bang description of our Universe according to which space itself was created and further expanded with objects being much closer in the past, and subsequently creating more space pushing galaxies apart, as the Universe was expanding.

*Dark matter*, mentioned above, is assumed to be non-baryonic which does not interact with the electromagnetic field, where recall that baryons consist of strongly interacting fermions such as the proton and the neutron. The need to assume the existence of dark matter in the Universe happened historically in the following manner. In 1932, the Dutch Jan Oort observed that the stars in our galaxy move too fast for the so-called visible matter it contains, and that more matter was needed than the visible matter, we are familiar with, to hold the stars from flying away into space

[1] You *don't* have to know general relativity and cosmology to follow the chapter. For the interested reader, detailed *pedagogical* treatments of the subject of general relativity, are given in Manoukian (2020, 2024, 2016) [26, 28, 29]. Also, detailed *pedagogical* treatments of modern cosmology are given in Manoukian (2020, 2024) [28, 29].

due to exceeding their escape velocity,[2] leading to the assumption of the existence of dark matter. This is in conformity with the predictions of others such as Lord Kelvin, Henri Poincaré, Knut Lundmark, and later with Fritz Zwicky.[3]

On the other hand, *dark energy*, also mentioned above, is a form of energy required for the description of an accelerating Universe, and is expressed in terms of a phenomenological constant, referred to as the cosmological constant. It is an energy density inherent to empty space, that is, the energy of the fabric of space itself. Einstein, about 1916,[4] introduced a cosmological constant to his field equations in order to describe a static rather than a dynamic universe. After the discovery of an expanding accelerating one, a cosmological constant was, now, introduced to take into account the acceleration of an expanding Universe. Such an acceleration was observed in the late 1990s by three researches, Saul Perlmutter, Brian P. Schmidt, Adam Reiss and their groups.[5]

The universe consists roughly of 5% matter, 27% dark matter and 68% dark energy.

The present largely accepted description of our Universe involves *inflation* in which it is assumed that the Universe experienced a gigantic exponential expansion within a fraction of a second after which the theory was considered to join the BB theory, prior of which the Universe was considered to be much smaller than the one assumed by the net BB theory in order to establish equilibrium in temperature consistent with later expansion of the Universe. It is important to dwell here on the aspect of the Universe being much smaller than the one assumed by the Big Bang one. The reason is that a very small Universe implies that its different parts of it are causally in contact so that uniformity of the temperature across it is initially established. This is the horizon problem of the BB.

Inflation was proposed in the early 1980s, by Alan H. Guth and with variations by others.[6]

Inflation led to *solve* some fundamental problems that arose in the Big Bang description. For example, it led to a self consistent description of the Universe as being *spatially* nearly flat as we observe it today, referred to as the flatness problem, as well as it was consistent with the high degree of isotropy of the so-called the cosmic microwave background across the sky to be studied below referred to as horizon problem, mentioned above. It also solved the problem as to why monopoles haven't been observed in nature, by showing that such an expansion must have diluted the concentration of monopoles quite significantly. Monopoles were, particularly, predicted by present grand unified gauge theories. Finally, present grand unified gauge theories also predict the violation of baryon number by some explicit processes. Such an enormous expansion would have almost nullified the probability that such

---

[2] Oort (1932) [36].

[3] Poincaré (1906) [41]; Lundmark (1930) [24]; Zwicky (1933) [51].

[4] Einstein (1915, 1916) [6–8].

[5] See, e.g., the (2012) Nobel Lectures of Perlmutte, Schmidt, and Riess [38, 42, 44].

[6] Guth (1981) [16]. See also Starobinsky (1980) [46], Kazanas (1980) [20], Linde (1982) [23], for pertinent variations of the theory.

processes may be reversed by physically separating the emerging particles, from the baryon conservation processes, to enormous distances making it highly unlikely that these particles would come together leading back to the initial state of a Universe with a conserved baryonic state. This is consistent with the fact that matter dominates over anti-matter in the Universe.

Finally, we come to the radiation we observe today, referred to as the *cosmic microwave background* (CMB) radiation all around us, relevant to the present chapter dealing with one of the most important radiation problems. The cosmic background radiation emerged when the Universe was about 380,000 years old—a number which we can actually calculate.[7]

In the 1940s and later, George Gamow and his collaborators realized that a Big Bang would produce intense radiation, and by the expansion of the Universe, such a radiation would cool down and remnants of this radiation may still exist today.[8]

Such a radiation was indeed discovered[9] in the mid 1960s by Arno Penzias and Robert Wilson, while trying to identify sources of noise in microwave satellite communications at Bell Laboratories in New Jersey, having an absolute temperature of the order of 3 K that came from all directions.

The radiation originated, when the Universe was about **380,000 years old**, which was cooled down sufficiently to about **3000 K**, when electrons and protons merged together, participated in forming neutral atoms, while photons, originally present after the Universe began its Big Bang expansion, became eventually mere spectators, as their chances of meeting charged particles diminished drastically, decoupled, and moved freely streaming through space ever since.[10]

Moreover, as the Universe expanded, the wavelength of light was stretched, that is, its frequency was reduced and red-shifted, giving rise to the present *microwave* background radiation observed, of black body spectrum of temperature $T_0 = 2.7255$ Kelvin,[11] involving about 410.79 photons per $\text{cm}^3$ which will be computed in the next section. Space expanded to the point that the wavelength is now only in the microwave region and may be detected by radio antennas.

The radiation seems the same in all directions around us on very large scales. The temperature variation $\delta T/T$, we see across the sky is quite small $\sim 10^{-5}$, taking into consideration of the Sun's motion relative to the CMB. That is, the temperature is nearly uniform.

[7] For the interested reader in such calculations, see, e.g., Manoukian (2020) [28], Chap. 86.

[8] See Gamow (1947, 1948, 1956) [12–14]; Gamow and Teller (1939) [15].

[9] Penzias and Wilson (1965) [37]. See also the earlier discovery by Andrew McKellar in 1941 [31].

[10] For the interested reader, for the relevant such technical details and derivations, see, e.g., Manoukian (2020) [28], Chap. 86.

[11] E.g, Fixsen (2009) [9].

A CMB anisotropy was first measured by the COBE satellite[12] in 1992, and many more accurate measurements of this anisotropy have been made later.[13]

In order to go into the details of the CMB, we need an equation which describes the time development of the expansion of the Universe described, in turn, by a *scaling factor* $R(t)$, and which at the present epoch, $t = t_0$, it is denoted by $R(t_0) \equiv R_0$. The dynamical equation describing this time development is referred to as a *Friedmann equation*,[14] and is given by

$$\left(\frac{\dot{R}(t)}{R(t)}\right)^2 = \frac{8\pi\,\mathrm{G_N}}{3}\left[\rho(t) + \frac{\Lambda c^2}{8\pi\,\mathrm{G_N}}\right] - \frac{kc^2}{R^2(t)}, \qquad (16.1.1)$$

where $\mathrm{G_N} = 6.6743 \times 10^{-11}\ \mathrm{m^3/kg\,s^2}$ denotes Newton's gravitation constant, $\rho(t)$ denotes the density of matter, baryonic and non-baryonic, as well of radiation, at time $t$, corresponding to the age of the Universe. $\Lambda$ denotes the cosmological constant, and $k$ is a measure of the spatial curvature of the Universe, such that

$$k = \begin{cases} +1 & \text{for positive constant curvature,} \\ 0 & \text{for a flat space,} \\ -1 & \text{for negative constant curvature.} \end{cases} \qquad (16.1.2)$$

If you draw a triangle (i.e., with three sides), in a space for which $k = 0$, the sum of the angles in the triangle is equal to 180°, i.e., you are dealing with an Euclidean (flat) space, while for $k = 1$, the sum of the angles $> 180°$, and finally for $k = -1$, the sum of the angles $< 180°$. For convenience, we have inserted the speed of light c, in Eq. (16.1.1).

The rate $\dot{R}(t)/R(t)$, with a dot referring to a time derivative, is referred to as the **Hubble expansion rate** of the Universe denoted by $H(t) = \dot{R}(t)/R(t)$, which, in particular, at the present epoch $t = t_0$, is given by

$$\frac{\dot{R}(t_0)}{R(t_0)} = H(t_0) \equiv H_0 = h \times \frac{1}{9.777752\ \mathrm{Gyr}}, \qquad \mathrm{Gyr} = 10^9\ \text{years}, \qquad (16.1.3)$$

where, in a standard notation, $h$, not to be confused with the Planck constant, is a scale factor for the Hubble expansion rate, and is a dimensionless parameter. The recent recommended value is $h = 0.674$.[15]

A derivation of the above dynamical equation for pedestrians is given here:

Consider a particle of mass $m$ attracted by a uniform sphere of matter of radius $Rr$ and mass density $\rho$ as shown here,

---

12 Smoot et al. (1992) [45].

13 Aghanim (2020) [1]; Planck Collaboration (2020, 2016) [39, 40]; Bennett et al. (2013) [2]; Kuo et al. [22]; Jones et al. (2006) [19]; Melchiorri et al. (2000); de Bernardis et al. (2000) [3]; Miller (1999) [33].

14 Friedmann (1922, 1924) [10, 11].

15 Navas et al. (PDG) (2024) [34].

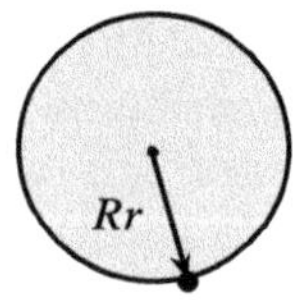


where $r$ here is taken to be *fixed*, and $R$, as a dimensionless scaling factor, which may, of course, *vary*.

The sum of kinetic energy of the particle and its potential energy is then given by

$$\frac{1}{2}mr^2\dot{R}^2 - \frac{Gm}{rR}\left(\frac{4\pi}{3}\rho r^3R^3\right), \tag{16.1.4}$$

where the dot means a derivative with respect to $t$ as usual. The total energy of the system $E$ is then given by

$$E = \frac{1}{2}mr^2\dot{R}^2 - \frac{Gm}{rR}\left(\frac{4\pi}{3}\rho r^3R^3\right). \tag{16.1.5}$$

Hence we may write

$$\left(\frac{\dot{R}(t)}{R(t)}\right)^2 = \frac{8\pi G}{3}\rho(t) - \frac{kc^2}{R^2(t)}, \tag{16.1.6}$$

where we have simply introduced a parameter $k$, arising from the above equation, defined by

$$-\frac{2E}{mr^2} = kc^2. \tag{16.1.7}$$

which is in conformity with the general relativistic case above in Eq. (16.1.1), up to the cosmological constant term, as a phenomenological parameter, for the description of an accelerating Universe.[16]

Consider light coming at us from a galaxy along the radial direction, with us at the center of coordinates. If $\lambda$ denotes the wavelength of light emitted and $\lambda_0$ of the wavelength of light determined upon receipt, the wavelength stretches with the expansion of the Universe as it cools, i.e., $\lambda \leq \lambda_0$ for $t \leq t_0$, implying a relation of the form $\lambda/\lambda_0 = R(t)/R(t_0)$, $t \leq t_0$, $\lambda \leq \lambda_0$. One sets $\lambda/\lambda_0 = 1/(1+z)$, where $z$ defines the *red-shift parameter*, $z_0 = 0$, at present. Upon the expansion of the Universe (i.e., more space), the density of matter $\rho_{\text{matt}}(t)$ decreases. On the other-hand although the density of radiation $\rho_{\text{rad}}(t)$ also decreases, it also decreases, in addition, due to the red-shift observed at the present time, i.e., the wavelength is

[16] From Eq. (16.1.7), one may formally infer that for a flat Universe, i.e., $k = 0$ in Eq. (16.1.2), that the energy $E = 0$. The often given argument that the Universe was created spontaneously with zero energy as a result of cancellation between the gravitational negative energy and the remaining one comes from here.

smaller at time $t$, or larger at the observation time $t_0$. That is, by setting $R(t_0) \equiv R_0$, we have

$$\rho_{\mathrm{matt}}(t) = \rho_{0\mathrm{matt}} \left(\frac{R_0}{R(t)}\right)^3, \quad \rho_{\mathrm{rad}}(t) = \rho_{0\mathrm{rad}} \left(\frac{R_0}{R(t)}\right)^3 (1+z) = \rho_{0\mathrm{rad}} \left(\frac{R_0}{R(t)}\right)^4. \tag{16.1.8}$$

That is, as the Universe expands, the radiation density dilutes faster than that of matter. The subscript "0" refers to present time values.

One defines mass density dimensionless parameters $\Omega$s by:

$$\Omega_{0[A]} \equiv \frac{8\,\pi\,\mathrm{G_N}}{3}\left(\frac{\rho_{0[A]}}{H_0^2}\right), \qquad \Omega_0 = \Omega_{0\mathrm{matt}} + \Omega_{0\mathrm{rad}} + \Omega_\Lambda, \tag{16.1.9}$$

$$\Omega_k \equiv -\left(\frac{kc^2}{H_0^2\, {R_0}^2}\right) = 1 - \Omega_0, \tag{16.1.10}$$

where $A = (\Lambda, \text{matt}, \text{rad})$, and $\rho_{0[\Lambda]} \equiv \rho_\Lambda \equiv \Lambda c^2/(8\pi \mathrm{G_N})$, $\rho_{\mathrm{matt}} + \rho_{\mathrm{rad}} = \rho$.

At the present time $t = t_0$, Eq. (16.1.1) may be expressed as

$$\frac{\dot{R_0}^2}{R_0^2} = H_0^2 = \frac{8\pi\,\mathrm{G_N}}{3}[\rho_{0\mathrm{matt}} + \rho_{0\mathrm{rad}} + \rho_\Lambda] - \frac{kc^2}{R_0^2}. \tag{16.1.11}$$

Equation (16.1.1) may be now, equivalently, rewritten in the convenient form[17]:

$$\dot{x} = \frac{H_0}{x}\big[x\,\Omega_{0\mathrm{matt}} + \Omega_{0\mathrm{rad}} + x^2\,\Omega_k + x^4\,\Omega_\Lambda\big]^{1/2}, \qquad x(t) = \frac{R(t)}{R_0}. \tag{16.1.12}$$

From Eq. (16.1.10), we note that for a flat space, i.e., for which $k = 0$, $\Omega_0 = 1$. Thus, the mass density parameter $\Omega_0$ *defines* a critical dimensionless mass density parameter corresponding to a spatially flat universe-density denoted by $\rho_{0\mathrm{crit}} = (3\,H_0^2)/(8\pi \mathrm{G_N})$, from Eq. (16.1.11).

In terms of the time variable, Eq. (16.1.12) gives

$$\mathrm{d}t = \frac{1}{H_0}\frac{x\,\mathrm{d}x}{[\,x\,\Omega_{0\mathrm{m}} + \Omega_{0\mathrm{rad}} + x^2\,\Omega_k + x^4\,\Omega_\Lambda\,]^{1/2}}. \tag{16.1.13}$$

The present numerical data is given by[18]

$$\Omega_{0\mathrm{matt}} = 0.315(7), \quad \Omega_{0\mathrm{rad}} \simeq 9.2 \times 10^{-5}, \quad \Omega_\Lambda = 0.685(7), \quad \Omega_k = 0.0007, \tag{16.1.14}$$

[17] Note that the expression within the square brackets in Eq. (16.1.12) equals 1 for $x = 1$.

[18] Navas et al. (PDG) [34].

where $\Omega_{0\text{rad}}$ includes ultra-relativistic neutrinos.[19]

Equation (16.1.13) may be also re-written as

$$\mathrm{d}t = \frac{R(t)}{H_0\, R_0} \frac{\mathrm{d}x}{[\, x\,\Omega_{0\text{m}} + \Omega_{0\text{rad}} + x^2\,\Omega_k \; + x^4\,\Omega_\Lambda \,]^{1/2}}, \tag{16.1.15}$$

In particular, the age of the universe at the present time $t_0$, may be simply written as

$$t_0 = \frac{1}{H_0}\int_0^1 \frac{x\,\mathrm{d}x}{[\, x\,\Omega_{0\text{m}} + \Omega_{0\text{rad}} + x^2\,\Omega_k \; + x^4\,\Omega_\Lambda \,]^{1/2}}. \tag{16.1.16}$$

where $R(t_0) = R_0$, and $1/H_0 = 14.507$ Gyr is given in Eq. (16.1.3) with $h = 0.674$. This gives $t_0 = 13.785$ billion years for the *age of the Universe* in agreement with the data,[20] computed earlier in my reference below.[21]

For a spatially flat Universe, consistent with a nearly one, the distance that light travels from the Big Bang at time $t = 0$ to a radial *coordinate* $r$ from the Bing Bang at the origin of a coordinate a system in time $t$, for which $R(t)/R_0 = 1/(1+z)$, is given by

$$r = \mathrm{c}\int_0^t \frac{\mathrm{d}t'}{R(t')}. \tag{16.1.17}$$

$r$ is a radial coordinate label, and doesn't take into account the expansion of the Universe. To obtain this, we must multiply the latter equation by $R(t)$, corresponding to the final time $t$, giving, what one calls, the *particle horizon*

$$r_{\text{hor}} = \mathrm{c}\,R(t)\int_0^t \frac{\mathrm{d}t'}{R(t')} = \mathrm{c}\frac{R(t)}{R_0 H_0}\int_0^{1/(1+z)} \frac{\mathrm{d}x}{[\, x\,\Omega_{0\text{m}} + \Omega_{0\text{rad}} \; + x^4\,\Omega_\Lambda \,]^{1/2}}, \tag{16.1.18}$$

$$= \frac{\mathrm{c}}{H_0\,(1+z)}\int_0^{1/(1+z)} \frac{\mathrm{d}x}{[\, x\,\Omega_{0\text{m}} + \Omega_{0\text{rad}} \; + x^4\,\Omega_\Lambda \,]^{1/2}}, \tag{16.1.19}$$

where we have used Eq. (16.1.15) to write the second line, and the definition of the red-shift parameter $z$, introduced above Eq. (16.1.8).

---

[19] This is determined in Chap. 84 of Manoukian (2020) [28], for the interested reader.

[20] Planck Collaboration et al. (2020) [40], Navas et al. (2024) (PDG) [34].

[21] Manoukian (2024) [29].

## 16.2 Remarkable Black Body Profile of the CMB Radiation and Photon Number Density

When the temperature of the Universe was much *hotter than* **3000 K**, it was too hot for electrons to be bound into atoms, the mean free path of photons, that is, the average distance that a photon travels in between scattering by two charged particles was small compared to the distance a photon would travel during the characteristic expansion time of the Universe for that period, and the photons were coupled to the electrons and protons, remembering that photons do not discriminate between negative and positive charges, they interact with both. As it got cooler, electrons and protons slowed down, merged together, and participated in forming neutral atoms.

One of the most remarkable aspect of the MCB radiation is its Black-Body spectrum which fits nearly *perfectly* with its theoretical form.[22] As a matter of fact the agreement with the black-body spectrum was so satisfying and needless to, say, important, that in 2006, John C. Mather and George F. Smoot, were awarded the Nobel Prize in physics for their discovery of the black-body form of the radiation as well as of the anisotropy of the cosmic microwave background radiation.[23]

Quite possibly, you have studied the black body spectrum in your elementary quantum mechanics course dealing with the ultraviolet catastrophe which led to the birth of the fundamental constant h of Planck. The energy, per unit volume, per unit frequency of radiation of photons with frequency $\nu$ is given by[24]

$$U_\gamma(\nu, T) = \frac{8\pi\,\mathrm{h}}{\mathrm{c}^3}\frac{\nu^3}{\exp[\mathrm{h}\nu/\mathrm{k_B}T] - 1}, \tag{16.2.1}$$

where we have used the letter $\gamma$ to denote the photon. Here $\mathrm{k_B}$ is the Boltzmann constant given by

$$\mathrm{k_B} = 8.617 \times 10^{-14}\,\mathrm{GeV/K} = 1.381 \times 10^{-23}\,\mathrm{J/K}, \tag{16.2.2}$$

where $1\,\mathrm{GeV} = 10^9\,\mathrm{eV}$, K is the Kelvin degree, and J denotes Joules, $1\,\mathrm{J} = 6.242 \times 10^9\,\mathrm{GeV}$.

The nature of the curve defined by $U_\gamma(\nu, T)$ versus $\nu$, is given in Fig. 16.1 for fixed temperature $T$. For convenience, we have inserted the speed of light c as well as the Planck constant h, instead of $2\pi$ where $\mathrm{h} = 2\pi\hbar$, in the above expression.

In the previous section, below Eq. (16.1.7), we have encountered the relation of the stretching of wavelength $\lambda(t) = \lambda_0(R(t)/R_0)$, $(t \leq t_0)$ as time $t$ increases up to the present time $t_0$, in the expanding Universe. Accordingly, if $\lambda_{\mathrm{dec}}$ denotes the wavelength emitted at the decoupling time $t_{\mathrm{dec}}$ of photons, i.e., $t_{\mathrm{dec}} < t_0$, then

---

22 For a derivation of the black body spectrum, see Manoukian (2020) [28], Chap. 82.

23 Mather and Smoot (2006) [30].

24 The reader who is interested in the derivation of this expression see Manoukian (2020) [28], Chap. 82.

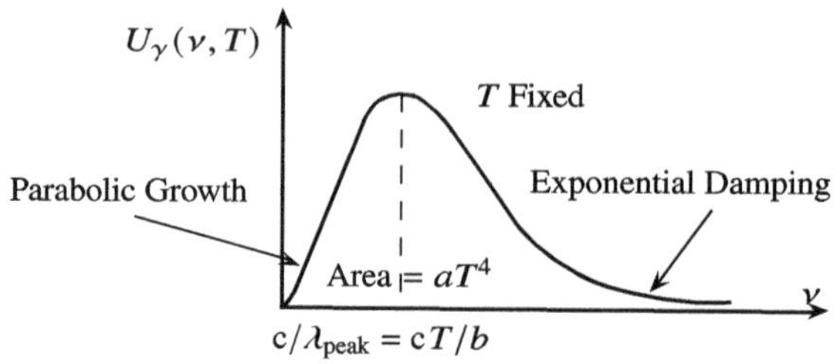


**Fig. 16.1** Here $U_\gamma(\nu, T)$ is energy per unit volume per unit frequency of radiation of a Black Body. Planck suggested that light can only be emitted in quantized form in units of frequency $\nu$, leading to exponential damping at high frequencies of $U(\nu, T)$, as shown above, and establishing the finiteness of the total energy per unit volume $\int_0^\infty d\nu\, U_\gamma(\nu, T) = a\,T^4$, $a = 4.723 \times 10^{-6}$ GeV/m$^3$K$^4$, referred to as the Stefan-Boltzmann Law. c denotes the speed of light, inserted here for convenience, and $b = 2.898 \times 10^{-3}$ m-K, where K stands for Kelvin degree. The peak corresponds to the familiar *Wien Displacement* Law. There exists a maximum energy of radiation at a given temperature

$$\frac{\lambda_{dec}}{\lambda_0} = \frac{R_{dec}}{R_0}, \quad \Rightarrow \quad \frac{\nu_0}{\nu_{dec}} = \frac{R_{dec}}{R_0} = \frac{1}{1 + z_{dec}}, \tag{16.2.3}$$

where $z_{dec}$ is the red-shift parameter at the decoupling time introduced before.

On the other hand, due to the expansion of space, we know from Eq. (16.1.8) that an energy density decreases by a multiplicative factor $(R_{dec}/R_0)^4$, by the time it reaches us. Hence the energy density $U_\lambda$, per unit frequency, decreases only by a multiplicative factor $(R_{dec}/R_0)^3$. That is, both $U_\gamma$ and $\nu^3$ decrease the same way

$$\frac{U_\gamma}{\nu^3} \quad \text{is invariant.} \tag{16.2.4}$$

What about the temperature $T$? To derive the corresponding expression for $T$, we first integrate $U_\gamma$ over $\nu$ to obtain an energy density. We introduce, in the process, the integration variable $x = h\nu/k_B T$, to obtain

$$U_\gamma(T) = \int_0^\infty U_\gamma(\nu, T)\, d\nu = \frac{k_B^4\, T^4}{\pi^2\, \hbar^3\, c^3} \int_0^\infty \frac{x^3\, dx}{e^x - 1} \equiv a\, T^4. \tag{16.2.5}$$

Using the integral

$$\int_0^\infty \frac{x^3\, dx}{e^x - 1} = \frac{\pi^4}{15}, \tag{16.2.6}$$

leads to an equation relating the total energy density of radiation and the temperature given by

$$U_\gamma(T) = a\,T^4, \qquad a = \frac{\pi^2\,k_B^4}{15\,\hbar^3\,c^3} = 4.723 \times 10^{-6}\,\text{GeV}/\text{m}^3\,\text{K}^4, \qquad (16.2.7)$$

known as the *Stefan-Boltzmann Law*.

The fact that the total energy density of radiation is proportional to $T^4$, it tells us that $T/T_0$ decreases by a multiplicative factor $R_{dec}/R_0$ as the radiation reaches us.

Thus from this, and Eqs. (16.2.3), (16.2.4), both[25]

$$\frac{U_\gamma}{\nu^3}, \quad \frac{\nu}{T} \quad \text{are invariant.} \qquad (16.2.8)$$

Thus we may infer the remarkable fact that Planck's law, given in Eq. (16.2.1), keeps its form from the time that the cosmic background emerged to the present epoch and its scales properly as just described.

With a black-body spectrum, with mean temperature[26] $T = 2.7255\,\text{K}$ at the present time (initially of 3000 K), the wavelength peak $\lambda_{peak}$ of the CMB is, according to the *Wien displacement* law, shown in the figure, is given by

$$\lambda_{peak} = \frac{2.898\ \text{mm}}{T/\text{K}} = \frac{2.898\ \text{mm}}{2.7255} \simeq 1.063\,\text{mm}, \qquad (16.2.9)$$

where $b = 2.898 \times 10^{-3}$ m-K (see Fig. 16.1).

Equation (16.2.3) also implies that

$$\frac{R(t_{dec})}{R_0} = \frac{2.7255}{3000} = \frac{1}{1100.72} \qquad (16.2.10)$$

in accordance with the tabulated result for $1 + z_{dec} = 1101$.[27] This remarkable result also tells us that the radial extension of the Universe at decoupling was about $1/1000$ of the size of the radial extension of the (visible) Universe at the present time.

The number density $N_\gamma(\nu, T)$, per unit frequency, of photons, is from Eq. (16.2.1) simply, by dividing by $h\nu$, given by

$$N_\gamma(\nu, T) = \frac{8\,\pi}{c^3}\,\frac{\nu^2}{\exp[\,h\nu/k_B T] - 1}. \qquad (16.2.11)$$

The number density, at a given temperature, of photons of all frequencies may be conveniently derived by introducing, in the process, the convenient variable $x = h\nu/k_B T$, as before, to obtain

---

[25] It is interesting to note for the keen reader, that from the invariance property of $R\nu$, discussed above Eq. (16.2.4), and the conservation of entropy, $RT$ is also invariant, as shown, e.g., in Manoukian (2020) [28], Eq. (81.28) in Chap. 81, taken together, also establish the invariance of $\nu/T$, separately.

[26] E.g. Fixsen (2009) [9].

[27] Navas et al. (PDG) (2024) [34].

$$N_\gamma(T) = \int_0^\infty N_\gamma(\nu, T)\,\mathrm{d}\nu = \frac{\mathrm{k}_\mathrm{B}^3\,T^3}{\pi^2\,\hbar^3\mathrm{c}^3}\int_0^\infty \frac{x^2\,\mathrm{d}x}{\mathrm{e}^x - 1}$$
$$\simeq 2.404\frac{\mathrm{k}_\mathrm{B}^3\,T^3}{\pi^2\,\hbar^3\mathrm{c}^3}, \qquad (16.2.12)$$

where we have used the integral

$$\int_0^\infty \frac{x^2\,\mathrm{d}x}{\mathrm{e}^x - 1} = 2\left[1 + \frac{1}{2^3} + \frac{1}{3^3} + \cdots\right]$$
$$\simeq 2[\,1.202\,] = 2.404. \qquad (16.2.13)$$

Accordingly, we have the following expression for the number density of photons at temperature $T$

$$N_\gamma(T) = 20.29 \times 10^6 \left(\frac{T}{K}\right)^3 \frac{1}{\mathrm{m}^3}. \qquad (16.2.14)$$

For example, for the **photon number density** $N_\gamma$ of CMB photons, at the present temperature $T = 2.7255\,\mathrm{K}$, is given by

$$N_\gamma(2.7255) = 410.79\ \text{photons/cm}^3, \qquad (16.2.15)$$

in conformity with well known tabulated result.[28]

Conversely, by introducing the fundamental constants h, c, and the Boltzmann constant $\mathrm{K_B}$, we may generate the dimensionless invariant expressions:

$$\frac{\nu}{T} \rightarrow \frac{\mathrm{h}\nu}{\mathrm{k_B}T}, \quad \text{and} \quad \frac{U_\gamma(\nu, T)}{\nu^3} \rightarrow \frac{\mathrm{c}^3 U_\gamma(\nu, T)}{\mathrm{h}\nu^3}, \qquad (16.2.16)$$

and with the energy density $U_\gamma$ expressed as a function of temperature and the frequency $\nu$ of radiation, as well as with the invariance properties, spelled out in Eq. (16.2.8), we may, on general grounds, explicitly write

$$U_\gamma(\nu, T) = \frac{\mathrm{h}\nu^3}{\mathrm{c}^3} F\left(\frac{\mathrm{h}\nu}{\mathrm{k_B}T}\right). \qquad (16.2.17)$$

Here $F$ is a dimensionless function of the variable given above. The nearly perfect fitting of the cosmic background radiation, leading to the Nobel prize mentioned earlier, turned out to be given simply by the remarkable expression

$$F\left(\frac{\mathrm{h}\nu}{\mathrm{k_B}T}\right) = \frac{8\pi}{\exp[\mathrm{h}\nu/\mathrm{k_B}T] - 1}, \qquad (16.2.18)$$

[28] Navas et al. (PDG) (2024) [34].

The CMB spectrum has been undoubtedly one of the most precisely measured black body spectrum.[29]

## 16.3 The CMB Radiation and the Power Spectrum

The CMB radiation is uniformly distributed across space to a high degree, regardless of where you look, and isotropic to an accuracy of $\delta T/T \sim 10^{-5}$ in temperatures variation.

The above means that the distribution of $[\delta T/T](\mathbf{n}) \equiv [T(\mathbf{n}) - <T>]/<T>$, in a convenient notation, is to a high degree the same for all directions specified by a unit vector $\mathbf{n}$. This, in turn, implies that if we expand the random variable, in question, $[T(\mathbf{n}) - <T>]/<T>$ in terms of the spherical harmonics, defined below, it is diagonal. In other words, regarding the coefficients $a^T_{\ell m}$ in the following expansion formula,

$$\frac{T(\mathbf{n}) - <T>}{<T>} = \sum_{\ell=0}^{\infty} \sum_{m=-\ell}^{\ell} a^T_{\ell m}\, Y_{\ell,m}(\mathbf{n}), \quad \mathbf{n} = (\sin\theta\cos\phi, \sin\theta\sin\phi, \cos\theta), \tag{16.3.1}$$

$$a^T_{\ell m} = \int \mathrm{d}\Omega\, \frac{T(\mathbf{n}) - <T>}{<T>}\, Y^*_{\ell,m}(\mathbf{n}), \quad \int \mathrm{d}\Omega\; Y_{\ell,m}(\mathbf{n}) Y^*_{\ell',m'}(\mathbf{n}) = \delta_{\ell'\ell}\,\delta_{m'm}, \tag{16.3.2}$$

the bilinear correlation of temperature fluctuations defined by $<a^{T*}_{\ell m} a^T_{\ell' m'}> = 0$, for $m \neq m'$ and/or $\ell \neq \ell'$, and may depend on $\ell$ but is independent of $m$, due to rotational invariance, as will be discussed below in Eq. (16.3.8).[30]

We adopt the following convention for the spherical harmonics defined as follows:

$$Y_{\ell,m}(\theta,\phi) = (\mathrm{i})^{m+|m|} \sqrt{\frac{(2\ell+1)(\ell-m)!}{4\pi(\ell+m)!}}\, P^m_\ell(\cos\theta)\, \mathrm{e}^{\mathrm{i}m\phi}, \tag{16.3.3}$$

$$P^m_\ell(x) = (-1)^m (1-x^2)^{m/2} \frac{\mathrm{d}^m}{\mathrm{d}x^m} P_\ell(x), \quad m \geq 0 \tag{16.3.4}$$

$$= \frac{(-1)^m}{2^\ell \ell!} (1-x^2)^{m/2} \frac{\mathrm{d}^{\ell+m}}{\mathrm{d}x^{\ell+m}} (x^2-1)^\ell, \quad m \geq 0, \tag{16.3.5}$$

$$P^{-m}_\ell = (-1)^m \frac{(\ell-m)!}{(\ell+m)!} P^m_\ell(x). \tag{16.3.6}$$

---

[29] White (1999) [47].

[30] It turns out to be more convenient to use the notation $Y_{\ell,m}$ rather than $Y^\ell_m$, especially when dealing with polarization aspects of CMB, as we will see in the next section.

and $Y_{0,0} = 1/\sqrt{4\pi}$. $P_\ell^m$ and $P_\ell$ are, respectively, the associated Legendre polynomials, and the Legendre polynomials.[31] With this convention, the spherical harmonics satisfy the following reality condition:

$$Y^*_{\ell,m}(\theta,\phi) = Y_{\ell,-m}(\theta,\phi). \tag{16.3.7}$$

The expectation value (average) of the random variable $\big\langle [T(\mathbf{n}) - <T>]/<T> \big\rangle$ is, by definition, zero. Statistically-isotropic sky means that all values of $m$, corresponding to a given $\ell$, are equivalent, i.e., there is no preferred orientation, and temperature correlations depend on angular separations and not orientation.

For a given angular variation $\theta$, a single spherical harmonic $Y_{\ell,m}$ corresponds to about $\ell \sim \pi/\theta$. By invoking rotational invariance, one may consider the bilinear correlation of temperature fluctuations, introduced above, which takes the form

$$< a^{T*}_{\ell m} a^T_{\ell' m'} >= C^{TT}_\ell \, \delta_{m'm} \, \delta_{\ell'\ell}, \qquad C^{TT}_\ell = \big\langle |a_{\ell m}|^2 \big\rangle, \tag{16.3.8}$$

$$\Big\langle \Big( \frac{T(\mathbf{n}) - <T>}{<T>} \Big) \Big( \frac{T(\mathbf{n}') - <T>}{<T>} \Big) \Big\rangle =$$
$$\sum_{\ell=0}^{\infty} \sum_{m=-\ell}^{\ell} C^{TT}_\ell \, Y_{\ell,m}(\mathbf{n}) \, Y^*_{\ell,m}(\mathbf{n}') = \sum_{\ell=0}^{\infty} C^{TT}_\ell \, \frac{(2\ell+1)}{4\pi} P_\ell(\mathbf{n}\cdot\mathbf{n}'), \tag{16.3.9}$$

where we have used an expansion formula in terms of Legendre polynomials $P_\ell$

$$\sum_{m=-\ell}^{\ell} Y_{\ell,m}(\mathbf{n}) Y^*_{\ell,m}(\mathbf{n}') = \frac{(2\ell+1)}{4\pi} P_\ell(\mathbf{n}\cdot\mathbf{n}'). \tag{16.3.10}$$

In particular for $\mathbf{n}' \equiv \mathbf{n}$, $P_\ell(1) = 1$, and we obtain the standard deviation of the statistic $(T(\mathbf{n}) - <T>)/<T>$:

$$\sqrt{\Big\langle \Big( \frac{T(\mathbf{n}) - <T>}{<T>} \Big)^2 \Big\rangle} = \sqrt{\sum_{\ell=0}^{\infty} C^{TT}_\ell \frac{(2\ell+1)}{4\pi}} \approx 10^{-5}, \tag{16.3.11}$$

indicating, in the process the rapid vanishing of the $C^{TT}_\ell$ with $\ell$ large, clearly observed experimentally.[32]

Under parity (e.g., left-right) transformation,

$$Y_{\ell,m}(\mathbf{n}) \to Y_{\ell,m}(-\mathbf{n}) = (-1)^\ell Y^m_\ell(\mathbf{n}), \tag{16.3.12}$$

[31] The associated Legendre polynomials and the Legendre polynomials have been also introduced in Chap. 13, Eqs. (13.5.18)–(13.5.20).

[32] Navas et al. (PDG) (2024) [34].

as directly follows from the definition of $Y_{\ell,m}(\theta, \phi)$ in Eq. (16.3.3), with the replacement $\theta \to \pi - \theta$, $\phi \to \phi + \pi$, in it under a parity transformation.

Isotropy of temperature then implies from Eq. (16.3.1), that under a parity transformation

$$a^T_{\ell,m} \ \to \ (-1)^\ell a^T_{\ell,m}. \tag{16.3.13}$$

The following reality conditions should be also noted:

$$Y^*_{\ell,m} \ = \ Y_{\ell,-m}, \ \text{ implies that } \ a^{T*}_{\ell,m} \ = \ a^T_{\ell,-m}, \ \ (C^{TT}_\ell)^* \ = \ C^{TT}_\ell. \tag{16.3.14}$$

The cosmic radiation is, however, not completely smooth and uniform, involving a small anisotropy in its temperature which results in a characteristic "grainy" pattern that varies with angular scale.[33] These temperature fluctuations reflect tiny fluctuations in the matter density. These tiny fluctuations provided seeds which grew into large structures as galaxies and cluster of galaxies formation later in the Universe, as it cooled down further, under the influence of gravitational attraction and, particularly, by the help of dark matter not being hindered by the pressure of the electromagnetic radiation.

We can almost "hear" a sound coming to an end from the remote past of the Universe. As if the Universe has stopped crying to announce that charged particles found their counterparts to form neutral atoms and finally photons where free. The sound was generated by density fluctuations from the inflation epoch as acoustic waves were produced, as a result of compression caused by the gravitational inward pressure on the cosmic fluid, consisting of photons coupled to electrons and protons, and the outward pressure set by radiation and the fluid to oppose such compressions with inherited periodic density fluctuations. These compressions followed by rarefaction of the cosmic fluid thus generating acoustic waves which were eventually brought to a halt when neutral atoms were predominantly formed, in an epoch referred to as the *recombination epoch*, and photons decoupled.

These acoustic waves have left their imprint at recombination, when neutral atoms formed, and their evidence is detected by the variations of observed densities reflecting slight temperature variations about a mean temperature.

Power spectrum is involved in studying the image of the Universe of about 380,000 years after the Big Bang. We consider sound waves that may have traveled, way before recombination, up to the decoupling time when the photon-baryon fluid stops oscillation and the photons were decoupled. The variation of the opposing pressure by photons with the corresponding variation of the mass density upon rarefaction of the cosmic fluid defines approximately the speed (squared) at which sounds propagate.[34]

That is, from the relation between pressure and the energy density $\rho$: $P = \rho c^2/3$, the (approximate) *speed of sound* $c_s$, in the photon baryon fluid, is easily worked out to be $c_s^2 = dP/d\rho$, or $c_s \approx c/\sqrt{3}$.

---

[33] Wright (2004) [49].

[34] Hu and White (1996) [18].

The distance that sound travels from the Big Bang to recombination is referred to as the *sound horizon* $\kappa_{\text{sh}}$. From Eq. (16.1.18), we may approximately take for the latter

$$\kappa_{\text{sh}} = \frac{\mathrm{c}}{\sqrt{3}} R_{\text{dec}} \int_0^{t_{\text{dec}}} \frac{\mathrm{d}t'}{R(t')}. \tag{16.3.15}$$

This is more conveniently rewritten in the form of Eq. (16.1.19), which for a flat Universe and the fact that for $t' < t_{\text{dec}}$, i.e., for $x < 1/1101$, we may neglect $x^4\Omega_\Lambda$, given in Eq. (16.1.13), considering the range of $x$ values just given, remembering from Eq. (16.2.10) $x_{\text{dec}} \simeq 1/1101$, leading to

$$\kappa_{\text{sh}} = \frac{\mathrm{c}}{\sqrt{3}H_0} \frac{R_{\text{dec}}}{R_0} \int_0^{1/1101} \frac{\mathrm{d}x}{[\Omega_{0\text{matt}}x + \Omega_{0\text{rad}}]^{1/2}} = 0.03612 \frac{\mathrm{c}}{H_0} \frac{R_{\text{dec}}}{R_0}, \tag{16.3.16}$$

where, $R_{\text{dec}}/R_0 \approx 1/1101, \Omega_{0\text{matt}} = 0.315$, $\Omega_{0\text{rad}} \simeq 9.2 \times 10^{-5}$ as given in Eq. (16.1.14).

The CMB radiation from the sky we measure today comes from a surface referred to as the *surface of last scattering* defining a region in space at which the decoupling event is estimated to have occurred.[35] The cosmic background radiation is coming at us from our right and left and from every direction in a symmetrical way. That is, the last scattering surface may be considered as a sphere with us, i.e., our galaxy, situated at the center.

Now we define the important distance used in cosmology, referred to as the angular distance, $D_{\text{ANG}}$. It is commonly, but not appropriately, referred to as the angular diameter distance.

To this end, in flat space, a source of small extension $\kappa$ at a large distance $D$ subtends an angle at the origin:

$\theta = \kappa/D$

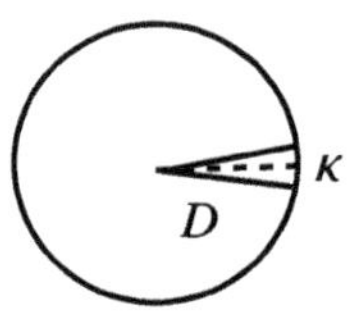


Surface of Last Scattering

conveniently depicted for additional clarity.

The circumference of a circle with a radial coordinate $r_{\text{o}}$ is, at a time $t$, given by $2\pi R(t) r_{\text{o}}$, where $t$ denotes the time at which light was emitted from the source. Hence one defines

$$\theta = \frac{\kappa}{R(t)\, r_{\text{o}}} = \frac{\kappa}{R_0\, r_{\text{o}}}(1+z), \quad \left(\frac{R_0}{R(t)} = (1+z)\right). \tag{16.3.17}$$

Upon comparing this equation with $\theta = \kappa/D$, one defines the angular distance by

[35] Nobel Laureates Mather and Smoot (2006) [30].

$$D_{\mathrm{ANG}} = \frac{R_0\, r_0}{1+z}. \tag{16.3.18}$$

We note that now $t > t_{\mathrm{dec}}$, we may neglect $\Omega_{\mathrm{rad}}$, but, on the other hand, $\Omega_\Lambda$ is a contributing factor, considering their magnitudes given in Eq. (16.1.14), and the range of $x$ values given below. The following expression then emerges for $D_{\mathrm{ANG}}$,

$$D_{\mathrm{ANG}} = \frac{\mathrm{c}}{H_0}\frac{R_{\mathrm{dec}}}{R_0}\int_{1/1101}^{1} \frac{\mathrm{d}x}{[\Omega_{0\mathrm{matt}}x + \Omega_\Lambda x^4]^{1/2}} = 3.1328\,\frac{\mathrm{c}}{H_0}\frac{R_{\mathrm{dec}}}{R_0}. \tag{16.3.19}$$

From Eqs. (16.3.16), (16.3.19), the *angular size of the sound horizon* is obtained to be,[36]

$$\theta_{\mathrm{sh}} = \frac{\kappa_{\mathrm{sh}}}{D_{\mathrm{ANG}}} = \frac{0.03612}{3.1328} = 0.01153\,\mathrm{rad} \simeq 0.66^\circ, \tag{16.3.20}$$

in conformity with the Planck Collaboration experimental fit to flat ΛCDM cosmology and of the tabulated result[37] which is given by 0.010410 rad. [Here CDM stands for cold dark matter.]

As we approach the recombination era from the remote past, the sound horizon becomes bigger from those of earlier times. The mode with the longest wavelength had time to compress to a maximum compression and stayed that way, when photons were about to decouple and gravity took over in a matter dominated Universe. Just at the moment of recombination, the scale of density of inhomogeneities is given by the sound horizon, whose subtended angle at Earth has been derived above. Maximum compression of a cosmic fluid corresponds to this large acoustic peak. The corresponding mode is referred to as the fundamental mode and is depicted in Fig. 16.2, as well as other lower acoustic peaks at even smaller angles.

To the experimentalists,[38] such peak values, shown for other angles as well, provide important information about the physical properties of the early universe, and the references just given provide useful information for studying their underlying intricacies for the interested reader.

---

[36] The angular size of the sound horizon was also computed in my previous work Manoukian (2020) [28], p. 532, Eq. (91.9).

[37] Plank Collaboration (2016, 2020) [39, 40]; Navas et al. (PDG) (2024) [34].

[38] Aghanim (2020) [1]; Planck Collaboration (2020, 2016) [39, 40]; Bennett et al. (2013) [2]; Kuo et al. [22]; Jones et al. (2006) [19]; Melchiorri et al. (2000) [32]; de Bernardis et al. (2000) [3]; Miller (1999) [33].

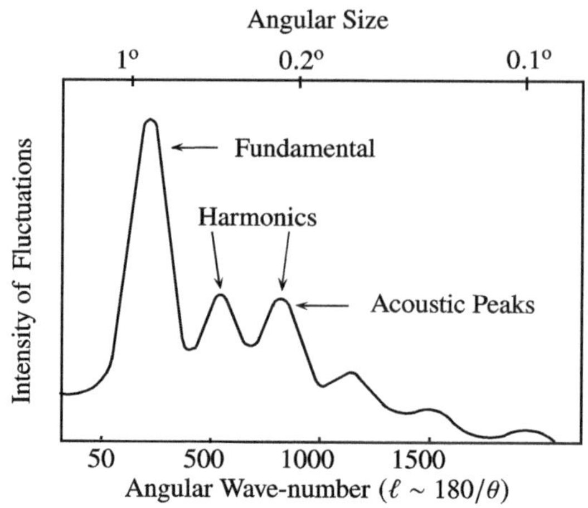


**Fig. 16.2** The figure, referred to as a the **angular power spectrum of temperature fluctuations of the CMB radiation**, formally depicts the acoustic peaks, corresponding to variations of observed densities and the slight temperature variations, with angular size $\theta$ and angular wave-number $\ell$. The first peak corresponds to a mode which has *compressed* once and with a maximum compression just before recombination giving rise to the highest peak. The first harmonic corresponds to a mode which entered the sound horizon earlier and had time to *decompress*, after a compression, resulting to a lower peak. The second harmonic had time to compress for a second time after rarefaction. Experimentalists represent the vertical axis by $\ell(\ell+1)C_\ell^{TT}/2\pi$. The peaks, including the lower peaks corresponding to even smaller angles, have been useful for the experimentalists to obtain information about the Universe and of the underlying parameters of the theory

The angular power spectrum has been precisely measured,[39] and is tabulated[40] as mentioned above. The measurement of the angle $\theta_{\rm sh}$, we have explicitly evaluated, is one of the most precisely determined parameters experimentally in cosmology.

## 16.4 CMB Polarization: Light Coming at Us from Our Right and Left

In the decoupling era, light is also expected to become polarized upon scattering, notably by electrons. We have already seen, e.g., in Chap. 13, Sect. 13.4 how a photon may acquire polarization upon scattering. Scattering of light by electrons is loosely referred to as Thomson scattering. As a matter of fact, it is predicted that the CMB radiation to be polarized at the level of about 5% of the temperature anisotropy.[41] In our study of the polarization aspects of the cosmic background radiation, we will, in the process, look at light coming at us from different directions from the last scatter-

[39] Planck Collaboration et al. (2016) [39].

[40] Navas et al. (PDG) (2024) [34].

[41] Hu and White (1996) [18].

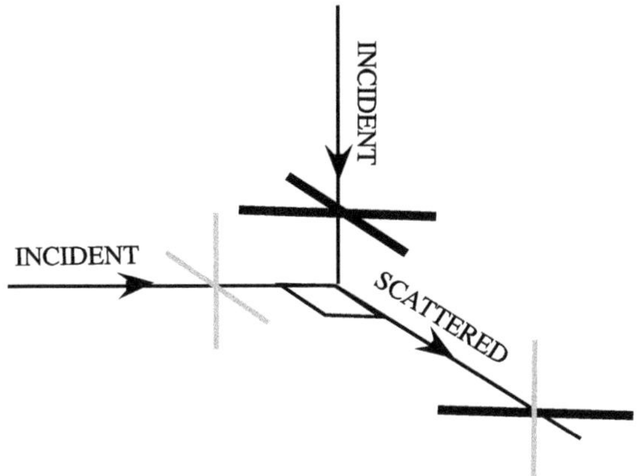


**Fig. 16.3** With incident light being far more intense from the vertical direction in comparison to the horizontal one, the analysis in Sect. 13.4 implies that the scattered light will be (practically) linearly polarized along the horizontal axis in a plane parallel to the plane of incidence

ing spherical surface.

To see how such polarization may develop in the cosmic background radiation upon scattering, consider Fig. 16.3 in which incident light, of significant different intensities, described in terms of two polarization directions, as shown in the figure, are scattered from an electron. According to our analysis in Sect. 13.4, light will be (essentially) observed linearly polarized along the horizontal axis in a plane parallel to the plane of incidence corresponding to the incident light of significantly more intense coming from the vertical axis direction.

Thus the cosmic background photons emitted in a given direction at last scattering to have a net polarisation requires the electrons on the last scattering surface to see an anisotropic distribution of incoming photons. The polarization is coming from Thomson scattering at $z \simeq 1100$ for the red shift parameter as derived in Eq. (16.2.10).

In addition to the polarization coming from Thomson scattering, polarization may arise from gravitational effects, such as gravitational waves as well as other related possibilities that will be mentioned below. There are two so-called modes of polarization, an $E$-mode due to the density fluctuations in the cosmic background radiation at the moment of decoupling, and another one referred to as $B$-mode expected to be due to gravitational effects. Conversion of $E$ modes to $B$ modes are also expected, again, due to the bending of light by gravity referred to as gravitational lensing,[42] which will be briefly discussed later in this section.

The cosmic background radiation is coming at us from our right and left and from every direction in a symmetrical way. That is, the last scattering surface may

[42] A reader who is interested in this aspect of gravitation referred to as gravitational lensing, see the pedagogical introduction in Manoukian (2024) [29], pp. 91–92.

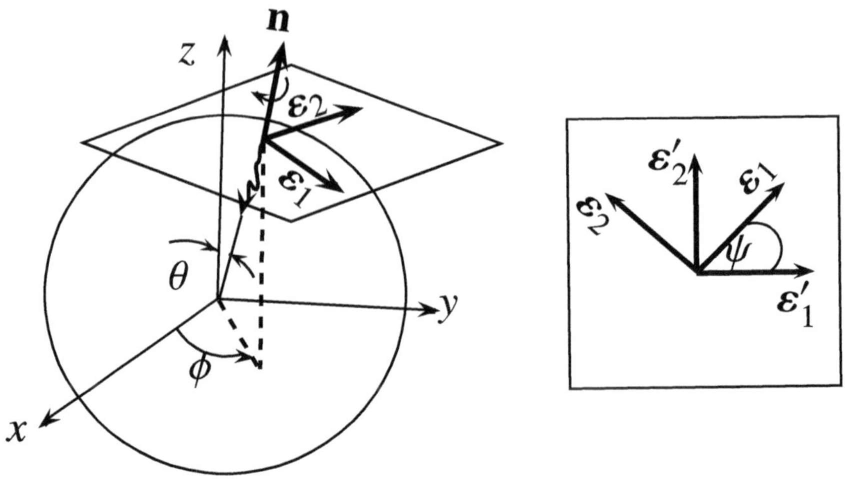


**Fig. 16.4** Consider a sphere, say, of radius $r$ centered at the origin of a given coordinate system. Carry out a c.c.w rotation of the $x, y, z$ coordinate system by angle $\phi$ around the $z-$axis. Then carry out a c.c.w rotation by an angle $\theta$ around the resulting new $y$-axis. The point on the sphere, initially at the North pole, is parametrized by the coordinate label $(r, \theta, \phi)$. One may introduce a tangent plane at the point of contact to the sphere where **n** is a unit vector perpendicular to the tangent plane at that point pointing the direction of the new $z-$axis, with resulting unit vectors $\boldsymbol{\varepsilon}_1$, $\boldsymbol{\varepsilon}_2$ in the tangent plane, arbitrariy identified. We set up an electric field **E** described in terms of polarization vectors $\boldsymbol{\varepsilon}_2, \boldsymbol{\varepsilon}_1$ as shown, where light, denoted by the zizag line, propagates toward the origin. Finally, we carry out a *c.w.* rotation, by an angle $\psi$, around the unit vector **n**, as shown. The angles $(\phi, \theta, -\psi)$ are referred to as *Euler angles* in the $z - y - z$ convention

be considered a sphere with us, i.e., our galaxy situated at the center. Such a sphere is shown in Fig. 16.4, with us situated at the origin, with a given coordinate system specified by $x, y, z$ axes, set up in our galaxy. Following the rotations described in the figure with respect to the angles $\theta$ and $\phi$, generating the unit vector **n**, we carry a c.w. rotation, around it, by an angle $\psi$. The following relations between the unit vectors are readily obtained from the figure.[43]

$$\boldsymbol{\varepsilon}_1 = \boldsymbol{\varepsilon}_1' \cos\psi + \boldsymbol{\varepsilon}_2' \sin\psi, \qquad \boldsymbol{\varepsilon}_2 = -\boldsymbol{\varepsilon}_1' \sin\psi + \boldsymbol{\varepsilon}_2' \cos\psi. \tag{16.4.1}$$

$$\mathbf{E} = E_1\, \boldsymbol{\varepsilon}_1 + E_2\, \boldsymbol{\varepsilon}_2 = E_1'\, \boldsymbol{\varepsilon}_1' + E_2'\, \boldsymbol{\varepsilon}_2'. \tag{16.4.2}$$

The intensity is defined by

$$I = |\mathbf{E}|^2 = |E_1|^2 + |E_2|^2. \tag{16.4.3}$$

[43] Some authors interchange the indices 1 and 2, thus simply interchanging the sign of our $Q$ below.

The so-called *Euler angles* are given here by $(\theta, \phi, -\psi)$. This is described in the Fig. 16.4 caption. The rotations described there are given by

First a c.c.w. rotation is made by an angle $\phi$ around the $z$ -axis.

A c.c.w. rotation is then made by an angle $\theta$ around the new $y$-axis, call it $y_1$.

Finally, a c.w. rotation is made by an angle $\psi$ around the new $z$-axis, along $\mathbf{n}$.

According, to the transformation rules described in Eq. (16.4.1) around the unit vector $\mathbf{n}$, we have directly from Eq. (16.4.2) that,

$$\begin{bmatrix} E_1' \\ E_2' \end{bmatrix} = \begin{bmatrix} \cos\psi & -\sin\psi \\ \sin\psi & \cos\psi \end{bmatrix} \begin{bmatrix} E_1 \\ E_2 \end{bmatrix}. \qquad (16.4.4)$$

The interesting CMB polarization physics may be entirely *described* in terms of two parameters $Q$, $U$, referred to as *Stokes parameters*, defined as follows:

$$Q = |E_1|^2 - |E_2|^2, \qquad U = E_1^* E_2 + E_1 E_2^*. \qquad (16.4.5)$$

The transformation rule of $(E_1, E_2)$ in Eq. (16.4.4), allows us to infer the transformation rule of the couple $(Q, U)$ to be given by

$$\begin{bmatrix} Q' \\ U' \end{bmatrix} = \begin{bmatrix} \cos 2\psi & -\sin 2\psi \\ \sin 2\psi & \cos 2\psi \end{bmatrix} \begin{bmatrix} Q \\ U \end{bmatrix}. \qquad (16.4.6)$$

The above, in particular, means the following transformation rules for the combinations $Q \pm \mathrm{i}U$, with $Q \to Q'$, $U \to U'$, as follow directly from the transformation rule in Eq. (16.4.6) by elementary algebra[44]

$$Q' \pm \mathrm{i}U' = \begin{cases} \mathrm{e}^{\pm \mathrm{i}2\psi}[Q \pm \mathrm{i}U] & \text{for a c.w. rotation by } \psi \\ \mathrm{e}^{\mp \mathrm{i}2\psi}[Q \pm \mathrm{i}U] & \text{for a c.c.w. rotation by } \psi \end{cases}. \qquad (16.4.7)$$

From such transformation rules, we may immediately conclude that for $\psi = \pi/4$:

$$[Q + \mathrm{i}U] \text{ is a spin 2 field and} \begin{cases} \text{a c.c.w rotation turns } (Q, U) \text{ into } (U, -Q) \\ \text{a c.w rotation turns } (Q, U) \text{ into } (-U, Q) \end{cases}. \qquad (16.4.8)$$

This is what one means by the *spin* of a field by its response under a rotation of a coordinate system. We also learn that the polarization described by $U$, in particular, may be *obtained* from the one described by $Q$ by a 45° c.c.w. rotation about the unit vector $\mathbf{n}$.

---

[44] With $(\boldsymbol{\varepsilon}_1, \boldsymbol{\varepsilon}_2, \mathbf{n})$, defining a right-handed system, makes $(\boldsymbol{\varepsilon}_2, \boldsymbol{\varepsilon}_1, -\mathbf{n})$ automatically a right-handed one.

Before going into the mathematical structure of $Q \pm \mathrm{i}U$ in studying further the CMB polarization, we already note that they cannot be expanded in terms of the spherical harmonics $Y_{\ell,m}$, as we have done in studying temperature fluctuations in Eq. (16.3.1), because the spherical harmonics $Y_{\ell,m}$, as we will see later, are of spin 0 fields.

Regarding polarization by Thomson scattering of CMB, we already learn, from the first Eq. (16.4.5), that

$$If\ |E_1|^2 \ll |E_2|^2,\ \text{a scattered light is essentially horizontally polarized,}$$
$$If\ |E_2|^2 \ll |E_1|^2,\ \text{a scattered light is essentially vertically polarized,}$$

horizontally/vertically, in reference to a coordinate system with $\boldsymbol{\varepsilon}_2$ specifying horizontal, and $\boldsymbol{\varepsilon}_1$ specifying vertical.

With the surface of the sphere representing the last scattering with us, at the orgin, we consider such light coming at us from our right and left. For the light coming at us from the right, consider the electric field described in terms of two polarization directions with $\alpha$ pointing out of the page, and $\beta$ perpendicular to both the direction of propagation of light and the $\alpha$-polarization axis:

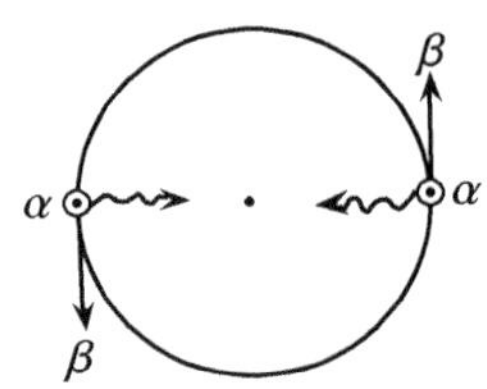


Now slide this combination, c.w. or c.c.w., invoking rotational invariance around our galaxy at the center, along the "circumference" to exactly the opposite end, i.e., to our left, having, necessarily, the direction of light pointing, again, toward us from the latter end. From this, what we learn rather intuitively, that one of the polarizations directions changes sign by going from right $\leftrightarrows$ left.

Accordingly, with the Stokes parameter $Q$, defined in Eq. (16.4.5), being quadratic in the field components, then as we go from right to left, it does not change sign, while the Stokes parameter $U$, defined also in Eq. (16.4.5), being expressed in terms of the product of the field components changes sign. A transformation from right to left, is what is meant by a parity transformation.[45] That is, under a parity transformation,

$$Q \to Q, \qquad U \to -U. \tag{16.4.9}$$

[45] For any point labeled by $(\theta, \phi)$ on the sphere, its reflection point on the sphere is given by $(\pi - \theta, \phi + \pi)$. We can always rotate our coordinate system at the center so that one of these points is on our right and the other one on our left invoking rotational symmetry.

As we will see later, this simple observation has important consequences experimentally. The intensity, of course, remains invariant as experiments also tell us.

A generalization of the spherical harmonics to which carry spin, with the spin spherical harmonics corresponding to the special case of spin 0, have been introduced by E. T. Newman and R. Penrose in 1966.[46] They are referred to as spin-weighted spherical harmonics. They are obtained from the spherical harmonics by application of spin raising and lowering operators.

From such an expansion of $(Q \pm \mathrm{i}U)$ in terms of spin-weighted spherical harmonics, one is able to transform $Q$ and $U$ into the $E$ and $B$ modes, mentioned earlier, *expressed* in terms of the angular wave number $\ell$ as done for temperature fluctuations in Eq. (16.3.1), much relevant to experiments on the CMB radiation and to the history of our Universe.

A spin-$s$ weighted spherical harmonics will be denoted by $Y^{(s)}_{\ell,m}(\mathbf{n})$, where

$$\mathbf{n} = (\sin\theta\,\cos\phi, \sin\theta\,\sin\phi, \cos\theta), \tag{16.4.10}$$

with $Y^{(0)}_{\ell,m}(\mathbf{n})$ denoting the ordinary spherical harmonics $Y_{\ell,m}(\mathbf{n})$.

An operation may be defined to raise the spin as follows

$$\begin{aligned}\eth Y^{(s)}_{\ell,m}(\mathbf{n}) &= \Big(s\cot\theta - \partial_\theta - \frac{\mathrm{i}}{\sin\theta}\partial_\phi\Big)Y^{(s)}_{\ell,m}(\mathbf{n})\\ &= \sqrt{(\ell-s)(\ell+s+1)}Y^{(s+1)}_{\ell,m}(\mathbf{n}),\end{aligned} \tag{16.4.11}$$

Now, according to the definition the right-hand side of the equation is of spin $(s+1)$. That is, if we apply the operation $\eth$ to the above expression one more time, the new multiplicative operator on the right hand side of the equation must be defined in terms of $(s+1)$ instead of $s$, i.e.,

$$\begin{aligned}(\eth)^2\,Y^{(s)}_{\ell,m}(\mathbf{n}) &= \Big((s+1)\cot\theta - \partial_\theta - \frac{\mathrm{i}}{\sin\theta}\partial_\phi\Big)\Big(s\cot\theta - \partial_\theta - \frac{\mathrm{i}}{\sin\theta}\partial_\phi\Big)\,Y^{(s)}_{\ell,m}(\mathbf{n})\\ &= \sqrt{(\ell-s)(\ell+s+1)}\,\Big((s+1)\cot\theta - \partial_\theta + \frac{\mathrm{i}}{\sin\theta}\partial_\phi\Big)\,\eth\,Y^{(s+1)}_{\ell,m}(\mathbf{n})\\ &= \sqrt{(\ell-s)(\ell+s+1)}\,\sqrt{(\ell-s-1)(\ell+s+2)}\,Y^{(s+2)}_{\ell,m}(\mathbf{n}).\end{aligned} \tag{16.4.12}$$

In particular, by choosing $s = 0$, we may infer that

[46] Newman and Penrose (1976) [35]. See also Wu and Yang (1976) [50], Torres del Castillo (2003) [43], Durrer (2021) [5].

$$
\begin{aligned}
&\Big(\cot\theta - \partial_\theta - \frac{\mathrm{i}}{\sin\theta}\partial_\phi\Big)\Big(-\partial_\theta - \frac{\mathrm{i}}{\sin\theta}\partial_\phi\Big)\, Y_{\ell,m}(\mathbf{n}) \\
&= \sqrt{(\ell)(\ell+1)}\,\Big(\cot\theta - \partial_\theta - \frac{\mathrm{i}}{\sin\theta}\partial_\phi\Big)\,\eth\, Y^{(1)}_{\ell,m}(\mathbf{n}) \\
&= \sqrt{(\ell)(\ell+1)}\,\sqrt{(\ell-1)(\ell+2)}\, Y^{(2)}_{\ell,m}(\mathbf{n}).
\end{aligned}
\tag{16.4.13}
$$

That is,

$$
Y^{(2)}_{\ell,m}(\mathbf{n}) = \sqrt{\frac{(\ell-2)!}{(\ell+2)!}}\Big(\cot\theta - \partial_\theta - \frac{\mathrm{i}}{\sin\theta}\partial_\phi\Big)\Big(-\partial_\theta - \frac{\mathrm{i}}{\sin\theta}\partial_\phi\Big)\, Y_{\ell,m}(\mathbf{n}),
\tag{16.4.14}
$$

where we have used

$$
\frac{1}{(\ell+2)(\ell+1)\ell(\ell-1)} = \frac{(\ell-2)!}{(\ell+2)!}.
\tag{16.4.15}
$$

Similarly, the operation to lower the spin of $Y^{(s)}_{\ell,m}$ is defined as follows

$$
\begin{aligned}
&\eth^* Y^{(s)}_{\ell,m}(\mathbf{n}) \\
&= \Big(-s\cot\theta - \partial_\theta + \frac{\mathrm{i}}{\sin\theta}\partial_\phi\Big) Y^{(s)}_{\ell,m}(\mathbf{n}) = -\sqrt{(\ell+s)(\ell-s+1)}\, Y^{(s-1)}_{\ell,m}(\mathbf{n}).
\end{aligned}
\tag{16.4.16}
$$

Another application of the operation to lower the spin again, gives

$$
\begin{aligned}
&(\eth^*)^2\, Y^{(s)}_{\ell,m}(\mathbf{n}) \\
&= \Big(-(s-1)\cot\theta - \partial_\theta + \frac{\mathrm{i}}{\sin\theta}\partial_\phi\Big)\Big(-s\cot\theta - \partial_\theta + \frac{\mathrm{i}}{\sin\theta}\partial_\phi\Big)\, Y^{(s)}_{\ell,m}(\mathbf{n}) \\
&= -\sqrt{(\ell+s)(\ell-s+1)}\,\Big(-(s-1) + \cot\theta - \partial_\theta + \frac{\mathrm{i}}{\sin\theta}\partial_\phi\Big)\, Y^{(s-1)}_{\ell,m}(\mathbf{n}) \\
&= \sqrt{(\ell+s)(\ell-s+1)}\,\sqrt{(\ell+s-1)(\ell-s+2)}\, Y^{(s-2)}_{\ell,m}(\mathbf{n}).
\end{aligned}
\tag{16.4.17}
$$

In particular by setting $s = 0$, the above equation gives,

$$
\begin{aligned}
&\Big(\cot\theta - \partial_\theta + \frac{\mathrm{i}}{\sin\theta}\partial_\phi\Big)\Big(-\partial_\theta + \frac{\mathrm{i}}{\sin\theta}\partial_\phi\Big)\, Y_{\ell,m}(\mathbf{n}) \\
&= \sqrt{(\ell)(\ell+1)}\,\sqrt{(\ell-1)(\ell+2)}\, Y^{-2}_{\ell,m}(\mathbf{n}),
\end{aligned}
\tag{16.4.18}
$$

or,

$$Y^{(-2)}_{\ell,m}(\mathbf{n}) = \sqrt{\frac{(\ell-2)!}{(\ell+2)!}}\Big(\cot\theta - \partial_\theta + \frac{\mathrm{i}}{\sin\theta}\partial_\phi\Big)\Big(-\partial_\theta + \frac{\mathrm{i}}{\sin\theta}\partial_\phi\Big)\, Y_{\ell,m}(\mathbf{n}). \tag{16.4.19}$$

To obtain the expression of $Y^{(\pm 2)}_{\ell,m}$, we may use the definition of the spherical harmonics, defined in Eq. (16.3.3)–(16.3.6), where in our convention, $Y^*_{\ell,m}(\theta,\phi) = Y_{\ell,-m}(\theta,\phi)$, as given in Eq. (16.3.7).

The spherical harmonics satisfy the following partial differential equation which is useful in relating various derivatives of it,

$$Y_{\ell,m} = \Theta_{\ell,m}(\theta)\mathrm{e}^{\mathrm{i}m\phi}, \text{ where} \tag{16.4.20}$$

$$\Big\{\frac{\mathrm{d}^2}{\mathrm{d}\theta^2} + \cot\theta\frac{\mathrm{d}}{\mathrm{d}\theta} + \Big[\ell(\ell+1) - \frac{m^2}{\sin^2\theta}\Big]\Big\}\Theta_{\ell,m}(\theta) = 0. \tag{16.4.21}$$

We note that in our convention, in particular, the following two properties of $Y_{\ell,m}(\mathbf{n})$,

$$Y^*_{\ell,m}(\mathbf{n}) = Y_{\ell,-m}(\mathbf{n}), \qquad Y_{\ell,m}(-\mathbf{n}) = (-1)^\ell Y_{\ell,m}(\mathbf{n}), \tag{16.4.22}$$

where $\mathbf{n}$ is given in Eq. (16.4.10), $\mathbf{n} \to -\mathbf{n}$ defining the parity (space reflection) transformation, with the angles $(\theta,\phi)$ transforming as

$$\theta \to \pi - \theta, \qquad \phi \to \phi + \pi. \tag{16.4.23}$$

Finally, the orthogonality property of the spherical harmonics should be noted

$$\int \mathrm{d}\Omega\; Y_{\ell,m}(\mathbf{n})\, Y^*_{\ell',m'}(\mathbf{n}) = \delta_{\ell'\ell}\,\delta_{m'm}, \qquad \mathrm{d}\Omega = \sin\theta\,\mathrm{d}\theta\,\mathrm{d}\phi. \tag{16.4.24}$$

From the derived expressions of the spin 2 weighted spherical harmonics $Y^{(\pm 2)}_{\ell,m}(\mathbf{n})$ in Eqs. (16.4.14), (16.4.19) in terms of $Y_{\ell,m}(\mathbf{n})$, and from a straightforward, though tedious, evaluation involving up to two derivatives with respect to $\theta$, the following useful expressions emerge for them

$$Y^{(\pm 2)}_{\ell,m}(\mathbf{n})) = \Big[\frac{(\ell-2)!}{(\ell+2)!}\Big]^{1/2}\big[a^{\pm}_{\ell m}(\theta)Y^m_\ell(\mathbf{n}) + b^{\pm}_{\ell m}(\theta)Y^m_{\ell-1}(\mathbf{n})\big], \tag{16.4.25}$$

$$a^{\pm}_{\ell m}(\theta) = \frac{2m^2 - \ell(\ell+1)}{\sin^2\theta} \mp 2m(\ell-1)\frac{\cot\theta}{\sin\theta} + \ell(\ell-1)\cot^2\theta, \tag{16.4.26}$$

$$b^{\pm}_{\ell m}(\theta) = 2\Big[\frac{(2\ell+1)}{(2\ell-1)}(\ell^2 - m^2)\Big]^{1/2}\Big(\pm\frac{m}{\sin^2\theta} + \frac{\cot\theta}{\sin\theta}\Big), \tag{16.4.27}$$

where $\ell \geq 2, -\ell \leq m \leq \ell$. The important step is to show that under rotation $Y^{(2)}_{\ell,m}(\mathbf{n})$, responds as a spin 2 object, and similarly for its complex conjugate. This step, which is, surprisingly, most often not carried out in the literature, is given in the Appendix in Sect. 16.5.

These functions satisfy the same orthogonality condition as the spherical harmonics in Eq. (16.4.24),

$$\int \mathrm{d}\Omega\ Y^{(\pm 2)}_{\ell,m}(\mathbf{n})\,(Y^{(\pm 2)}_{\ell',m'}(\mathbf{n}))^* \ =\ \delta_{\ell'\ell}\,\delta_{m'm}. \qquad (16.4.28)$$

From the reality condition of $Y_{\ell,m}$ in Eq. (16.3.7), we may refer from Eqs. (16.4.25)–(16.4.27), the following condition for $(Y^{(\pm 2)}_{\ell,m})^* \ =\ Y^{(\mp 2)}_{\ell,-m}$. Also under parity transformation, $\theta \to \pi - \theta,\ \phi \to \phi + \pi$, we know from Eq. (16.3.12) that $Y_{\ell,m}(\mathbf{n}) \to (-1)^\ell\, Y_{\ell,m}(\mathbf{n}),\ Y_{\ell-1,m}(\mathbf{n}) \to (-1)^{(\ell-1)}\, Y_{\ell-1,m}(\mathbf{n})$. Also, from Eqs. (16.4.26), (16.4.27), $a^{\pm}_{\ell,m} \to a^{\mp}_{\ell,m}$, and $b^{\pm}_{\ell,m} \to -b^{\mp}_{\ell,m}$, as a consequence of the facts that $\sin\theta \to \sin\theta$, and $\cos\theta \to -\cos\theta$.

Accordingly, we may infer that

$$\big(Y^{(\pm 2)}_{\ell,m}\big)^* \ =\ Y^{(\mp 2)}_{\ell,-m}, \quad \text{and under parity transformation}\ \ Y^{(\pm 2)}_{\ell,m} \ \to\ (-1)^\ell Y^{(\mp 2)}_{\ell,m}. \qquad (16.4.29)$$

Since the intensity $I$ in Eq. (16.4.3) is invariant under both transformations above, we may scale the stokes parameters $Q$ and $U$ in Eq. (16.4.5) as

$$\mathbb{Q} = \frac{Q}{I}, \qquad \mathbb{U} = \frac{U}{I} \qquad (16.4.30)$$

and by recalling the transformation under a c.w. rotation in Eq. (16.4.7), we may readily expand the spin-2 functions $\mathbb{Q} \pm \mathrm{i}\mathbb{U}$ in terms of the 2-weighted spherical harmonics:

$$(\mathbb{Q} + \mathrm{i}\,\mathbb{U})(\theta,\phi) \ =\ \sum_{\ell=2}^{\infty}\sum_{m=-\ell}^{\ell} c_{\ell,m}\, Y^{(2)}_{\ell,m}(\theta,\phi), \qquad (16.4.31)$$

$$(\mathbb{Q} - \mathrm{i}\,\mathbb{U})(\theta,\phi) \ =\ \sum_{\ell=2}^{\infty}\sum_{m=-\ell}^{\ell} c^*_{\ell,m}\, Y^{(-2)}_{\ell,-m}(\theta,\phi), \qquad \big(Y^{(2)}_{\ell,m}\big)^* = Y^{(-2)}_{\ell,-m} \qquad (16.4.32)$$

$$=\ \sum_{\ell=2}^{\infty}\sum_{m=-\ell}^{\ell} c^*_{\ell,-m}\, Y^{(-2)}_{\ell,m}(\theta,\phi), \qquad (16.4.33)$$

where the last equality follows by using the fact that one is summing over both signs of $m$.

Using the orthogonality relation of the spin weighted spherical harmonics in Eq. (16.4.28), the coefficients $c_{\ell,m}$ are then given by

$$c_{\ell,m} = \int \mathrm{d}\Omega\, (\mathbb{Q} + \mathrm{i}\,\mathbb{U})(\theta, \phi)\, (Y^{(2)}_{\ell,m}(\theta, \phi))^*, \tag{16.4.34}$$

$$c^*_{\ell,-m} = \int \mathrm{d}\Omega\, (\mathbb{Q} - \mathrm{i}\,\mathbb{U})(\theta, \phi)\, (Y^{(-2)}_{\ell,m}(\theta, \phi))^*. \tag{16.4.35}$$

Under a parity transformation, $(Q + \mathrm{i}U)' = (Q - \mathrm{i}U)$, as given in Eq. (16.4.9), $I' = I$, being quadratic in the component fields, and $(Y^{(2)}_{\ell,m})' = (-1)^\ell Y^{(-2)}_{\ell,m}$, given in Eq. (16.4.29). Accordingly, under a parity transformation

$$c_{\ell,m} \to (-1)^\ell \int \mathrm{d}\Omega\, (Q - \mathrm{i}U)\big(Y^{(-2)}_{\ell,m}\big)^* \equiv (-1)^\ell\, c^*_{\ell,-m}, \tag{16.4.36}$$

using Eq. (16.4.35).

Therefore, we may define the coefficients

$$\begin{aligned} a^E_{\ell,m} &= \frac{1}{2}\big(c_{\ell,m} + c^*_{\ell,-m}\big), \qquad \big(a^E_{\ell,m}\big)^* = a^E_{\ell,-m}, \\ a^B_{\ell,m} &= \frac{\mathrm{i}}{2}\big(c^*_{\ell,-m} - c_{\ell,m}\big), \qquad \big(a^B_{\ell,m}\big)^* = a^B_{\ell,-m}, \end{aligned} \tag{16.4.37}$$

which, according Eq. (16.4.36), have the following transformation rules under a parity transformation:

$$a^E_{\ell,m} \to (-1)^\ell\, a^E_{\ell,m}, \qquad a^B_{\ell,m} \to (-1)^{(\ell+1)}\, a^B_{\ell,m}. \tag{16.4.38}$$

This allows us to equivalently rewrite Eq. (16.4.31) as ,

$$(\mathbb{Q} + \mathrm{i}\,\mathbb{U})(\theta, \phi) = \sum_{\ell=2}^{\infty} \sum_{m=-\ell}^{\ell} (a^E_{\ell,m} + \mathrm{i}\, a^B_{\ell,m})\, Y^{(2)}_{\ell,m}(\theta, \phi). \tag{16.4.39}$$

In the Appendix Sect. 16.5, in Eq. (16.5.21) and below it, we establish the important property of the spin 2 behavior of $Y^2_{\ell,m}$, but now for a c.c.w. rotation by the angle $\psi$, that

$$Y^2_{\ell,m} \to \mathrm{e}^{-\mathrm{i}2\psi} Y^2_{\ell,m}, \quad \text{and for } \psi = \pi/4, \tag{16.4.40}$$

$$Y^2_{\ell,m} \to -\mathrm{i}\, Y^2_{\ell,m}. \tag{16.4.41}$$

While from Eq. (16.4.8), again for a c.c.w. rotation by $\psi = \pi/4$, which is equally true for $(\mathbb{Q} + \mathrm{i}\,\mathbb{U})$, that

$$(\mathbb{Q} + \mathrm{i}\,\mathbb{U}) \to (\mathbb{U} - \mathrm{i}\,\mathbb{Q}), \tag{16.4.42}$$

to be used on the left-hand side of Eq. (16.4.39).

That is, under a c.c.w. rotation, Eq. (16.4.39) becomes

$$(\mathbb{U}-\mathrm{i}\mathbb{Q})(\theta,\phi) = \sum_{\ell=2}^{\infty}\sum_{m=-\ell}^{\ell}(a^B_{\ell,m}-\mathrm{i}\,a^E_{\ell,m})\,Y^{(2)}_{\ell,m}(\theta,\phi), \tag{16.4.43}$$

which brings us back to Eq. (16.4.39) establishing, as expected, consistency of the underlying transformation properties, and showing, in the process, in particular, transformation of $E$ modes to $B$ modes.

To consider the pairwise correlations between temperature fluctuations and the $E$ and $B$ polarization modes, we first record right here the appropriate properties of the relevant coefficients $a^T_\ell$, $(a^E_\ell,\ a^B_\ell)$, derived in Eqs. (16.3.13)/(16.3.14), Eqs. (16.4.37)/(16.4.38), respectively. They are given by

$$(a^T_{\ell,m})^* = a^T_{\ell,-m},\quad (a^E_{\ell,m})^* = a^E_{\ell,-m}\quad (a^B_{\ell,m})^* = a^B_{\ell,-m}, \tag{16.4.44}$$

$$a^T_{\ell,m} \to (-1)^\ell\, a^T_{\ell,m},\quad a^E_{\ell,m} \to (-1)^\ell\, a^E_{\ell,m},\quad a^B_{\ell,m} \to (-1)^{\ell+1}\, a^B_{\ell,m}, \tag{16.4.45}$$

corresponding, respectively, to their reality conditions, and their parity transformations.

Accordingly, by invoking rotational invariance, the pairwise bilinear correlations between temperature fluctuations, the $E$, and $B$ polarization modes, are given by

$$\big\langle (a^T_{\ell,m})^* a^T_{\ell',m'}\big\rangle = C^{TT}_\ell\,\delta_{\ell,\ell'}\delta_{m,m'},\quad \big\langle (a^T_{\ell,m})^* a^E_{\ell',m'}\big\rangle = C^{TE}_\ell\,\delta_{\ell,\ell'}\delta_{m,m'}, \tag{16.4.46}$$

$$\big\langle (a^E_{\ell,m})^* a^E_{\ell',m'}\big\rangle = C^{EE}_\ell\,\delta_{\ell,\ell'}\delta_{m,m'},\quad \big\langle (a^B_{\ell,m})^* a^B_{\ell',m'}\big\rangle = C^{BB}_\ell\,\delta_{\ell,\ell'}\delta_{m,m'}, \tag{16.4.47}$$

and

$$\big\langle (a^T_{\ell,m})^* a^B_{\ell',m'}\big\rangle = 0,\qquad \big\langle (a^E_{\ell,m})^* a^B_{\ell',m'}\big\rangle = 0. \tag{16.4.48}$$

In particular we learn, that due to the even-odd parity transformations of the coefficient of temperature fluctuations and the $B$ mode polarizations, and the even-odd parity transformations of the coefficient of the $E$ mode and the B mode polarizations, their two point correlations vanish. As a matter of fact any observation of a non-vanishing of the correlation between these pairs will be clear indication of parity violation. The coefficients $C^{TT}_\ell$, already considered in the previous section, $C^{TE}_\ell$, and $C^{EE}_\ell$ are real. This is obvious for $C^{TT}_\ell$, and $C^{EE}_\ell$, while

$$C^{TE}_\ell\,\delta_{\ell,\ell'}\delta_{m,m'} = \big\langle (a^T_{\ell,m})^* a^E_{\ell',m'}\big\rangle = \big\langle a^T_{\ell,-m}(a^E_{\ell',-m'})^*\big\rangle = (C^{TE}_\ell)^*\delta_{\ell,\ell'}\delta_{m,m'}. \tag{16.4.49}$$

Measurement of pairs of correlations between temperature fluctuations and polarization are recorded in terms the coefficients $C^{TT}_\ell$, $C^{TE}_\ell$, $C^{EE}_\ell$, $C^{BB}_\ell$. From an expansion of $(Q \pm \mathrm{i}U)$ in terms of spin-weighted spherical harmonics, one was able to transform $Q$ and $U$ into the $E$ and $B$ modes, mentioned earlier, *expressed* in terms of the angular wave number $\ell$ as done for temperature fluctuations in Eq. (16.3.1), much relevant to experiments on the CMB radiation and to the history of our Universe. The dependence being solely on $\ell$, and independent of $m$, arises by invoking rotational invariance. These modes are considered next.

As we have seen earlier, referring to Fig. 16.3, in the beginning of the section, the $E$-mode polarisation arises from Thomson scattering of photons off of the electrons just before the electrons combined with the nuclei, after which the photons became merely spectators on their way to us. Moreover, in order that these photons emitted at last scattering to have a net polarisation required the electrons on the last scattering surface to see an anisotropic distribution of incoming photons.[47]

In a time evolution process, as light passes through the gravitational fields of massive objects of galaxies, they bend its trajectory, through a process referred to as gravitational lensing, distorting its polarization and converting $E$-modes into $B$-modes, in conformity with the transformation of the Stokes parameter $Q$ to the Stokes parameter $U$, described in Eq. (16.4.8).[48] *Thus, quite importantly, this provides information about the distribution of matter in the universe.*

In gravitational lensing, simply, the deflection of light in the gravitational field a massive object, may, e.g., cause a source emitting the light to have several images upon deflection as well as lead to other interesting phenomena.[49]

Other gravitational effects may also generate $B$-modes. For example, the enormous expansion, within a fraction of a second, of the Universe at inflation, would have produced gravitational waves as ripples in spacetime, with their inherited tensorial character,[50] may have also generated $B$-modes again by twisting the polarization of the photons. This, however, has not yet been detected, as such. Their production would have given evidence for the existence of gravitational waves produced during inflation, and, if detected, may provide additional information on this early stage of the Universe. Of course, gravitational waves have now been independently detected,[51] about 100 years after Einstein's prediction.[52]

From an expansion of $(Q \pm \mathrm{i}U)$ in terms of spin-weighted spherical harmonics, one is able to transform $Q$ and $U$ into the $E$ and $B$ modes, mentioned earlier, *expressed* in terms of the angular wave number $\ell$ as done for temperature fluctuations in Eq. (16.3.1), much relevant to experiments on the CMB radiation and to the history of our Universe.

As bilinear correlation of temperature fluctuations $C_\ell^{TT}$ in Eq. (16.3.8), we have the generated the bilinear correlations involving the $E$ and $B$ modes of polarization, $C_\ell^{TE}$, $C_\ell^{EE}$, $C_\ell^{BB}$ in Eqs. (16.4.42), (16.4.43). We have also seen that by invoking

---

[47] The $E$-mode was successfully detected in (2002) by the DASI (Degree Angular Scale Interferometer) Team: Kovac et al. [21].

[48] For the detection of the expected conversion of $E$-modes to $B$-modes, by gravitational lensing from SPT (South Pole Telescope) see: Hanson et al. (2013) [17].

[49] A reader who is interested in this aspect of gravitation, referred to as gravitational lensing, see the pedagogical introduction in Manoukian (2024) [29], pp. 91–92.

[50] For a precise treatment of the tensorial nature of the polarization of gravitation and the 45° relative orientations of planes of polarizations sates of gravitational waves, see Manoukian (2016b) [27].

[51] See, e.g., Castelvecchi and Witze (2016) [4], and pertinent references to the original discoverers within.

[52] Einstein (2016) [8].

parity transformations, that no such bilinear correlations exist between temperature and the $B$-mode, and no correlation exists between the two modes $E$ and $B$. An observation of a non-vanishing of the correlation between the pairs $(TB)$, $(EB)$, will be clear indication of parity violation.

As mentioned earlier, the first measurement of a polarization signal was in 2002[53] coming from Thomson scattering at $z \simeq 1100$. While the conversion of $E$ mode to the $B$ mode by gravitational lensing was detected in 2013.[54]

The correlations factors $\ell(\ell+1)C_\ell^{TE}/2\pi$ and $\ell(\ell+1)C_\ell^{EE}/2\pi$ exhibit oscillations as functions of $\ell$ in a similar way as for the temperature fluctuations $\ell(\ell+1)C_\ell^{TT}/2\pi$ given in Fig. 16.2 in the previous section, but with much lower amplitudes, where remember that the angles $\theta$ at the peaks of such oscillations correspond to angular wave-numbers $\ell$ given roughly by $\ell \sim 180/\theta$. That is, in particular, polarizations anisotropies are relatively small and that is why it took longer to detect. For the reader interested in the experimental results, the relevant graphs have been clearly tabulated by the "Planck Collaboration".[55] The expected amplitudes of $C_\ell^{BB}$ are very small, and their studies are more challenging.

## 16.5 Appendix: Spin Content of the Spin-Weighted Spherical Harmonics

Undoubtedly, the reader has been impatiently waiting to see that the spin associated with the spin weighted spherical harmonics $Y_{\ell,m}^{(2)}$ given in Eqs. (16.4.14), (16.4.25) is actually 2, and similarly for its complex conjugate $Y_{\ell,-m}^{(-2)}$. That is, under a rotation, as described in Fig. 16.4, they respond in the say way as $Q \pm iU$ in Eq. (16.4.7) for $s = \pm 2$. Unfortunately, and rather surprisingly, this step is often not carried out in many presentations of spin-weighted spherical harmonics in the literature.

We use some basic properties of the theory of angular momentum in quantum mechanics to establish this essential property mentioned above.

In Fig. 16.4, bellow Eq. (16.4.3), we have defined how the angles $(\phi, \theta, -\psi)$, referred to as *Euler angles*, are defined in the $z - y - z$ convention.

They were spelled out in the caption of the figure, as well as bellow Eq. (16.4.3). In more details, the rotations are given in three steps as follows:

1. A c.c.w rotation around the $z-$axis by an angle $\phi$. The $x-$axis then changes, say, to $x_1$, the $y-$axis becomes, say, $y_1$, while the $z-$axis remains the same.

2. A c.c.w. rotation around the $y_1-$axis by an angle $\theta$. The $x_1-$axis becomes, say, $x_2$, the $y_1-$axis doesn't change, while the $z-$axis becomes, say, $z_1$.

3. A c.w. rotation around the $z_1-$axis by angle $\psi$. The $x_2-$axis becomes, say, $x_3$, the $y_1$ one becomes, say, $y_2$, while the $z_1$ doesn't change.

The resulting angles $(\theta, \phi, -\psi)$ are the Euler angles in the $z - y - z$ convention.

---

[53] Kovac et al. [21].

[54] Hanson et al. (2013) [17].

[55] Planck Collaboration et al. (2020) [40].

Denote by $R_z(\phi)$, e.g., the operation for carrying out a c.c.w. rotation around the $z-$axis by an angle $\phi$. That is, the overall rotation $\mathcal{R}$ operation, involving the three rotations, corresponding to the *appropriate* order just defined, is given by:

$$\mathcal{R} = R_{z_1}(-\psi)\, R_{y_1}(\theta)\, R_z(\phi), \tag{16.5.1}$$

defining the rotation in the $z - y - z$ convention.

Here note that a c.w. rotation around the $z_1-$axis is the same, analytically, as a c.c.w rotation around the $z_1-$axis by $-\psi$, in the tangent plane in Fig. 16.4, and hence the minus sign in $R_{z_1}(-\psi)$. Note also that, with, e.g., $R_z^{-1}(\phi)$ denoting the inverse of the rotation around $z-$axis, i.e., a c.w. rotation, by an angle $\phi$, we have

$$R_z^{-1}(\phi)\, R_z(\phi) = I, \tag{16.5.2}$$

where $I$ denotes the identity operation.

By inspection, from Fig. 16.4,

$$R_{y_1}(\theta) = R_z(\phi) R_y(\theta) R_z^{-1}(\phi), \tag{16.5.3}$$

$$R_{z_1}(-\psi) = R_{y_1}(\theta) R_z(-\psi) R_{y_1}^{-1}(\theta). \tag{16.5.4}$$

That is,

$$\begin{aligned} R_{z_1}(-\psi) R_{y_1}(\theta) &= R_{y_1}(\theta) R_z(-\psi) R_{y_1}^{-1}(\theta)] R_{y_1}(\theta) \\ &= [R_z(\phi) R_y(\theta) R_z^{-1}(\phi)] R_z(-\psi). \end{aligned} \tag{16.5.5}$$

Using the obvious fact of the commutativity of a rotation about the same axis,

$$R_z(-\psi) R_z(\phi) = R_z(\phi) R_z(-\psi), \tag{16.5.6}$$

the overall rotation $\mathcal{R}$, given in Eq. (16.5.1), simplifies to an expression involving rotations around the *original* axes[56]:

$$\begin{aligned} \mathcal{R} &= R_{z_1}(-\psi) R_{y_1}(\theta) R_z(\phi) \\ &= [R_z(\phi) R_y(\theta) R_z(-\psi)]. \end{aligned} \tag{16.5.7}$$

The operation $R_z(\phi)$ operating on states $|\ell, m>$ is given by the well known expression

$$R_z(\phi) = \mathrm{e}^{-\mathrm{i}\phi L_z}, \tag{16.5.8}$$

and hence

---

[56] This derivation going from the expression of the rotation defined in Eq. (16.5.1) to the final expression given in Eq. (16.5.7) is, surprisingly, not carried out in many textbooks.

$$\mathcal{R} = \mathrm{e}^{-\mathrm{i}\phi L_z}\mathrm{e}^{-\mathrm{i}\theta L_y}\mathrm{e}^{\mathrm{i}\psi L_z}, \tag{16.5.9}$$

where, here, $L_z$, $L_y$ (scaled by $1/\hbar$) are the familiar angular momentum operators you studied in your introductory course in quantum mechanics, satisfying the simple relations, when applied to $|\ell, m>$, given by

$$L_z|\ell, m> = m\,|\ell, m>, \tag{16.5.10}$$

$$L_y|\ell, m> = -\frac{\mathrm{i}}{2}\sqrt{(\ell-m)(\ell+m+1)}|\ell, m+1> + \frac{\mathrm{i}}{2}\sqrt{(\ell+m)(\ell-m+1)}|\ell, m-1>. \tag{16.5.11}$$

Consider, in particular, the matrix element

$$\langle \ell, m|\mathcal{R}|\ell, -s\rangle = \Big\langle \ell, m\Big|\,\mathrm{e}^{-\mathrm{i}\phi L_z}\mathrm{e}^{-\mathrm{i}\theta L_y}\mathrm{e}^{\mathrm{i}\psi L_z}\,\Big|\ell, -s\Big\rangle \equiv D^{\ell}_{m,-s}(\phi, \theta, \psi), \tag{16.5.12}$$

where $[D^{\ell}_{m,m'}(\phi, \theta, -\psi)]$ is referred to as a Wigner $D$- matrix.[57]

From Eq. (16.5.10), involving $L_z$ as a multiplicative operators, we may immediately write

$$\Big\langle \ell, m\Big|\,\mathrm{e}^{-\mathrm{i}\phi L_z}\mathrm{e}^{-\mathrm{i}\theta L_y}\mathrm{e}^{\mathrm{i}\psi L_z}\,\Big|\ell, -s\Big\rangle = \mathrm{e}^{-\mathrm{i}m\phi}\mathrm{e}^{-\mathrm{i}s\psi}\big\langle \ell, m\big|\,\mathrm{e}^{-\mathrm{i}\theta L_y}\,\big|\ell, -s\big\rangle. \tag{16.5.13}$$

This suggests to introduce, in turn, the following matrix element

$$d^{\ell}_{m,-s}(\theta) = \big\langle \ell, m\big|\,\mathrm{e}^{-\mathrm{i}\theta L_y}\,\big|\ell, -s\big\rangle. \tag{16.5.14}$$

With the action of the $L_y$ operator explicitly given in Eq. (16.5.11) on states $|\ell, m>$, the expressions in the above equation, involving the last rotation by the angle $\psi$, gives the following remarkable result, in particular, for $s = 2$

$$\begin{aligned} \mathrm{e}^{-\mathrm{i}m\phi}\, d^{\ell}_{m,-2}(\theta)\,\mathrm{e}^{-\mathrm{i}2\psi} &= \sqrt{\frac{4\pi}{2\ell+1}}\, Y^{(-2)}_{\ell,-m}(\theta, \phi)\,\mathrm{e}^{-\mathrm{i}2\psi} \\ &= \sqrt{\frac{4\pi}{2\ell+1}}\,\big(Y^{(2)}_{\ell,m}\big)^*(\theta, \phi)\,\mathrm{e}^{-\mathrm{i}2\psi}, \end{aligned} \tag{16.5.15}$$

where $(Y^{(-2)}_{\ell,-m})^* = Y^{(2)}_{\ell,m}$, given in Eq. (16.4.29), and $Y^{(2)}_{\ell,m}$ is given in Eq. (16.4.25), and a similar expression for any integer $s$.

For spin $s = 0$, we simply have

$$\mathrm{e}^{-\mathrm{i}m\phi}\, d^{\ell}_{m,0}(\theta) = \sqrt{\frac{4\pi}{2\ell+1}}\, Y^*_{\ell,m}(\theta, \phi). \tag{16.5.16}$$

[57] Wigner (1951) [48].

An expression which is more familiar in the literature.

Upon taking the complex conjugate of Eq. (16.5.15), we obtain,

$$e^{im\phi}\,(d^{\ell}_{m,-2}(\theta))^{*}\,e^{i2\psi} = \sqrt{\frac{4\pi}{2\ell+1}}\,Y^{(2)}_{\ell,m}(\theta,\phi)\,e^{i2\psi}. \tag{16.5.17}$$

For $\psi = 0$, this gives

$$e^{im\phi}\,(d^{\ell}_{m,-2}(\theta))^{*} = \sqrt{\frac{4\pi}{2\ell+1}}\,Y^{(2)}_{\ell,m}(\theta,\phi). \tag{16.5.18}$$

That is, under no rotation by the angle $\psi$,

$$Y^{(2)}_{\ell,m}(\theta,\phi) = \sqrt{\frac{2\ell+1}{4\pi}}\,e^{im\phi}\,(d^{\ell}_{m,-2}(\theta))^{*}. \tag{16.5.19}$$

while under rotation by the angle $\psi$, as described for $(Q + iU)$ in Eq. (16.4.7), Sect. 16.4, we have from Eqs. (16.5.17), (16.5.19),

$$Y^{(2)}_{\ell,m}(\theta,\phi) \to \sqrt{\frac{2\ell+1}{4\pi}}\,e^{im\phi}\,(d^{\ell}_{m,-2}(\theta))^{*}e^{i2\psi}, \tag{16.5.20}$$

that is

$$Y^{(2)}_{\ell,m} \to Y^{(2)}_{\ell,m}\,e^{i2\psi}, \tag{16.5.21}$$

which is the defining equation of a spin 2 object, and where recall that we have considered a c.w. rotation by an angle $\psi$, which, operationally, is equivalent to a c.c.w. rotation by an angle $-\psi$. Therefore, for a c.c.w. by an angle $\psi$, simply replace $\psi$ by $-\psi$ in Eq. (16.5.21) above. A similar analysis is carried out for the complex conjugate of $Y^{(2)}_{\ell,m}$. Note that $d^{\ell}_{m,-s}(\theta)$, defined in Eq. (16.5.14), involves a *finite* number of operations, given through Eq. (16.5.11), as a consequence of the facts that for a given $\ell$, $|s| \leq \ell$, $|m| \leq \ell$.

## References


1. Aghanim, N., et al. (2020). Planck 2018 results.VI. Cosmological parameters. *Astronomy & Astrophysics 641, A6* 1-67.
2. Bennett, C.L. et al. (2013). Nine-year Wilkinson Microwave Anisotropy Probe (WMAP) observations: Final maps and results. *The Astrophysical Journal Supplement Series. 208 (2)*, 20, (54pp).
3. de Bernardis, P. et al. (2000) A flat Universe from high-resolution maps of the Cosmic Microwave Background Radiation. *Nature, 404*, 955-959.

4. Castelvecchi, D, & Witze, A. (2016). Einstein's gravitational waves found at last.*Nature News 2016*, 19361.
5. Durrer, R. (2021). *The cosmmic microwave background*. Cambridge: Cambridge University Press.
6. Einstein, A. (1915). Die feldgleichungen der gravitation. *Preussischen Akademie der Wissenschaften zu Berlin, Sitzungsberichte, 1915*, 844-847.
7. Einstein, A. (1916). Die grundlage der allgemeinen relativitätstheorie. *Annalen der Physik, 49*, 769-822.
8. Einstein, A. (1916). Näherungsweise integration der feldgleichungen der gravitation. *Preussischen Akademie der Wissenschaften (Berlin), Sitzungsberichte*, 688-696.
9. Fixsen, D. J. (2009). The temperature of the Cosmic Microwave Background. *The Astrophysical Journal, 707 (2)*, 916–920.
10. Friedmann, A. (1922). *Über die krümmung des raumes. Zeitschrift für Physik. 10*, 377–386. [English translation in: Friedman, A. (1999). On the curvature of space. *General Relativity and Gravitation, 31*, 1991-2000.]
11. Friedmann, A. (1924).Über die möglichkeit einer welt mit konstanter negativer krümmung des raumes. *Zeitschrift für Physik, 21*, 326–332. [English translation in: Friedmann, A. (1999). On the Possibility of a World with constant negative curvature of space. *General Relativity and Gravitation, 31*, 2001-2008.]
12. Gamow, G. (1947). Expanding Universe and the origin of the elements. *Physical Review, 70*, 572-573.
13. Gamow, G. (1948). The evolution of the Universe. *Nature, 162*, 680-682.
14. Gamow, G. (1956). The physics of the expanding Universe. *Vistas in Astronomy, 2*, 1726-1732.
15. Gawow, G., & Teller, E. (1939). Thee expanding Universe and the origin of the grand nebulae. *Nature, 143*, 116-117.
16. Guth, A. (1981). Inflationary universe: A possible solution to the horizon and flatness problems. *Physical Review, D 23*, 347-356.
17. Hanson, D., et al. (2013). Detection of $B$-mode polarization in the Cosmic Microwave Background with data from the South Pole Telescope. *Physical Review Letters 111, (14)*, 141301. [arXiv:1307.5830]
18. Hu, W., & White, M. (1996). Acoustic signatures in the Cosmic Microwave Background, *Astrophysical Journal, 471*, 30-51.
19. Jones, W. C., et al. (2006) A measurement of the angular power spectrum of the CMB temperature anisotropy from the 2003 flight of the BOOMERANG. *Astrophysical Journal, 647*, 823-832.
20. Kazanas, D. (1980). Dynamics of the universe and spontaneous symmetry breaking. *Astrophysical Journal, 241*, L59–63.
21. Kovac, E.M., et al. (2002). Measurement of polarization with the degree angular scale interferometer. *Nature 420*, 763–771
22. Kuo, C. L., et al. (2007). Improved measurement of the CMB power spectrum with ACBAR. *Astrophysical Journal, 664*, 687-701.
23. Linde, A. D. (1982). A new inflationary universe scenario: A possible solution of the horizon, flatness, homogeneity, isotropy and primordial monopole problems. *Physics Letters, B 108*, 389-393.
24. Lundmark, K. (1930). Über die Bestimmung der Entfernungen, Dimensionen, Massen und Dichtigkeit fur die nächstgelegenen anagalacktischen Sternsysteme. *Meddelanden Fran Lunds Astronomiska Observatorium Serie I, 125*, 1–13.
25. Manoukian, E. B. (2006). *Quantum theory: A wide spectrum*. Dordrecht: Springer.
26. Manoukian, E. B. (2016). *Quantum field theory II: Introductions to quantum gravity, supersymmatry and string theory*. Switzerland: Springer.
27. Manoukian, E. B. (2016). $45^{\text{o}}$ Relative orientations of planes of polarizations sates of gravitational waves and the graviton. *International Journal of Theoretical Physics, 55*, 4878-4884.
28. Manoukian, E. B. (2020). *100 years of fundamental theoretical physics in the palm of your hand: Integrated technical treatment*. Switzerland: Springer.

29. Manoukian, E. B. (2024). *How energy considerations have shaped our fundamental modern theories of physics: Theory and key historical moments*. Switzerland: Springer.
30. Mather, J.C., & Smoot, G. F. (2006). Cosmic Microwave Background Radiation anisotropies: Their discovery and utilization. *Nobel Lecture. Nobel Foundation, NobelPrize.org*.
31. McKellar, A. (1941). Molecular lines from the lowest states of diatomic molecules composed of atoms probably present in interstellar space. *Publications of the Dominion Astrophysical Observatory, Vancouver, B.C., Canada, 7*, 251-272.
32. Melchiorri, A., et al. (2000). A Measurement of $\Omega$ from the North American Test Flight of Boomerang. *The Astrophysical Journal Letters. 536 (2)*, L63–L66.
33. Miller, A. et al. (1999). A measurement of the angular power spectrum of the Cosmic Microwave Background from $\ell = 100$ to 400. *Astrophysical Journal, 524*, L1-L4.
34. Navas, S., et al. (2024). Particle Data Group. *Physical Review D 110*, 030001.
35. Newman, E. T., & Penrose, R. (1966). Note on the Bondi-Metzner-Sachs Group, *Journal of Mathematical Physics, 7 (5)*, 863–870.
36. Oort, J.H. (1932). The force exerted by the stellar system in the direction perpendicular to the galactic plane and some related problems. *Bulletin of the Astronomical Institutes of the Netherlands, 6*, 249–287.
37. Penzias, A. A., & Wilson, R. W. (1965). A measurement of excess antenna temperature At 4080 Mc/s. *Astrophysical Journal Letters, 142*, 419-421.
38. Perlmutter, S. (2012). Measuring the acceleration of the cosmic expansion using Supernovae. *Reviews of Modern Physics, 84*, 1127-1149.
39. Planck Collaboration et al. (2016). Planck 2015 results. XIII. Cosmological parameters. *Astronomy & Astrophysics, 594*, A1.
40. Planck Collaboration et al. (2020). Planck 2018 results. I. Overview and the cosmological legacy of Planck. *Astronomy & Astrophysics. 641*, A1.
41. Poincaré, P. (1906). La Voie lactée et la théorie des gaz. [The Milky Way and the theory of gases]. *Bulletin de la Société Astronomique de France, 20*, 153–165.
42. Riess, A. G. (2012). My path to the accelerating Universe. *Reviews of Modern Physics, 84*, 1165-1175.
43. Torres del Castillo, G. F. (2003). *3-D spinors, spin-Weighted Functions and their applications*. Boson: Birkhäuser.
44. Schmidt, B. P. (2012). Acceleration expansion of the Universe through observations of distant Supernovae. *Reviews of Modern Physics, 84*, 1151-1163.
45. Smoot, G. F. (1992). Structure in the COBE differential Microwave Radiometer first-year maps. *Astrophysical Journal Letters, 396*, L1.
46. Starobinsky A. A. (1980). A new type of isotropic cosmological models without singularity. *Physics Letters B, 91*, 99-102.
47. White, M. (1999). Anisotropies in the CMB. *Proceedings of the Los Angeles Meeting, DPF 99. UCLA*. arXiv:astro-ph/9903232.
48. Wigner, E. P. (1951). *Group theory and its application to the quantum mechanics of atomic spectra*. New York: Academic Press.
49. Wright, E.L. (2004). Theoretical overview of Cosmic Microwave Background anisotropy. In W. L. Freedman (ed.): *Measuring and modeling the Universe. Carnegie Observatories Astrophysics Series*. Cambridge: Cambridge University Press. p. 291. arXiv:astro-ph/0305591.
50. Wu, T. T., & Yang, C. N. (1976). Dirac monopole without strings: monopole harmonics, *Nuclear Physics B, 107 (3)*, 365–380.
51. Zwicky, F. (1933). Die rotverschiebung von extragalaktischen nebeln. *Helvetica Physica Acta 6*, 110-127.

# Index





E. B. Manoukian, *Radiation*,
https://doi.org/10.1007/978-3-032-21506-2

GPSR Compliance
The European Union's (EU) General Product Safety Regulation (GPSR) is a set of rules that requires consumer products to be safe and our obligations to ensure this.

If you have any concerns about our products, you can contact us on

ProductSafety@springernature.com

In case Publisher is established outside the EU, the EU authorized representative is:

Springer Nature Customer Service Center GmbH
Europaplatz 3
69115 Heidelberg, Germany

www.ingramcontent.com/pod-product-compliance
Ingram Content Group UK Ltd.
Pitfield, Milton Keynes, MK11 3LW, UK
UKHW020936180926
475692UK00003B/59

* 9 7 8 3 0 3 2 2 1 5 0 5 5 *